SPRINGER TRACTS
IN MODERN PHYSICS

Ergebnisse
der exakten Natur-
wissenschaften

Volume **58**

Editor: G. Höhler

Editorial Board: P. Falk-Vairant S. Flügge J. Hamilton
F. Hund H. Lehmann E. A. Niekisch W. Paul

Springer-Verlag Berlin Heidelberg GmbH 1971

Manuscripts for publication should be adressed to:

G. HÖHLER, Institut für Theoretische Kernphysik der Universität, 75 Karlsruhe 1, Postfach 6380

Proofs and all correspondence concerning papers in the process of publication should be addressed to:

E. A. NIEKISCH, Kernforschungsanlage Jülich, Institut für Technische Physik, 517 Jülich, Postfach 365

ISBN 978-3-662-15590-5   ISBN 978-3-540-36505-1 (eBook)
DOI 10.1007/978-3-540-36505-1

Contents

Survey of Cosmology

Is "Our World" Implied by Thermal Equilibrium in the Hadron Era?*

W. KUNDT

Contents

Introduction

Until a few years ago, "Cosmology" meant the search for global solutions of some relativistic field equations, or the study of their properties. Which had better be called "world geometry". Today's Cosmology means the *physics beyond laboratory scale*, both in space and time. Its main tasks are to explain the existence, and history of all the objects in the sky, ranging (in size) from cosmic particles (including photons) through dust, planets, stars and star clusters to galaxies, and galaxy clusters.

If one trusts the simplest existing spacetime theory, namely Einstein's with vanishing cosmological constant Λ, there is (still) a host of *admissible cosmological models* as we do not know a complete set of initial or boundary conditions. However, the observed isotropy of the background

---

* Dedicated to Mr. *E. Witt* for his 60th birthday.

radiation implies isotropic geometry (wherever it holds), and the generality is thereby restricted (approximately) to the class of Robertson-Walker models. Redshift and mass density measurements then tell us that the 3-space curvature ε/R^2 satisfies $-\dot{R}^2/R^2 c^2 \lesseqgtr \varepsilon/R^2 \lesssim 0$, with zero as a suggestive value; (this curvature measures the energy density of the substratum). Adopting $\varepsilon = 0$, one is left with essentially a one parameter class of cosmological models depending on the (present) mean ratio of baryon number over photon number. Omnès has endeavoured to derive this surviving parameter from the assumption of charge symmetry (zero baryon number), as a stochastic phenomenon.

The purpose of this article is to discuss in greater detail why the *Einstein-de Sitter model* (considered above) with $0 = \Lambda = \varepsilon = Q = B = L_e = L_\mu$ is indeed a very suggestive one, ($Q =$ charge, $B =$ baryon number, $L_j =$ lepton numbers), and to sketch how *thermodynamics* in this model can explain our present observations. This survey will deviate in parts from other publications, mainly in the treatment of baryon separation, and its consequences.

1. Coarse Grained World Geometry

1.1 Isotropy

The 2.7 deg *background radiation* has not yet revealed any intensity anisotropy, or polarisation: isotropy of 0.1 % has been established down to apertures of 1°, 1 % down to 2′, 15 % down to 10″; and polarization has been excluded at the 1 % level; compare *Partridge* (1969), and also *Wolfe* (1969). The existence of a test particle field with an isotropic velocity distribution (with respect to the cosmic substratum) implies an isotropic world geometry, see *Ehlers, Geren, Sachs* (1968), or *Kundt* (1968), Section 4.1. When we invoke the cosmological or Copernican principle, the background radiation is such an isotropic test particle field (of zero rest mass), and our neighbouring world geometry is (approximately) of the Robertson-Walker type.

Of course isotropy is violated within the solar system, or within our galaxy: Isotropy is claimed for the *coarse grained* universe whose grains are about 10^8 light years in diameter. Smaller domains show a fine structure which will be discussed in Chapter 3.

How far back in time is isotropy a "good" symmetry? The isotropy of the background radiation down to quite small apertures tells us that our "*surface of last scattering*" (of photons) has a rather smooth temperature and velocity distribution, i.e. is rather *homogeneous*. Inhomogeneities (and resulting anisotropies) could only be imagined at much earlier times, before an era that is able to smooth out such inhomogeneities, geometrically, or through particle fluxes. Particle motions are, however, restricted to cosmologically small scales because velocities have to stay inside the light cone; and even inside much narrower cones if the mean free path is small. How far the light cones spread depends upon the world geometry at early times; we come back to this point. Suffice it to say that within the simplest cosmological models, particle smoothing of inhomogeneities via pressure is restricted to the scale of galactic masses. And that an inhomogeneous early universe is imaginable but by no means necessary. The homogeneity of the surface of last scattering speaks against an inhomogeneous past.

The line element of a Robertson-Walker spacetime can be written

$$ds^2 = R^2(t)\, d\sigma^2 - c^2 dt^2 \tag{1}$$

where the "*expansion factor*", or "world radius" $R(t)$ is the only unspecified function, and where $d\sigma^2$ describes a 3-space of constant curvature ε. For uniqueness, ε can be gauged to equal ± 1 or 0. In this gauge, $R(t)$ is the 3-space curvature radius of the universe for $\varepsilon \neq 0$.

For given mean energy density $\mu = : (c^2/8\pi G)\mu^*$, ($G$ = Newton's constant), there are a number of further *characteristic cosmic lengths*:

gravitational or Schwarzschild radius of the universe $\qquad m := \mu^* R^3/3c^2$

trapping length of the universe $\qquad l := c(3/\mu^*)^{1/2} = R(R/m)^{1/2}$

maximal spatial extent of past light cone $\qquad L(t_0) := \max_t \left(R(t) \int_t^{t_0} d\sigma \right).$

m is constant throughout dust periods (pressure = 0); l is the radius of a homogeneous matter ball which is just inside its Schwarzschild throat (horizon); it equals essentially one over the square root of the Ricci scalar. l and L are defined for all values of ε, and are presently nearly equal to R and ct. Finally, $l \leq R \leq m$ for $\varepsilon = +1$. These remarks show that even in the spatially flat case ($\varepsilon = 0$) there are preferred lengths which take over the role of the "world radius", and which define "pockets" of the world in a physically significant way. Cosmology deals with only one such pocket.

1.2 Big Bang

When one solves the Einstein field equations for a Robertson-Walker metric, he finds a *singular beginning* at some finite time when matter density, pressure, and velocity gradients become infinite. Is this so-called "big bang" a peculiar feature of the isotropic (*Friedmann*) world models?

There are several answers "no" of differing strength and generality. They have been surveyed by *Kundt* (1968), *Penrose* (1968), *Hawking* and *Sciama* (1969), *Hawking* and *Penrose* (1970). Let us repeat two of them: 1) A big bang as defined above is common to all spatially homogeneous models. 2) When spacetime has no symmetry, the strongest "*singularity theorem*" that takes account of our experimental situation reads as follows: Spacetime cannot be geodesically complete (to the past) if a) the energy inequality holds: $\mu + 3p + \Lambda c^4/4\pi G \geqq 0$, ($\Lambda =$ cosmological constant, isotropic pressures assumed), and if b) there are no closed timelike worldlines. It would be nice to know if geodesic incompleteness implies unbounded matter density.

Can the *energy inequality* (in the last singularity theorem) be violated? A negative Λ could do so, and can not (yet) be ruled out by observations. As μ and p increase towards the past (at least for some time), the strongest violation of the energy inequality would occur in the present. A negative Λ would be required for a closed universe such that light can get "around the world". Contrary to my earlier belief, there seems to be no hint at an experimental verification of this kind; nor does a non-vanishing Λ seem to be required by any other observation. Moreover, the important role played by the energy inequality speaks in favour of a vanishing Λ.

The energy inequality could also be violated by exceedingly *negative pressures* as suggested by *Libby* and *Thomas* (1969) for densities just above nuclear density when the Pauli principle is circumvented by cold pressing of new particles. In this state the mutual attraction of baryons could dominate their hard core repulsion, and during two orders of magnitude in the density the pressure would be negative. A violation of $\mu + 3p \geqq 0$ appears, however, to be marginally excluded. Moreover, this equation of state does not apply to cosmological situations where nuclear densities occur at extreme temperatures only.

The foregoing considerations reassure us that our world cannot have existed forever, and that the Friedmann models should not be too unrealistic even for a description of the early universe. With one possible exception raised by *Misner* (1969a): the Friedmann models possess a particle horizon. A *particle horizon* is a future light cone with vertex at the past boundary. Its existence means that the big bang is spacelike. Which implies that for each pair of particles there is a finite minimum time before which they cannot interact; so dissipative effects are restricted to limited scales.

The *existence* of a particle horizon is a difficult property to check, so that even among the class of spatially homogeneous world models it seems not yet certain whether copies without a horizon exist; compare *Misner* (1969b), *Belinsky* and *Khalatnikov* (1970), and *Mac Callum* (1971).

More generally, if the past boundary has a causal structure it can be piecewise spacelike, null, or timelike, and it can be at finite, or infinite distance; compare *Schmidt's* and *Seifert's* contributions to the Gwatt seminar, to appear (1971) in GRG. Friedmann spacetimes have a spacelike past boundary at finite distance; they describe one *connected big bang*. Models with null or timelike boundaries would allow for more effective mixing.

1.3 Topological Peculiarities

The Friedmann worlds are topologically R^4. Who guarantees that more complicated topologies are not realized in our actual universe? For instance, there is a countable infinity of different 3-*spaces of constant curvature*; compare the contribution by *Ellis* to the Gwatt seminar. Such possibilities are in principle observable but hardly in practice. And they would barely influence the local physics, unless their scale was of comparable (i.e. small) size.

What can be said about the topology of our *past light cone?* A natural cosmological hypothesis appears to be *Hawking's* stable causality: there are no closed causal lines even when the local light cones are varied. Stable causality is implied by global hyperbolicity. It is equivalent to the existence of a global time coordinate; which helps in discussing possible topological peculiarities. Could they be of importance to our thermal history? All we will need in the next chapter is – looking backwards in time – a smooth shrinking of 3-space, i.e. a locally Friedmann behaviour.

What about a *black hole*, i.e. matter of subcosmic size fallen inside its gravitational radius? From far away it looks like a static worldtube; can such a tube cut through our past light cone? The answer is "no": black holes hide behind future pointing horizons which lie in the future of "our" past light cone; they "cover" part of the future boundary.

1.4 Dynamics

After a lengthy discourse on how good the isotropic approximation may be for times before the last scattering (of photons), we now return to the isotropic Robertson-Walker spacetimes, and restrict all further consideration to them.

In this section we want to discuss the *field equations*. They read:

$$R_{ab} - \frac{R}{2} g_{ab} - \Lambda g_{ab} = \frac{8\pi G}{c^2} T_{ab}. \tag{2}$$

We will see in the next chapter that throughout all eras, cosmic matter can be safely approximated by a mixture of ideal gases. With the possible exception of neutrinos, these gases will be at rest relative to each other (because of the high collision rates at early times), so that their 4-*momentum tensor* reads $(c = 1)$:

$$\left.\begin{aligned} T_{ab} &= (\mu + p)\, u_a u_b + p g_{ab} \\ \mu &= \sum_j \mu_j, \, p = \sum_j p_j, \, u_a u^a = -1 \end{aligned}\right\} . \tag{3}$$

Moreover, the background radiation confirms that u_a, (the center-of-mass velocity of the substratum), is tangent to the preferred timelike congruence of the isotropic models: $u_a = t_{,a}$.

Energy densities μ_j and pressures p_j satisfy *equations of state* of the form

$$\left.\begin{aligned} \mu_k &= \mu_k(T_k, p_j) \\ p_k &= p_k(T_k, p_j) \end{aligned}\right\} \tag{4}$$

which follow from statistical mechanics (grand canonical distributions and mass-action law). For example, the photon gas satisfies $p_\gamma = \mu_\gamma/3 \sim T_\gamma^4$, and cold (pressurefree) matter satisfies $p_m = 0$.

The time-component of the reduced Bianchi identities now reads:

$$u_a T^{ab}_{\;\;;b} = 0 \;\Leftrightarrow\; 0 = d(\mu R^3) + p d(R^3). \tag{5}$$

It states energy balance⋆ during the expansion. Whenever the cosmic expansion is reversible – which it is to a good approximation, for all times large compared with 10^{-16} sec say, despite burning stars, exploding and collapsing objects – this equation states *conservation of entropy*. (Entropy growth is mainly due to the inhomogeneities, but a tiny contribution comes from bulk viscosity of matter: even an ideal gas has a non-zero bulk viscosity in between pre-relativistic and ultra-relativistic temperatures, compare *Stewart* (1969).) For cold matter, Eq. (5) reduces to mass conservation; for a relativistic gas it says $\mu R^4 = \text{const}$.

Finally we get Friedmann's expansion equation from Einstein's field equations:

$$\dot{R}^2 - \frac{\mu^* - \Lambda c^2}{3} R^2 = -\varepsilon c^2 \tag{6}$$

which expresses again an *energy balance* when multiplied by one half the mass inside a small cell: Conserved is the Newtonian energy inside this cell as evaluated by an observer at distance $R(t)$, (as though he was in Euclidean space), whereby $\mu - \Lambda c^4/8\pi G$ is the active energy density,

---

⋆ Energy is not conserved because its gravitational potential part is left out.

and $-\varepsilon/2$ is the (constant) energy by rest energy ratio. The expansion is unhalted iff $\varepsilon \leq 0$.

Let us *summarize*: 1) Coarse grained cosmology is described by the variables t, R, T_j, p_j, μ_j as functions of each other. The equations of state (4) relate p_j, μ_j to as many temperatures T_k as there are non-interacting subsystems. These temperatures T_k are related to R via the energy law (5). R and t are related via (6). 2) Entropy is nearly conserved in an isotropic world; its weak growth is due to a) bulk viscosity and b) deviations from homogeneity. 3) Free parameters are Λ, ε, and the present values of R, T_j, μ_j. $\Lambda = 0$ is compatible with observations, and so is $\varepsilon = 0$ as so to speak a lower limit to the permitted energy by rest energy ratio; cf. *Rindler* (1969). All the other parameters can in principle be derived from a thorough analysis of our past; which will be done in the next chapter.

One more remark referring to the *growth of entropy*, to the *heat death* of this world, and to the arrow of time – subjects on which textbooks contain most confusing paragraphs: The entropy in a comoving volume element of a Friedmann world is practically constant and positive; (it is a function of the photon temperature alone). How can life develop in a homogeneous gas of constant entropy? Because a gravitating gas of constant density is not in its state of maximal entropy (though in a state of relatively maximal entropy if its mass is below the Jeans mass, see third chapter): Its state of maximal entropy is reached when it collapses to one black hole. The "heat death" of this world is gravitational collapse!

Numerically we find the following estimate: Today the universe contains roughly one nucleon in 10^8 photons (or one nucleon in more than 10^9 photons if the submillimeter background is cosmic and as high as recently suggested, cf. remarks below Eq. (39)), which means that matter contributes roughly a fraction $3 \cdot 10^{-7}$ to the total *entropy*. When stars collapse, an amount of the same order of magnitude is produced, and the total entropy increases roughly by three parts in 10^7 (only). Nevertheless, all life and structure in this world owes its existence to this small fractional increase. Which is not related to the cosmic arrow of time (redshift instead of blueshift): If we lived in a model with $\varepsilon = 1$, entropy would still increase in the contracting phase due to local collapse and mixing.

In other words: the fact that the thermodynamic *arrow of time* agrees with the *cosmic arrow of time* is accidental. On the other hand, a coincidence of the thermodynamic arrow of time with the electrodynamic one follows from the hypothesis of local equilibrium in the early universe (i.e. no correlations at "minus infinity").

2. Thermal History

2.1 Eras

In this chapter we restrict all considerations to Friedmann models, and if necessary (for definiteness) to the Einstein- de Sitter model with *vanishing parameter values:* $0 = \Lambda = \varepsilon = Q = B = L_e = L_\mu$.

We first of all convince ourselves that the cosmic expansion has been nearly *quasi static:* The most obvious particles in the universe are photons, electrons, and protons. Their scattering cross section σ over electrons is on the order of the Thomson cross section: $\sigma_T = 4\pi r^2$, where r is the classical electron radius: $r = e^2/mc^2 \doteq 3 \cdot 10^{-13}$ cm. The mean time τ between collisions is: $\tau = 1/n\sigma v$, with $n =$ density of targets, and $v =$ velocity of projectiles. τ has to be compared with the characteristic cosmic expansion time $t_{ch} = R/\dot{R} = (3/\mu^*)^{1/2}$. The result reads:

$$\frac{\tau}{t_{ch}} = 10 \left(\frac{\mu}{\mu_0} \right)^{1/2} \frac{\mu_{t0}}{\mu_t} \frac{c}{v} \tag{7}$$

where μ is the total energy density, μ_t is the energy density of the targets (i.e. electrons or protons), and an index 0 denotes their present values. As long as electron pairs cannot be produced thermally, the number densities of electrons and protons are equal, and nearly inversely proportional to the (comoving) volume, so that Eq. (7) simplifies to

$$\frac{\tau}{t_{ch}} = 10 \left(\frac{R}{R_0} \right)^{3/2} \frac{c}{v} \tag{8}$$

whereby $c/v \approx (mc^2/3kT)^{1/2}$ (also) decreases with decreasing R (except during reheating). Which means that collisions have been frequent in the near past (when compared with the cosmic time scale), and have been overwhelmingly frequent at early times.

When we state that the cosmic expansion has been quasi static in the past, we claim not only that *certain* local processes have been frequent but that all components of the substratum have been in local equilibrium. This forces us to consider *neutrinos,* as the least interacting particles (to our present knowledge). According to *Bahcall* (1964), their mean time between collisions over thermal electrons or muons respectively is on the order of

$$\tau_\nu \approx \frac{1}{c\sigma_\nu} \left(\frac{\hbar c}{kT} \right)^3 \left(\frac{mc^2}{kT} \right)^2, \tag{9}$$

where m is the electron or muon rest mass, $\sigma_\nu \sim m^2$, and $\sigma_{\nu_e} = 2 \cdot 10^{-44}$ cm^2. We divide by the cosmic time $t_{ch} = R/\dot{R} = (3/\mu^*)^{1/2}$ as above where for high temperatures, μ is slightly larger than Planck's photon energy density. As a result, the ratio

$$\frac{\tau_v}{t_{ch}} \approx 10 \left(\frac{m_e c^2}{kT} \right)^3 \tag{10}$$

(for both electron and muon neutrinos) drops quickly below unity for temperatures just above thermal electron pair creation. In other words: when temperatures are above thermal electron pair creation, electron neutrinos have frequent collisions with electrons; when temperatures reach (even) thermal muon pair creation, muon neutrinos collide rapidly. Below thermal electron creation temperatures, however, neutrinos are *collisionfree* (decoupled); we return to this point below Eq. (30).

Whilst they decouple, and even thereafter, neutrinos can *damp out* a large amount of *inhomogeneity* in the universe; (viscosity is inversely proportional to the mean free path); compare *Misner* (1968). Whether or not they have done so is (still) an open question. The effect must not be overestimated because of its *short range* (solar scale during decoupling, horizon scale after decoupling). Nevertheless, the quasi static character of the cosmic expansion might be weakly violated by the presence of (decoupled) neutrinos.

The quasi static character is *violated* by element formation: chemical abundances in the primordial plasma cannot relax to their thermal equilibrium values. Moreover, baryon pair annihilation may have been hampered by a preceding phase transition in which baryons and anti-baryons separated into droplets. Finally, the formation of stars and star clusters is not fast compared with the cosmic expansion because otherwise there would be nothing left but black holes; (an intermediate state would be black holes plus radiation).

When we speak of a *nearly quasi static* expansion, we weigh the non equilibrium processes by their relative violation of entropy conservation. Nevertheless, as stressed above, it is this very violation which has produced mankind.

Let us *summarize:* the main local feature of a coarse grained universe is adiabatic cooling of a near equilibrium mixture of charged and neutral gases, a temperature bath.

A *temperature bath* has characteristic *phases:* At low temperatures it contains only radiation, electromagnetic, neutrino, and gravitational. A second phase starts when thermal energies suffice for electron pair creation (the lightest non zero rest mass particles known). Whenever thermal energies reach the energy of a new particle species, this species will join in. So that one gets as many plus one phases of the temperature bath as there are elementary particles of different rest mass, as long as these rest masses are widely spaced. When their spacing shrinks below

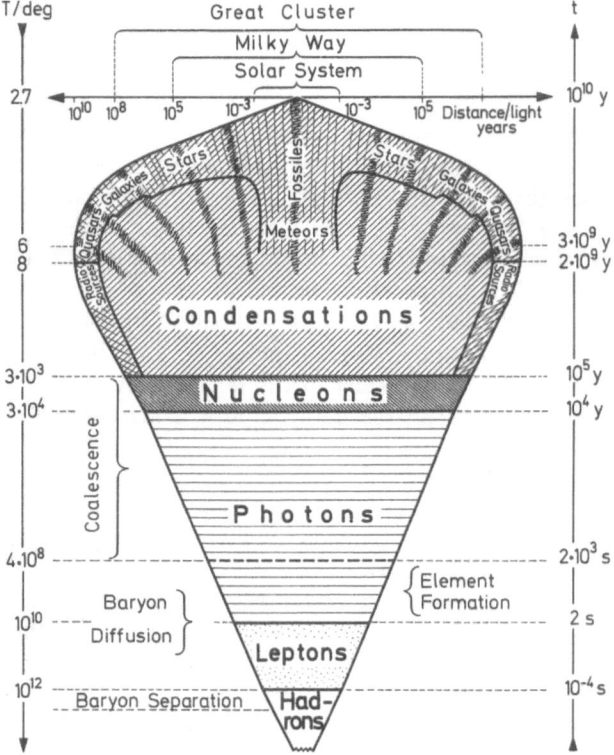

Fig. 1. 2-section through our past light cone in a nonlinear scale, whereby 3-space expansion is qualitatively accounted for. The hatched spacetime domains in the era of condensations are directly accessible to observation. Fine grained density contrast is symbolically marked

the width of a phase transition – which it practically does at the nucleon threshold – a new phase sets in: the hadron phase. This phase will be the ultimate one (towards high temperatures) unless the level density of elementary particle rest masses decreases again – which could only happen above 10 GeV, according to present experiments – or unless a new theory takes over.

We have found, coarsely, three phases of a temperature bath: the radiation phase, the lepton phase, and the hadron phase. The universe will traverse these temperature phases in reverse order if these has been a hot big bang – a hypothesis for which we have given strong support. This leads to three *eras* in our *cosmic history*. The last of these three shows important suberas: first there dominate photons in the energy balance, and Friedmann's equation is that of a light cosmos. During cooling, the energy density of the surviving nucleons eventually overtakes that of the photon, and the universe turns metrically into a dust universe. Another important aspect is whether or not the universe is opaque (to photons): When the universe becomes transparent, i.e. when the mean free path of photons increases beyond the horizon scale, local density fluctuations are no more stabilized by photon pressure, and condensations can contract. This happens about when the plasma recombines (so that the photons find no more electrons to scatter).

We therefore distinguish between the *five eras* of hadrons, leptons, photons, nucleons, and condensations. The first four of them are called after the particle species dominating in the energy balance; the last era is also energetically dominated by nucleons, but matter condensations are the prominent local feature. Photon and nucleon era are often combined into radiation era, the eras of nucleons and condensations into matter era. Note that in low density models ($\mu/\mu_{crit} < 0.06$, "crit" standing for the critical Einstein-de Sitter value), opacity stops before nucleons dominate so that the fourth era does not exist.

Fig. 1 shows a section through "our" *backward light cone* in a non-linear scale. The striped areas are under observational control. The curves during the era of condensations symbolize flowlines of condensed matter.

2.2 Equations of State

We now treat our thermal history in greater detail. As discussed above, *thermal equilibrium distributions* should be an excellent approximation for thermally created particles, at least above threshold.

At each temperature, we will deal with a *mixture* of particle species which can each be approximately treated as an *ideal gas*. This because plasma corrections to the free energy are of relative order $\alpha : = (d/3\lambda_D)^3$ where $d : = n^{-1/3} =$ mean particle separation, and where $\lambda_D : = (kT/4\pi ne^2)^{1/2}$ is the Debye length of the plasma. We find $\alpha \approx (\hbar c/10^2 kTd)^{3/2}$, which yields $\alpha \approx 3 \cdot 10^{-8}$ for $T = 3 \cdot 10^5$ deg and corresponding (electron) number density $n_e = 10^{10}$ cm^{-3}. α is nearly constant for temperatures below electron pair creation. Above threshold, kT in λ_D has to be replaced by the relativistic Fermi energy $kT_F = \pi\hbar c n^{1/3}$, and we get $\alpha \approx 3 \cdot 10^{-4}$; which should, however, not be taken too seriously because the expression for α only retains approximate validity. In any case, plasma corrections appear to be negligible to first order.

For an ideal (relativistic quantum) gas of particle rest mass m we have the following expressions for *particle number N, energy U, pressure p,* and *entropy S* in a *volume V*:

$$
\left.
\begin{aligned}
n &:= \frac{N}{V} = A(T) \cdot I^{11}\left(\frac{mc^2}{kT}, \eta, \varepsilon\right) \\
\frac{\mu}{kT} &:= \frac{U}{kTV} = A(T) \cdot I^{21}\left(\frac{mc^2}{kT}, \eta, \varepsilon\right) \\
\frac{3p}{kT} &= A(T) \cdot I^{03}\left(\frac{mc^2}{kT}, \eta, \varepsilon\right) \\
\frac{S}{kV} &= A(T)\left(I^{21} + \frac{1}{3} I^{03} - \eta I^{11}\right)
\end{aligned}
\right\}
\tag{11}
$$

with: $\varepsilon = \pm 1$ for $\left\{\begin{matrix} \text{fermions} \\ \text{bosons} \end{matrix}\right\}$, $A(T) := \frac{w}{2\pi^2}\left(\frac{kT}{\hbar c}\right)^3$,

$$
w := \text{spin statistical weight} = \left\{\begin{matrix} 1 & \text{for neutrinos} \\ 2 & \text{for photons} \\ 2s+1 & \text{for spin } s, m \neq 0 \end{matrix}\right\},
$$

$$
I^{ab}(\gamma, \eta, \varepsilon) := \int_{\gamma}^{\infty} \frac{x^a(x^2 - \gamma^2)^{b/2}\, dx}{e^{x-\eta} + \varepsilon}.
$$

The *chemical potential*, or *free enthalpy per particle g* satisfies $g = \eta kT$, (*Gibbs-Duhem*). For $\eta = \text{const}$, the integrals I^{ab} are exponentially small for $\gamma \to \infty$ (i.e. low temperatures); for $\eta = 0$, $\gamma \to 0$ they rapidly converge towards

$$
I^{11}(0,0,1) = \frac{3}{2}\zeta(3), \quad I^{21}(0,0,1) = I^{03}(0,0,1) = \frac{7\pi^4}{120},
$$

$$
I^{11}(0,0,-1) = 2\zeta(3), \quad I^{21}(0,0,-1) = I^{03}(0,0,-1) = \frac{\pi^4}{15},
\tag{12}
$$

with $\zeta(3) = 1.202$. When T is fixed, the only free parameter is η; it is uniquely determined by n.

More generally, when we are given a mixture of reacting gases each described by expressions of the above kind, we have to minimize the Gibbsian free enthalpy G at constant p, T in order to determine the parameters η_j of the jth species: $d_{p,T} G = 0$ where

$$
G(T, p_j, N_j) = \sum_j N_j \cdot g_j(T, p_j).
\tag{13}
$$

Remember that $g := u + pv - Ts$ where small letters refer to one particle, and that $0 = \sum_j N_j d_{p,T} g_j$ because of $p = \sum_j N_j p_j$. As a consequence, we have

$$
0 = \sum_j d_{p,T} N_j \cdot g_j.
\tag{14}
$$

Here the differentials $d_{p,T}N_j$ are proportional to the stoechiometric numbers $v_j^{(k)}$ describing the kth reaction: $d_{p,T}N_j = v_j^{(k)}d\alpha_k$, $0 \leq \alpha_k \leq 1$: whence

$$0 = \sum_j v_j^{(k)}g_j, \quad 1 \leq k \leq r;\tag{15}$$

r counts the number of linearly independent reactions. When r is smaller than the number s of different components (species), there must be $c = s - r$ conservation laws

$$0 = \sum_j (\alpha_j^{(l)}N_j + B_j^{(l)}), \quad 1 \leq l \leq c.\tag{16}$$

Eqs. (15) and (16) form s equations for the s unknown parameters η_j which determine all thermodynamic quantities of the reacting ideal gas mixture *(mass action law)*.

Let us start by considering *photons*. They form a one component system with no particle conservation ($v = 1$), hence their chemical potential vanishes ($\eta = 0$). Next consider *nucleons* (or electrons). They can decay into photons: $(v_j) = (1, 1, -x)$, (where j stands for baryons, antibaryons, and photons in succession), and Eq. (15) yields: $0 = \eta_B + \eta_{\bar{B}}$. On the other hand, baryon number is conserved: $0 = N_B - N_{\bar{B}} - B$. If charge symmetry holds ($B = 0$), $N^B = N_{\bar{B}}$ implies $\eta_B = \eta_{\bar{B}}$ so that $\eta_B = \eta_{\bar{B}} = 0$. If charge symmetry is weakly violated (as has been assumed for the universe before the work of Omnès), the potential $\eta_B = -\eta_{\bar{B}}$ is a function of temperature determined by Eq. (16), which vanishes for $\gamma_B \to 0$.

As a good approximation, therefore, we can treat the substratum of the early universe as a mixture of ideal gases with *vanishing chemical potentials*. Eqs. (11), (12) then show that $n/A(T)$ approximates a step function with jumps at the rest masses m_j of the components, $\left(n := \sum_j n_j\right)$, that the energy per particle above threshold is about $u_j = 3,15\,kT$ for fermions, $u_j = 2.7\,kT$ for bosons, and that the entropy per particle above threshold approximates a constant: $s_j \approx 4\,k$ for fermions, $s_j \approx 3,5\,k$ for bosons. We also see that for *adiabatic processes* in between thresholds, const $= S \sim VA(T) \sim VT^3 \sim N$; which for the universe means $RT = $ const, and particle number conservation.

Before we draw further conclusions, let us consider the *hadronic phase* in which thresholds lie nearly dense on the temperature scale. Replacing summation over rest masses by integration, we now have for the total particle density

$$n = \sum_j n_j \to A(T) \sum_\varepsilon \int_0^\infty dm\, \varrho_\varepsilon(m)\, I^{11}\left(\frac{mc^2}{kT}, 0, \varepsilon\right),\tag{17}$$

where $\varrho_\varepsilon(m)$ is the spectral density of fermion or boson rest masses respectively, and where the spin statistical weight w in $A(T)$ has been set equal to one (it is taken care of by $\varrho_\varepsilon(m)$). Corresponding formulae hold for the other thermodynamic densities listed in Eqs. (11).

Their evaluation needs a knowledge of the *spectral densities* $\varrho_\varepsilon(m)$ of hadrons. Empirically

$$\varrho(m) := \varrho_+(m) + \varrho_-(m) = a(m^2 + m_h^2)^{-5/4} e^{mc^2/kT_h} \qquad (18)$$

gives a good description at least up to 11 GeV rest energies, with $a = 2.6 \cdot 10^4 \, (\text{MeV}/c^2)^{3/2}$, $m_h c^2 = 0.5$ MeV, $T_h = 1.8 \cdot 10^{12}$ deg; (the index h stands ambiguously for "highest", "hadronic" or "Hagedorn"; we must avoid *Hagedorn*'s index 0 which in cosmology denotes present values).

Hagedorn (1965, 1968, 1970) has derived the *asymptotic validity* of Eq. (18) for $m \to \infty$ under the following *assumptions:*

1. All hadrons form an ideal gas (their interaction is described by the presence of all the other hadrons), whose energies are canonically distributed in Fock space.

2. Particle numbers are variable; (conservation laws can be eluded e.g. by pair creation).

3. The number of different hadrons (= mass spectral density) approaches asymptotically the number of their excited states (= energy spectral density) up to a polynomial factor.

Assumption 1) is common to this whole section; ass. 2) is equivalent with vanishing chemical potentials. Ass. 3) is *Hagedorn*'s fireball bootstrap which implies formula (18) (with $m_h = 0$), in which a and T_h are unspecified parameters. Approximate values for these two parameters can be obtained on comparison with a model of distinguishable particles within a Fermi cell V_h (= hadronic interaction volume): $a \approx 15 m_\pi^{3/2}$, $V_h := (\hbar/m_\pi c)^3$, $kT_h \approx m_\pi c^2$; we relegate all proofs to an appendix. See also *Huang* and *Weinberg* (1970).

In order to insert the spectral density (18) into Eq. (17), it is necessary to realize that ε in I^{11} can be replaced by zero without large error; the result reads (see appendix):

$$n \approx \frac{1}{V_h} \ln \frac{T_h}{4\Delta T}, \quad \Delta T := T_h - T \ll T_h. \qquad (19)$$

n grows infinite for $T \to T_h$ = highest temperature.

All the other *thermodynamic densities* of interest can be obtained from (19) by differentiation and linear combination when one verifies that $N \approx \ln Z(1)$, Z being the *grand canonical partition sum*. They read in the limit of infinite energy density $(T \to T_h)$:

$$\left. \begin{aligned} \frac{\mu}{kT} &\simeq \frac{1}{V_h} \frac{T_h}{\Delta T} &\simeq \frac{S}{kV} \\[2mm] \frac{p}{kT} &\simeq \frac{1}{V_h} \ln \frac{T_h}{4\Delta T} &\simeq n \end{aligned} \right\} \qquad (20)$$

As a consequence, a change of hadronic state is *isentropic* iff at constant energy, and it satisfies

$$\frac{V}{V_h} \sim \frac{\varDelta T}{T_h}. \tag{21}$$

The ratio of pressure and energy density approaches zero for $T \to T_h$ (as it does for $T \to 0$). This can be easily understood: For increasing temperature, the number of thermally created particles grows only logarithmically whereas their rest mass grows inversely as $\varDelta T$; energy is stored away in increasingly heavy particles with increasingly non relativistic velocities.

Quantitatively one obtains for the average *magnitude of momentum p*:

$$\frac{\langle |\boldsymbol{p}| \rangle c}{kT} \approx 2 \left(\frac{2T_h}{\varDelta T} \right)^{1/2} \left(\ln \frac{T_h}{4 \varDelta T} \right)^{-1}, \tag{22}$$

and for the average *energy u per particle* (from Eqs. (20)):

$$\frac{u}{kT} = \frac{\mu}{nkT} \simeq \frac{T_h}{\varDelta T} \left(\ln \frac{T_h}{4 \varDelta T} \right)^{-1}, \tag{23}$$

so that:

$$\frac{v}{c} = \frac{\langle |\boldsymbol{p}| \rangle c}{u} \simeq 2 \left(\frac{2 \varDelta T}{T_h} \right)^{1/2} \xrightarrow[T \to T_h]{} 0. \tag{24}$$

The *thermal de Broglie wavelength* $\lambda := h/\langle |\boldsymbol{p}| \rangle$ shrinks with increasing T according to Eq. (22). We compare it with the average particle distance $d := n^{-1/3}$:

$$\frac{\lambda}{d} = \pi \left(\frac{\varDelta T}{2 T_h} \right)^{1/2} \left(\ln \frac{T_h}{4 \varDelta T} \right)^{4/3}, \tag{25}$$

and see that hadrons show increasingly non quantum statistical behaviour ($\lambda \ll d$) for temperatures $\varDelta T/T_h \ll 10^{-2}$. Phase transitions (such as baryon-antibaryon separation) are only possible if $\lambda \gtrsim d$, i.e. at the lower end of hadron temperatures.

2.3 Local Cosmology (or: General Relativistic Thermodynamics)

We are now ready to pursue the program outlined in Section 1.4, namely to find t, R, T_j, n_j, μ_j as functions of each other throughout our history. The essential hypothesis we make is that of a *hot big bang:* In the beginning there was a high concentration of thermalized energy. More precisely this hypothesis is needed beginning with (down from) lower hadronic temperatures, when collision times were about twenty orders of magnitude shorter than evolution times (due to isotropic models).

It seems of no relevance to know the states of matter before such an early equilibrium era because all the information of such states is (statistically) lost.

Let us then start with the *hadron era*. As outlined in the last section, a good approximation to the energy density is (cf. Eqs. (20), (21)):

$$\mu = \frac{kT_h}{V_h}\frac{T_h}{\Delta T} = \frac{2kT_h}{V_h}\left(\frac{R_h}{R}\right)^3 \tag{26}$$

with $R = R_h$ for $T = T_h/2$; and Friedmann's Eq. (6) with $\Lambda = 0 = \varepsilon$

$$\frac{\dot{R}}{R} = \left(\frac{\mu^*}{3}\right)^{1/2}, \quad \mu^* := \frac{8\pi G}{c^2}\mu, \tag{27}$$

integrates to:

$$t = 2(3\mu^*)^{-1/2} = t_h\left(\frac{R}{R_h}\right)^{3/2} = t_h\left(\frac{2\Delta T}{T_h}\right)^{1/2} \tag{28}$$

with

$$t_h := \left(\frac{(\hbar c)^3}{12\pi G}\right)^{1/2}\frac{c}{(kT_h)^2} \approx 10^{-4} \text{ sec}.$$

The scale parameter R_h (for the expansion factor R) is arbitrary; we will express it below in terms of its present value. The hadron era ends when $2\Delta T \approx T_h$, i.e. at a temperature of 10^{12} deg, 10^{-4} sec after the big bang, with a particle number density n_h on the order of

$$n_h \approx \frac{1}{V_h} = \left(\frac{kT_h}{\hbar c}\right)^3 \approx 10^{39} \text{ cm}^{-3} = 1 \text{ fm}^{-3}. \tag{29}$$

Note that the Einstein-de Sitter model makes absolute predictions which only involve constants of nature!

When the temperature drops below $T_h/2 = 10^{12}$ deg, hadrons quickly clear the stage. How fast they annihilate depends on whether their abundances manage to stay in thermal equilibrium during the expansion, or whether e.g. baryon-antibaryon separation hampers their disappearance; we come back to this point in the next section. In any case, annihilation will be fast enough to make baryons an energetically small component below $T_h/2$. We now enter the lepton era.

During the *lepton era*, muons, electrons, and their neutrinos share the energy with photons. There is no more a simple dependence of the total energy density on temperature. Rather we have a stepwise pumping of energy from the leptons to photons during their annihilation. Moreover, the neutrinos decouple during, or shortly after the annihilation of their leptons, and retain thereafter temperatures of their own. Let us treat each step as though it was discontinuous as a function of R.

Before muon pair annihilation, the entropy in a fixed volume V reads (cf. Eqs. (11)):

$$\frac{3}{4} S \approx \frac{1}{T} [U_\gamma + 2(U_e + U_{\nu_e} + U_\mu + U_{\nu_\mu})] ; \tag{30}$$

the factor two in front of energies takes care of antiparticles; for the time being we assume charge symmetry. After muon pair annihilations, μ-neutrinos retain thermal contact with the surviving leptons almost as long as e-neutrinos do, see *De Graaf* (1970). We therefore have after muon disappearance:

$$\frac{3}{4} S \approx \frac{1}{T} [U_\gamma + 2(U_e + U_{\nu_e} + U_{\nu_\mu})] ; \tag{31}$$

and after electron pair annihilation:

$$\frac{3}{4} S \approx \frac{U_\gamma}{T_\gamma} + \frac{2U_{\nu_e}}{T_{\nu_e}} + \frac{2U_{\nu_\mu}}{T_{\nu_\mu}} . \tag{32}$$

Entropy is approximately conserved throughout so that during annihilation, successive entropy expressions can be equated. For the energy of thermal particles we have $U \sim T^4$, whereby according to Eqs. (12):

$$U_+/U_- \approx I^{21}(0, 0, +)/I^{21}(0, 0, -) = 7/8 .$$

Temperatures of decoupling particles are conserved. This leads to the *temperature* ratios:

$$\frac{T_{\nu_\mu}}{T_\gamma} \lesssim \frac{T_{\nu_e}}{T_\gamma} = \left(\frac{4}{11} \right)^{1/3} , \tag{33}$$

and to the *total energy density* μ_{rad} after both muon and electron pair annihilation:

$$\mu_{\text{rad}} \approx \mu_\gamma \left\{ 1 + \frac{7}{4} \left(\frac{4}{11} \right)^{4/3} \right\} \approx \mu_\gamma \cdot 1.45 . \tag{34}$$

On this occasion, let us determine the *neutrino number densities* after decoupling. From Eqs. (12) we get for thermal particles:

$$n_+/n_- = I^{11}(0, 0, +)/I^{11}(0, 0, -) = 3/4 ,$$

and $n \sim T^3$ plus Eqs. (33) imply:

$$\frac{2n_{\nu_\mu}}{n_\gamma} \lesssim \frac{2n_{\nu_e}}{n_\gamma} = \frac{3}{11} . \tag{35}$$

Zero rest mass particles satisfy the adiabatic expansion law $V T^3 = \text{const}$, so that $T \sim R^{-1}$. The total energy density satisfies $\mu \sim T^4$

in between thresholds, and is continuous during the stepwise pair annihilations when certain temperatures jump. As a consequence, we have
$\mu \sim R^{-4}$ in the *lepton era*, and Friedmann's Eq. (27) integrates to

$$t - t_h = \frac{1}{2} \left(\frac{3}{\mu^*} \right)^{1/2} \left[1 - \left(\frac{R_h}{R} \right)^2 \right] \sim R^2 - R_h^2 . \qquad (36)$$

When we ignore the small integration constants t_h and R_h, and express
μ^* in terms of the photon temperature called T (without index), we get
from Eqs. (11), (12):

$$\frac{t}{\text{sec}} \approx c(T) \left(\frac{10^{10} \deg}{T} \right)^2 \approx \left(\frac{R}{R_l} \right)^2, \; c(T) := \left(\frac{T}{T_{\nu_\mu}} \right)^2, \qquad (37)$$

with an adaptable parameter R_l, and with a correction factor $c(T)$
which ranges between 1 and 2.5 taking care of pair annihilation. Comparison of Eqs. (28) and (37) yields $R_l \approx 10^2 R_h$.

The lepton era ends at about 2 sec when the temperature drops below
10^{10} deg; (below 10^{10} deg, electron pairs disappear). For the remaining
photon gas we still have $\mu \sim T^4$ so that formulae (37) remain valid. Had
all nucleons and electrons annihilated, this would be the end of the story.

However we know that our present universe contains more energy
in the form of nucleons than of photons. Two explanations offer themselves: Either total annihilation is prohibited by a small non-zero baryon
number per comoving volume which God has wisely chosen, or baryon
pair decay is imperfect due to separation into baryon and antibaryon
clouds. The latter possibility is in any case there, and all we have to
do is believe *Omnès*, or else sit down and work out for ourselves the
most complicated separation and annihilation mechanism that gives rise
to *fine grained baryon number fluctuation*.

Whichever theory is correct, baryons are practically conserved in
the *matter era* which starts when the (rest) energy density of the surviving
baryons bypasses that of the cooling photon gas. Thereafter, conservation
of mass expresses itself in the form $\mu V = \text{const}$, or $\mu \sim R^{-3}$, and Friedmann's Eq. (27) gives (again ignoring small constants of integration):

$$\frac{t}{t_0} \approx \left(\frac{R}{R_0} \right)^{3/2} = \left(\frac{T}{T_0} \right)^{-3/2}, \qquad (38)$$

where T is again the photon temperature satisfying $RT = \text{const}$, and
where an index 0 denotes the present value.

The *present age* t_0 (of the universe), and *background temperature* T_0
are not independent if one believes in the Einstein-de Sitter model, and

in a predictable *nucleon to photon ratio* ξ after the photon era

$$\xi := \frac{n_N}{n_\gamma} = \frac{\mu_N}{\mu_\gamma} \frac{2.7\, kT_0}{m_N c^2} \approx 10^{-8}. \tag{39}$$

We borrowed the value of ξ from observation:

$$\mu^* = 3H^2, \quad H_0^{-1} := (R/\dot{R})_0 = 1.3 \cdot 10^{10} \text{ years}$$

(compare *Sandage*, 1968, 1969) imply $\mu_0 = 10^{-29}\, c^2$ g cm^{-3}; and $T_0 = 2.7$ deg corresponds to $\mu_{\gamma_0} = 4 \cdot 10^{-34}\, c^2$ g/cm^3. The value of ξ is rather uncertain: Hubble's parameter H_0 is not better known than to 30%, and recent measurements of the background radiation produced up to 10^2 times the intensity of a 2.7 deg blackbody for wavelengths below 1 mm; compare *Shivanandan et al.* (1968), *Houck* and *Harwit* (1969), *Muehlner* and *Weiss* (1970), *McNutt* and *Feldman* (1970); *Petrosian et al.* (1969), *Smith* and *Partridge* (1970), *Peebles* (1970b, c), the latter four for theoretical discussion. If this radiation turned out to be cosmic rather than galactic, ξ could fall as low as 10^{-10}.

Let us *match* Eqs. (37) to their successors (38). The linking temperature T_{eq} is defined by the equality $\mu_N = \mu_\gamma$ of nucleon and photon energy density which occurs for

$$3kT_{eq} = \xi m_N c^2 \tag{40}$$

whence: $T_{eq} \approx 3 \cdot 10^4$ deg (with the uncertainty given by ξ), and: $t_{eq} \approx 10^4$ years (from (37)). Matching to Eqs. (38) yields $t_0 = (T_{eq}/T_0)^{3/2}\, t_{eq} \approx 1.2 \cdot 10^{10}$ years for $T_0 = 2.7$ deg, which is in excellent agreement with $t_0 = 2/3\, H_0 \approx 0.9 \cdot 10^{10}$ years as obtained directly from Eq. (38). Notice that for given ξ, there is just one parameter necessary to describe our location in spacetime: the present age t_0, or the present background temperature T_0! Finally we get $2R_0 \approx 10^{10}\, R_l \approx 10^{12}\, R_h$. Let us *collect results*:

$$\left. \begin{aligned}
\frac{t}{t_h} &= \left(\frac{2\Delta T}{T_h}\right)^{1/2} &&= \left(\frac{R}{R_h}\right)^{3/2}, \; 0 < t \le t_h = 10^{-4} \text{ sec, (hadrons)} \\
\frac{t}{\text{sec}} &= c(T)\left(\frac{10^{10} \text{ deg}}{T}\right)^2 &&= \left(\frac{R}{R_l}\right)^2, \; t_h \le t \le t_{eq} = 10^4 \text{ y,} \binom{\text{leptons}}{\text{photons}} \\
\frac{t}{t_0} &= \left(\frac{T_0}{T}\right)^{3/2} &&= \left(\frac{R}{R_0}\right)^{3/2}, \; t_{eq} \le t \le t_0 = 10^{10} \text{ y, (matter).}
\end{aligned} \right\} \tag{41}$$

After these rather positive results, let us ask ourselves if we have possibly overlooked a number of *alternatives* e.g. by insufficient knowledge of elementary particle physics. The first question concerns neutrinos: suppose there are more than *two kinds* of *neutrinos*, x say. In this case,

2*

the total energy density has been higher during the lepton and photon era: $c(T)$ in Eqs. (37), or (41) has to be multiplied by roughly $(7/(5 + x))^{1/2}$, and as a consequence, t_{eq} and t_0 shrink by the same factor. This conflicts with measurements of the age of our galaxy which is taken to be $0.7 \cdot 10^{10}$ years, see *Dicke* (1969). We conclude that x must be smaller than 10^2, in which case $x = 2$ is no more unplausible.

Next we ask if the *lepton number* densities are zero. A blackbody radiation of 2.7 deg contains $4.0 \cdot 10^2$ cm^{-3} photons. Eqs. (35) then imply that there are at present 55 neutrinos of each kind per cm^3 but only $4 \cdot 10^2 \, \zeta = 0.6 \cdot 10^{-5}$ electrons (and nucleons) per cm^3. This under the assumption of charge symmetry. It shows that at present, almost all leptons are neutrinos. Reaction equations of the form (15) tell us that the sum of the chemical potentials of neutrino-antineutrino pairs must vanish. A non zero lepton number would imply a non vanishing difference of such potentials, whence $\eta_\nu = -\eta_{\bar{\nu}} \neq 0$, and the component with the positive η would have a large energy density μ_ν of the asymptotic size

$$\mu_\nu(\eta)/\mu_\nu(0) \simeq 15\eta^4/4\pi^4 \approx \eta^4/26 .$$

$\mu_\nu(0)$ is of the order $0.2 \, \mu_\gamma \approx 10^{-34} \, c^2\text{g cm}^{-3}$, compare formula (34). $\mu(\eta)$ must not exceed the critical density $10^{-29} \, c^2 \, \text{g cm}^{-3}$ (in order to keep the universe older than our galaxy). This implies $\eta < 40$, or a chemical potential smaller than 10^{-2} eV. Up to now there is no evidence for a non zero lepton number.

What makes us believe that the total energy density μ^* equals the *critical density* $3H^2$ even though we can at best discover 10% of it in the sky (via the virial theorem applied to great clusters)? It is the fact that matter can withhold from our telescopes if it is in the state of either ionized hydrogen (near $3 \cdot 10^5$ deg; [compare *Oort* (1970)]), black holes or dead galaxies [compare *Peebles* (1969a)], gravitational radiation, or neutrinos; see also *Kafka* (1970). And it is our conviction that nature is simple.

On the other hand, the fact that matter density cannot be much larger than the critical density (in order to keep the universe older than our galaxy) allows us to set upper bounds to *possible neutrino rest masses* which for the muon neutrino are far more stringent than laboratory estimates. We have found above that the present number densities of neutrinos are around $n_\nu \approx 50$ cm^{-3}. When one divides the critical energy density $10^{-29} \, c^2 \, \text{g cm}^{-3} = 0.6 \cdot 10^4$ eV cm^{-3} by n_ν, he obtains the upper bound $m_\nu < 2 \cdot 10^{-4} \, m_e$.

Are there further particles in the universe of which we do not know because of their low abundance, or small cross section? Stable heavy particles with small cross section *(quarks)* will partially decouple before temperatures fall below their creation threshold, and thus have higher than thermal abundances. Following *Novikov* and *Zeldovic* (1967), we get a present quark to photon ratio of the order

$$\frac{n_q}{n_\gamma} \approx \frac{n_{\bar{q}}}{n_\gamma} \approx \frac{1}{2.4 \cdot A(T_\mathrm{h}) \cdot \sigma \cdot v \cdot t_\mathrm{h}} \approx \left(\frac{Gm_q^2}{\hbar c} \right)^{1/2} \approx 10^{-18}, \qquad (42)$$

where $\sigma \approx (h/m_q c)^2$ is a typical quark cross section, $v \approx c$ is a typical particle velocity during decoupling, and t_h is the hadronic time given in Eq. (28). This formula results when one integrates the most important reaction equations $(q + q \leftrightarrow N + \bar{q}, q + \bar{q} \leftrightarrow \text{mesons})$ through decoupling, whereby the main effect happens right after decoupling: slowness of reactions reduces an exponential (thermal) decay to an algebraic one. Note that Eq. (42) predicts as much quark as there is gold in the universe! Up to now there is no ultimately convincing evidence for the existence of quarks, compare *Chu et al.* (1970).

More important *non equilibrium processes* are baryon annihilation, and the formation of the elements; we treat them in the next section. Before, let us mention the following recent cosmological literature: *Zeldovic* and *Novikov* (1967), *Harrison* (1968), *Dautcourt* and *Wallis* (1968), *Sciama* (1969), *Kundt* (1969), *Zeldovic* (1970).

2.4 Baryon Separation and Element Formation

We come back to the important question whether or not the universe is charge symmetric. Or, more specifically, whether or not baryon pairs managed to escape annihilation by means of separation into distinct droplets (of like particles). If they had kept thermal equilibrium with photons, they would have disappeared in proportion to $\gamma^{3/2} e^{-\gamma}$, cf. Eq. (11), unless there had been a relative surplus of 10^{-8} baryons present in the hadron era (so that their chemical potential did not exactly vanish).

In agreement with *Omnès* (1969, 1970), and *Schatzmann* (1970), we are going to support *baryon number zero*. Our numerical results will be similar to those of the mentioned authors but their mechanisms and criteria could not always be adopted.

Baryon pair annihilation is hampered by different cooperating mechanisms which give rise to three phases, or periods: those of *separation*, *diffusion*, and of *coalescence*. Separation takes place at the end of the hadron era due to a phase transition of the second kind (demixing). Separation has a short range due to lack of time ($3 \cdot 10^{-4}$ cm corresponding to 10^5 g droplets): At the end of the hadron era (10^{-4} sec), temperatures drop below pair creation threshold, and demixing is followed by its converse: diffusion. During the diffusion period there is strong annihilation of pairs, but again time is short: Before rms-baryon-concentrations can drop to their present value $\zeta_0 = 10^{-8}$, surface tension caused by the annihilation products at the boundaries (between matter and anti-matter) overcomes the baryon pressure. With this we enter the coalescence period: surface tension shovels and reshovels matter droplets (of initial mass 10^9 g) until they have fused into at least galactic masses, with only small losses by annihilation at the boundaries. The actual losses during the coalescence period – about a fraction 10^{-2} survives – depend on the surviving concentration after diffusion in such a way that the total losses are independent of the initial concentration: the less is left the less annihilates (as long as the baryon concentration has remained above the present one). This stabilizes the result. We now discuss the details.

1. *Separation Period*. The intuitive idea of baryon-antibaryon sepa-ration just above thermal creation threshold is that like baryons attract each other whereas unlike ones don't. For a phase transition it is necessary that the de Broglie wave length of a particle be (at least) as large as the average particle spacing. According to Eq. (25) this happens for $\Delta T/T_h \gtrsim 10^{-2}$; (our asymptotic formulae have to be improved for $\Delta T \approx T_h$). Another necessary condition (in the present case) is that relative baryon number fluctuations within the range of interaction grow unity; in the appendix we shall verify this condition for the same temperature region. On the other hand, it is sufficient for proving a phase transition to establish a *minimum* of the *free enthalpy* for relative baryon numbers $b_j \neq 0$, where of course $b_j = b_j(T)$. *Omnès* claims that indeed $b_j \neq 0$ for $T > T_c$ where T_c corresponds to thermal energies of order $2kT_h (= 3 \cdot 10^8$ eV); his latest proof is based on a result by *Dashen, Ma*, and *Bernstein* (1970).

Let us belief his proof, which is at present far out of laboratory reach; (in my eyes, it is the only weak link in the deductive chain). We can then estimate the typical *emulsion size* r_h at the end of the hadron era – we often speak of "droplets of radius r_h": This size r_h is the rms-distance that a baryon can random walk during the separation period; the latter essentially coincides with the late stage of the hadron era; (cosmic time is always viewed logarithmically!). With a diffusion coefficient $D = \lambda v/3$, $\lambda \approx 1$ fm, $v \approx c$, and $r = (Dt)^{1/2}$ for diffusion, we have

$$r_h \approx (\lambda c t_h/3)^{1/2} \approx 3 \cdot 10^{-4} \text{ cm},\qquad(43)$$

which should be considered an order of magnitude estimate (of the emulsion size at the end of the separation period). In order to keep annihilation low, r_h had better not be much smaller than this value.

When we use one third the particle number density (29), and a proton mass $m_p = 2 \cdot 10^{-24}$ g, a ball of radius (43) contains $m_p n_h 4\pi r_h^3/9 = 10^5$ g baryons as pronounced above. Eventually, only a fraction

$\xi_0 = 10^{-8}$ of these baryons will survive so that the *droplets* contain 10^{-3} g of present matter. A galaxy needs some 10^{47} of these droplets.

2. *Diffusion.* Immediately after separation, temperatures fall below pair creation threshold, and pairs annihilate wherever they meet. Diffusion smears out what has been separated before, but only through *random walk distances*

$$d(t) = \left[R^2(t) \int_{t_h}^{t} R^{-2}(t') D(t') \, dt' \right]^{1/2} \tag{44}$$

where $D(t) = \lambda_N v_N/3$ is the time dependent nucleon diffusion coefficient, λ = mean free path, v = velocity.

The *rms*-distance d needs a *derivation*. It is the result of diffusion in an expanding medium in Euclidean space. Let n be the (nucleon) number density, and $R(t)$ the expansion factor (= world radius) so that the space metric reads $R^2(t) \sum_{\alpha} (dx^{\alpha})^2$. The temporal change of n obeys, according to Gauss' theorem:

$$-\partial_t(nR^3) = R\partial_{\alpha}(D\partial^{\alpha}n) . \tag{45}$$

For $\partial_{\alpha}D = 0$ we find the elementary solution:

$$n(x, t) = cd^{-3} \exp \{-(Rx/2d)^2\} \tag{46}$$

with $x^2 := \sum_{\alpha} (x^{\alpha})^2$, and $d = d(t)$ given by Eq. (44), q.e.d.

d is the new *emulsion size*. It grows with time, hence the dominant contribution comes during the late stage of diffusion which falls into the early photon era. At that time, there are almost no neutrons left because thermal energies are smaller than the neutron-proton energy difference. Diffusion of protons is controlled by collisions with (almost) thermal (relativistic) electrons so that their mean free path is of the order $\lambda_p \approx 1/n_e\sigma_p \approx 1/n_\gamma\sigma_T$ with Thomson's cross section

$$\sigma_T = 4\pi \left(\frac{e^2}{m_e c^2} \right)^2 \approx 10^{-24} \text{ cm}^2 . \tag{47}$$

Consequently, $D = \lambda_p v_p/3 \approx v_p/3n_\gamma\sigma_T$,

$$v \approx (3kT/m)^{1/2} , \tag{48}$$

$$n_\gamma \approx \frac{1}{4} \left(\frac{kT}{\hbar c} \right)^3 \tag{49}$$

(compare Eqs. (11), (12)), and from (44) and $T \sim t^{-1/2}$ we get

$$d(t) = 5 \cdot 10^{-5} \text{ cm } (t/t_h)^{9/8} \tag{50}$$

for the emulsion size in the diffusion period.

The *annihilation factor*, or *nucleon concentration* ξ after diffusion is the product of its initial value in the separated droplets (depending on the degree of demixing) times a factor describing the random excess of matter over antimatter (or vice versa) due to imperfect cancellation. The former is near unity whilst the latter equals $1/\sqrt{N}$ when N is the number of initial droplets in the diffusion volume $4\pi d^3/3$. These initial droplets – if they had kept their individuality – would have grown (in radius) in proportion to $R \sim t^{1/2}$ so that

$$\xi \approx \left(\frac{r_h R}{d R_h}\right)^{3/2} \approx 15 \cdot \left(\frac{t}{t_h}\right)^{-15/16}. \tag{51}$$

Satisfactorily, ξ stays above 10^{-8} for $t < 10^{10} \, t_h = 10^6$ sec. (Actually, the last statement is false when we use the correct diffusion coefficient D which rises markedly during electron pair annihilation. However, coalescence takes over just in time to make this correction superfluous). We will soon see that diffusion is stopped much earlier by the annihilation pressure along the boundaries between matter and antimatter.

For this it is necessary to calculate the *number* of *pair annihilations* per area and time between matter and antimatter. A tin-shaped domain containing a piece of boundary is approximately described by two semi-infinite regions separated by the plane $x = 0$. Far on the left there are only antiprotons; far on the right there are only protons. They approach each other via diffusion, and annihilate each other in a slab containing $x = 0$. When annihilation products and non-uniform motion are ignored, the particle densities $n_\pm$ obey the balance eqs.*

$$(\partial_t \mp v\partial_x - D\partial_x^2)\, n_\pm = -\alpha n_+ n_- \tag{52}$$

where the diffusion coefficient D and the proton drift velocity $\pm v$ towards $x = 0$ have been assumed constant, $\alpha = \langle \sigma_a | v_+ - v_- | \rangle = \sigma_a \sqrt{2}\, v_p$ is the annihilation coefficient, and σ_a is the annihilation cross section. We are looking for a stationary solution with boundary values $n_\pm(\pm\infty) = n_0$, $n_\pm(\mp\infty) = 0$ for suitable eigenvalue v.

Omnès shows that there is only one *eigen solution* which reads

$$n_\pm = n_0 \frac{\exp(\pm 2x/b)}{4\,\mathrm{ch}^2(x/b)} \tag{53}$$

with

$$\left.\begin{array}{l} b = (24D/\alpha n_0)^{1/2} \\ v = (\alpha D n_0/6)^{1/2} \end{array}\right\}; \tag{54}$$

---

* Eqs. (52) do not describe the actual physical situation in which the center-of-mass velocity v of the plasma vanishes, and in which baryon currents are driven by diffusion plus convection. A more realistic treatment is needed. (I am thankful to *J. Peebles* for stressing this point.)

b is a measure of the thickness of the annihilation zone. The number of annihilations v per area and time is

$$v = \alpha \int_{-\infty}^{\infty} n_+ n_- \, dx = \alpha b n_0^2 / 12 \,. \qquad (55)$$

We will apply these formulae primarily during the coalescence period when proton diffusion is controlled by (charge neutrality and) electron collisions over thermal photons. In that case $D \approx \lambda_e v_e / 3$, $\lambda_e = 1/n_\gamma \sigma_e$, and we get from Eqs. (54), (55), with n_0 replaced by $n_p = : \xi n_\gamma$:

$$\left. \begin{array}{l} b \approx \dfrac{2}{n_\gamma \sigma_e} \left(\dfrac{\sigma_e}{\sigma_a} \dfrac{v_e}{v_p} \dfrac{1}{\xi} \right)^{1/2} \\[3mm] v \approx 0{,}3 \, c \left(\dfrac{\sigma_a}{\sigma_e} \dfrac{v_e v_p}{c^2} \xi \right)^{1/2} \\[3mm] v = n_p v \end{array} \right\} \qquad (56)$$

These formulae are valid after electron pair annihilation, but are a good approximation already during diffusion.

Next we need the *surface tension* exerted by the annihilation products. On annihilation, a baryon pair produces charged and neutral pions. The charged pions decay into electrons and neutrinos via muons; the neutral pions decay into photons. As a result, the pair energy is eventually converted into neutrinos (50%), high energy photons (33%. average energy $= 1.8 \cdot 10^8$ eV), and electrons (17%, average energy $= 10^8$ eV). These decay products transfer momentum to other particles approaching the boundary. When the boundary is curved, the density of decay products is higher on the concave side than on the convex side. As a result there acts a pressure trying to decrease the curvature. This pressure grows with the mean free path of the annihilation products as long as the latter stays below the emulsion size. Consequently, neutrinos with their long mean free path — compare Eq. (9) — exert no net force (when isotropically emitted). Electrons, on the other hand, have a short mean free path compared with photons. The main effect therefore comes from the *high energy photons*.

We calculate the *surface tension p_s* exerted by the decay photons on a spherical piece of boundary of radius r. There are v annihilations per area and time, each producing 5/3 photons of average momentum $m_p c/5$. When λ_γ is their mean free path, they convey momentum to a neighbouring sphere at distance $\lambda_\gamma \cos\vartheta$ where ϑ is the angle enclosed with the boundary normal. Their momentum transfer normal to the boundary per time and area of the neighbouring sphere is (for $\lambda_\gamma \ll r$):

$$\frac{1}{2} \frac{5}{3} v \left(\frac{m_p c}{5} \cos\vartheta \right) \frac{r^2}{(r + \lambda_\gamma \cos\vartheta)^2} \,, \qquad (57)$$

and p_s is the difference of two such expressions for $\vartheta \gtrless \pi/2$ averaged over ϑ:

$$p_s = \frac{vm_pc}{6} \left\langle \frac{\cos\vartheta}{(1 - \lambda_\gamma \cos\vartheta/r)^2} - \frac{\cos\vartheta}{(1 + \lambda_\gamma \cos\vartheta/r)^2} \right\rangle, \qquad (58)$$

or for small λ_γ/r:

$$p_s \approx \frac{2}{9} vm_pc \frac{\lambda_\gamma}{r}. \qquad (59)$$

This last formula is basic to all further discussion; it yields a valid★ order of magnitude estimate of p_s up to $\lambda_\gamma \lesssim 2r$.

We return to the question: when is *diffusion stopped* by surface tension of the annihilation products? The condition is easy to state: p_s must be comparable with the kinetic nucleon pressure $p = n_p kT$ inside a matter droplet. From Eqs. (56) and (59) we get

$$\frac{p_s}{p} = \frac{m_pc^2}{15\,kT} \frac{\lambda_\gamma}{r} \left(\frac{\sigma_a}{\sigma_e} \frac{v_e v_p}{c^2} \xi \right)^{1/2}. \qquad (60)$$

During diffusion, r stands for the emulsion size d found in (50). λ_γ, the mean free path of $1.8 \cdot 10^8$ eV photons, is controlled by pair production on collision with electrons, and is given by

$$\lambda_\gamma \approx \frac{30}{n_e \sigma_e}, \quad \sigma_e \approx \frac{2}{3} \sigma_T. \qquad (61)$$

The annihilation cross section σ_a is estimated by *Schatzmann* to be

$$\sigma_a \lesssim \pi \left(\frac{\hbar}{m_\pi c} \right)^2 \left(\frac{c}{v_p} \right)^2 = 3 \cdot 10^{-26} \text{ cm}^2 \left(\frac{c}{v_p} \right)^2. \qquad (62)$$

As a result we get during diffusion:

$$\frac{p_s}{p} = 0.6 \cdot 10^{-8} \left(\frac{T_h}{T} \right)^{7/4} \xi_e^{-1} \xi_p^{1/2} = 3 \cdot 10^{-8} \left(\frac{T_h}{T} \right)^{13/16} \gamma^{-3/2} e^\gamma \qquad (63)$$

where $n_e =: \xi_e n_\gamma$, $n_p =: \xi_p n_\gamma$ (i.e. the former ξ is now called ξ_p), where use has been made of Eqs. (51) and (11) for the annihilation factors ξ, and where $\gamma := m_e c^2/kT$ as above. The condition $p_s \doteq p$ now determines a critical temperature T_a at which diffusion changes into coalescence; by iteration it is quickly evaluated:

$$\gamma_a = 15 \Leftrightarrow T_a = 4 \cdot 10^8 \text{ deg} \Leftrightarrow t_a = 1.5 \cdot 10^3 \text{ sec}, \qquad (64)$$

and

$$(\xi_e)_a = 3 \cdot 10^{-5}, (\xi_p)_a = 4 \cdot 10^{-6}, d_a = 8 \cdot 10^3 \text{ cm}. \qquad (65)$$

★ A more correct limit of validity is $\lambda_\gamma \lesssim r$.

Let us *repeat:* at the critical temperature T_a, surface tension catches up with nucleon pressure, and from thereon the boundaries between matter and antimatter act like strained membranes that tend to decrease their curvature, i.e. to grow. λ_γ and the annihilation length b stay small compared with the emulsion size, which means that annihilation is a surface rather than a volume effect. t_a falls into the photon era. At t_a, the proton concentration is still more than 10^2 times the present one; it has been almost reached by the electron concentration. After t_a, surface tension is strong enough to move the enclosed matter around, and to make like droplets fuse. We thereby enter the coalescence period.

3. *Coalescence Period.* We have to convince ourselves that surface tension is indeed strong enough to eventually *fuse* at least 10^{36} *droplets.* To this end we compare the *velocity* of a droplet, acted upon by p_s, with the velocity needed to exchange a droplet with its neighbour during the characteristic cosmic time t_{ch} (introduced in (7)). When p_s acts on matter plus radiation within a cylindrical volume of height r, the resulting acceleration (ignoring viscosity) is $a = p_s c^2/\mu r$. From Eqs. (59), (56) we get for the velocity ratio of interest during lepton and photon era:

$$\frac{at_{ch}^2}{r} = \frac{2vm_p c^3 \lambda_\gamma t_{ch}^2}{9\mu_\gamma r^3} = 2 \cdot 10^{34} \, \xi_e^2 \xi_p^{3/2} \frac{T}{T_a} \left(\frac{\lambda_\gamma}{r}\right)^3 \tag{66}$$

where use has been made of

$$t_{ch}^2 = \frac{3}{\mu^*} = \frac{3c^2}{8\pi G\mu_\gamma}. \tag{67}$$

At the end of the photon era we have from (66):

$$\left(\frac{at_{ch}^2}{r}\right)_{eq} = 10^2 \left(\frac{\lambda_\gamma}{r}\right)^3 \tag{68}$$

which is (still) bigger than unity for an emulsion size $r \simeq 5\,\lambda_\gamma$. (Notice that the emulsion size was called r during separation, d during diffusion, and r again during coalescence.) Throughout the photon era, therefore, surface tension is strong enough to shovel droplets around such that the emulsion size r can grow with λ_γ (the needed velocities are subsonic). The ratio λ_γ/r is determined by the two requirements: $p_s \gtrless p$, and: $a \cdot t_{ch}^2 \gtrless r$. Initially the first inequality is stronger, and λ_γ/r rises from about 1/8 to about 2 (see (60)), then falls for a while in proportion to T until the second inequality takes over and forces it up again to its limiting value 2.

This limiting value is reached near the end of the nucleon era. For determining the exact time (temperature), we have to modify Eq. (66) in three ways: 1) the cosmic time t_{ch} is shortened by a higher energy density than that of the photons (used in (67)), 2) the mass of the droplets resides now in the protons (rather than photons) and 3) recombination of the cosmic plasma increases the proton diffusion coefficient D used in v: proton diffusion is henceforth controlled by neutral hydrogen collisions over thermal photons. As a result, the exponent of the temperature factor in (66) changes first from 1 to 3, and near recombination from 3 to 1; (the recombination temperature is $T_{rec} = 4 \cdot 10^3$ deg). Growth of the emulsion stops at a temperature T_e for which $at_{ch}^2 = r = \lambda_\gamma/2$. Estimates starting from Eq. (68) yield: 10^3 deg $\lesssim T_e \lesssim 3 \cdot 10^3$ deg; let us adopt $T_e = 3 \cdot 10^3$ deg. We then find (cf. (61))

$$r_e = \frac{1}{2}(\lambda_\gamma)_e = 3 \cdot 10^{21} \text{ cm}. \tag{69}$$

For temperatures just below T_e, surface tension is (still) huge compared with the baryon pressure but too small for moving matter drops; as a result, the *emulsion size r freezes in*, and cosmic expansion achieves a present value

$$r_0 = r_e \frac{R_0}{R_e} = 3 \cdot 10^{24} \text{ cm} = 3 \cdot 10^6 \text{ light years}. \tag{70}$$

A present coarse grained volume of radius r_0 contains nucleons of mass $4\pi r_0^3 \mu_{N0}/3 = 10^{45}$ g $= 10^{12} M_\odot$, (if $\xi_p = 10^{-8}$ which we will soon try to prove). In other words: the emulsion grows by coalescence to a galactic scale. This mass scale has been determined up to a factor 10^3, corresponding to an uncertain factor 3 in T_e. If our tentative value is correct, every other galaxy is made of antimatter (on the average). If not, we may even have clusters made of the same kind of matter! Observation speaks against the former, cf. *Reinhardt* (1970, 71).

Remains a determination of the total *amount of* baryon *annihilation* (described by ξ_p). The annihilation rate $-\dot{N}_p$ along a boundary of size A is: $-\dot{N}_p = vA$. Whence for the density n_p in cubic cells of constant volume $A \cdot r/3$

$$-\dot{n}_p = 3\frac{v}{r} = 0.9\frac{n_p c}{r}\left(\frac{\sigma_a}{\sigma_e}\frac{v_e v_p}{c^2}\xi_p\right)^{1/2}, \tag{71}$$

where use has been made of our model (56). The emulsion size r is obtained from Eq. (60) for $p_s = p$:

$$\frac{r}{\lambda_\gamma} = \frac{m_p c^2}{15\,kT}\left(\frac{\sigma_a}{\sigma_e}\frac{v_e v_p}{c^2}\xi_p\right)^{1/2} = 4 \cdot 10^3\,\xi_p^{1/2}\frac{T_a}{T}. \tag{72}$$

When this is inserted into (71) we get

$$-\dot{n}_p = qn_p^2\frac{T}{T_a}\frac{\xi_e}{\xi_p}, \quad q := 2 \cdot 10^{-19} \text{ cm}^3 \text{ sec}^{-1}. \tag{73}$$

Up to now we have ignored cosmic expansion. Adding its effect on density decrease to the last formula we find

$$-\dot{n}_p = q n_p^2 \frac{T}{T_a} \frac{\xi_e}{\xi_p} + 3 \frac{\dot{R}}{R} n_p,$$ (74)

or, multiplying by $(R/R_a)^3$, and using $R \sim T^{-1} \sim t^{1/2}$:

$$\partial_t \left(n_p^{-1} \left(\frac{T}{T_a} \right)^3 \right) = q \left(\frac{t_a}{t} \right)^2 \frac{\xi_e}{\xi_p},$$ (75)

whence

$$n_p = \left(\frac{T}{T_a} \right)^3 \left[q t_a \int_1^{t/t_a} x^{-2} \frac{\xi_e}{\xi_p} dx + (n_p^{-1})_a \right]^{-1}.$$ (76)

The integrand in the brackets contains the concentration ratio ξ_e/ξ_p which drops exponentially from 8 to 1. It can be ignored with small error, and the integral converges to one as $(1 - t_a/t)$. The constant of integration $(n_p^{-1})_a$ is negligible after fractions of t_a. Which means that whatever nucleon desity we start out with, we quickly end up with the *frozen density*

$$n_p = \left(\frac{T}{T_a} \right)^3 (q t_a)^{-1}.$$ (77)

We are interested in the *asymptotic nucleon concentration* $\xi := n_p/n_\gamma$. With the photon density (49), and with the data (64), (73) we get from (77):

$$\xi = \frac{4}{q t_a} \left(\frac{\hbar c}{k T_a} \right)^3 = 2 \cdot 10^{-12}.$$ (78)

This value is three to four orders of magnitude smaller than wanted; (the larger values obtained by *Omnès* and *Schatzmann* are due to errors in the calculation). However, it has been obtained with too crude a model: We have used a stationary annihilation model in flat spacetime, and obtained in (56) a proton drift velocity v (of order $c\xi_p^{1/2}/2$) which at $t = t_a$ is about 30 times larger than the proton diffusion velocity (of order $(D/t)^{1/2} \sim T^{-13/4}$, compare Eq. (44))! Moreover, we have neglected the pressure of the annihilation products which (likewise) tends to decrease the nucleon density near the annihilation boundary. This means that our values obtained for v, ν, p_s, and p are something like a factor 10^2 too high at the beginning of the coalescence period. Secondly, the critical ratio p_s/p, (taken equal to one in Eqs. (64), (65), (72)), ought to be lowered by a similar factor because surface tension fights pressure differences, not pressures, and Eq. (66) shows how strong it is. As a result, q in Eq. (78) drops by the square of this factor. More time has to be spent before a result like $\xi = 10^{-8}$ (or: $\xi \ll 10^{-8}$) can be claimed as an order of magnitude statement.

When we calculated the amount of baryon annihilation in Eqs. (71) to (77), we used the emulsion size following from (60). Which is correct at the beginning of the coalescence period but not towards its end, when (66) takes over. After recombination, the annihilation rate (55) rises considerably due to strong diffusion, and relatively small emulsion size. The resulting radiation can be partially thermalized for photon temperatures $\geqq 3 \cdot 10^2$ deg. It may amount to more than 10% of the nucleons in energy, corresponding to about 10^2 times the energy density of the background radiation! This would solve the present *submillimeter radiation puzzle* mentioned below Eq. (39); compare *Peebles* (1970b, c).

Even more annihilation would take place at temperatures below $3 \cdot 10^2$ deg. If it is true that every other galaxy is made of antimatter, we should see a lot of *annihilation radiation* in the sky; compare *Steigman* (1969), *Jones* and *Jones* (1970). *Schatzmann* (1970) finds, however, that at (the present) low densities, stochastic magnetic fields build up parallel to the boundaries of matter and antimatter, and reduce the annihilation rate ν considerably. He also shows that magnetic pressure is strong enough to make cosmic clouds of opposite matter bounce (rather than penetrate). More than 10^4 of such collisions may have occurred within our past light cone; their redshifted radiation could explain the observed γ-ray *intensity peak* at 1 MeV.

With these remarks we leave the problem of baryon separation, and discuss the related problem of *element formation*. Element formation is a non equilibrium process because otherwise there would now only be iron in the world, except in very high density stars. There is the unsettled question whether the elements have been cooked in stars, or in the cosmic plasma. The latter can only have happened near 10^9 deg, when deuterons are no more thermally disintegrated whilst neutrons have not yet decayed. Cosmic element formation therefore starts at the end of the lepton era, and essentially ends after the neutron decay time of 11 minutes (compare *Taylor*, 1968). This interval coincides roughly with the intersection of the baryon diffusion period and the photon era.

The resulting *chemical abundances* depend critically upon densities, temperature, and time. If baryon pairs annihilated according to thermal equilibrium, we know from *Wagoner, Fowler* and *Hoyle* (1967) the production rates of all the lightest elements up to Al^{26}; (they integrated the most relevant 144 reaction equations). Only helium is produced at a rate comparable with its present abundance; all the other elements would have to be cooked in stars. If, on the other hand, baryon separation has taken place as described above, we deal with much higher densities similar to the ones considered by *Wagoner et al.* in bouncing massive stars. As a result, heavy elements are likewise produced at appreciable rates; compare *Wagoner* (1969). *Amiet* and *Zeh* (1968) and *Zeh* (1970) have pointed out that the cosmic abundances of all the heavy nuclei can be gained in three steps the first of which happens at nucleon densities $\mu_N \approx 2 \cdot 10^{10} \, c^2 \, \text{g/cm}^3$ and temperatures $kT \lesssim 2{,}5 \cdot 10^5$ eV; this condition is close to what is realized during baryon diffusion. However, the simultaneous presence of matter and antimatter complicates the situation. Under these circumstances, *Schatzmann* (1970) has pointed out that the final hydrogen to helium ratio depends critically on the initial baryon to antibaryon ratio which in turn is related to the emulsion size r_h after separation. More detailed calculations will have to be done before this important question can be ultimately decided. See also *Cameron* (1970).

3. Fine Grained Cosmology

Fine grained structure comes into existence when stochastic homogeneity is violated. *Baryon-antibaryon separation*, if true, destroys homogeneity on scales below the emulsion size, and thereby opens the chapter. After recombination of the cosmic plasma, a lot more fine structure is introduced through *gravitational instability:* Planets, stars, star clusters, galaxies, and galaxy clusters condense out of the rather homogeneous cosmic substratum. They possess peculiar velocities, angular momentum and magnetic fields, surrounding gas, dust, and plasma. Stellar (thermal) and synchrotron radiation produce *fine grained anisotropy*. The formerly near equilibrium hadronic gas prepares to breed mankind!

3.1 Growth of Density Fluctuations

We ask for an explanation of the observed *mass spectrum* of cosmic condensations. A first step will be to study the (gravitational) growth of density fluctuations irrespective of how initial fluctuations came into being.

This problem can be attacked by *linear perturbation* of an isotropic world model (at least for not too early times): Density perturbations $\delta\mu$ can be linearly superimposed when the curvature radius they produce is large compared to their body radius, i.e. when their body radius r is large compared to their gravitational radius. In symbols:

$$G \; \delta\mu \; r^2/c^2 \ll 1 \,. \tag{79}$$

A third way of expressing this condition is that gravitational binding energies be small compared to rest energies. The condition is satisfied for all objects in the sky we know except neutron stars, hyperon stars (if they exist), and black holes. Even for the latter the condition is satisfied as long as we stay far enough away from them, because their far field is identical with that of normal stars plus possibly (weak) gravitational radiation. Consequently, when we study the growth of density fluctuations with $\delta\mu/\mu \lesssim 1$ we can trust linear perturbation of the coarse grained world geometry.

Is $\delta\mu/\mu \lesssim 1$ a *valid hypothesis* at least before recombination? It should certainly be valid on mass scales up to $10^6 M_\odot$ which are in the range of the homogenizing photon pressure. For larger mass scales we learn from observation that at recombination, $\delta\mu/\mu \lesssim 1\%$ should be a fair estimate. This because galaxies would be denser than they are, ($\mu_{\text{gal}} \approx 10^{-23} c^2 \text{ g cm}^{-3}$), if their density contrast at recombination reached 10%: they would have stopped expanding at too early times. On the other hand, a contrast $\lesssim 10^{-4}$ at recombination would be too small for present densities, and for the rather developed (stellar) structure.

We are interested in the fate of a small amplitude *density contrast* $\delta\mu/\mu$ throughout cosmic history. In this section we ignore stabilizing or driving pressures. The problem reduces to a small perturbation of a Friedmann universe, or, in the long wavelength limit, to a small perturbation of the function $R(t)$ governed by *Friedmann*'s expansion Eq. (6). For $\varepsilon \ll 1$, $\Lambda = 0$, $c = 1$, the solutions of Eqs. (6), (5) read

$$\left. \begin{aligned} \frac{R}{m} &= \left(\frac{3}{2}\frac{t-t_0}{m}\right)^{2/3}\left\{1+O\left[\varepsilon\left(\frac{t-t_0}{m}\right)^{2/3}\right]\right\} \quad \text{for} \quad p=0 \\ \frac{R}{s} &= \left(\frac{t-t_0}{s}\right)^{1/2}\left\{1+O\left[\varepsilon\frac{t-t_0}{s}\right]\right\} \quad \text{for} \quad p=\frac{\mu}{3} \end{aligned} \right\} \quad (80)$$

with $m := \mu^* R^3/3 = \text{gravitational radius}$, and $s := 2(\mu^* R^4/3)^{1/2} \sim \text{entropy}^{2/3}$. The first line describes dust, the second line radiation. Both solutions depend on three parameters: a scale parameter m (s) measuring comoving entropy, ε measuring the energy by rest energy ratio, and t_0 measuring the time of big bang.

When varying these three parameters, we obtain for $t_0 = 0$:

$$\left. \begin{aligned} \delta_m R/R &\sim m^{-1} \sim R^{-3} \\ \delta_\varepsilon R/R &\sim t^{2/3} \sim R \\ \delta_{t_0} R/R &\sim t^{-1} \sim R^{-3/2} \end{aligned} \right\} \quad \text{for} \quad p=0, \quad (81)$$

$$\left. \begin{aligned} \delta_s R/R &\sim s^{-1} \sim R^{-2} \\ \delta_\varepsilon R/R &\sim t \quad\;\; \sim R^2 \\ \delta_{t_0} R/R &\sim t^{-1} \sim R^{-2} \end{aligned} \right\} \quad \text{for} \quad p=\frac{\mu}{3}. \quad (82)$$

Variations of ε describe fluctuations of thermodynamic character; they *grow* algebraically with time. The other variations *diminish* with time. They may or may not have been important at early epochs; simplicity of the initial state suggests the latter.

The density contrast $\delta\mu/\mu$ behaves as $\delta R/R$ because μ obeys a power law in R for both dust and radiation. We therefore see that density fluctuations grow in proportion to R during the hadron and matter era, and in proportion to R^2 during lepton and photon era. This *growth law* holds identically on all scales that are large compared to reaction scales; especially it holds on all scales exceeding the horizon. The earlier we start with a certain fluctuation level, the larger are the fluctuations today. When we want to derive a statement like "$\delta_k\mu/\mu = 1\%$ at recombination and wave number k" (k corresponding to a certain mass scale), we must, among others, know the fluctuation level at some early time for comoving wave number k. The following sections pursue this goal.

3.2 Dynamical Restrictions

The growth Eqs. (81), (82) were derived for scales beyond the range of interaction. Matter of size $10^j M_\odot$ comes within the *horizon* when $(4\pi/3)\,\mu(2ct)^3 = 10^j M_\odot c^2$, which happens at

$$t_H = 5 \cdot 10^{-8+2j/3} \text{ years}. \tag{83}$$

For galaxies, $t_H \approx 1$ year. After t_H, *photon pressure* causes the density contrast to oscillate until the photon diffusion length has increased to the scale considered. Thereafter, oscillation changes into exponential decay. The situation changes again at recombination when the photon mean free path increases beyond the horizon scale: From now on, all fluctuations can grow unhampered until molecular, or nuclear reactions stabilize via radiation pressure; which happens for $\delta\mu/\mu \gg 1$.

Right now we have oversimplified the picture. A *thorough treatment* of the fate of *density fluctuations* after they have come within the horizon needs a distinction between adiabatic and isothermal fluctuations – essentially only the isothermal components survive – and a distinction between photon and nucleon dominated eras. From the long list of contributions to this subject we mention a few (by no means representative) examples: *Peebles* (1968a), (1969c), *Peebles* and *Dicke* (1969), *Peebles* and *Yu* (1970), as well as the elaborate survey article by *Rees* (1970), and the two shorter reviews by *Rees* and *Sciama* (1969). Exhaustive references can be found in *Rees* (1970). The variety of observed objects in the sky, and their possible regularities are discussed in *Arp* (1966), *Ryle* (1968); *Peebles* (1968b), *Yu* and *Peebles* (1969).

In this section we are going to discuss the *Jeans criterion* which determines the *minimum mass* unstable against gravitational collapse in the presence of stabilizing pressures. This minimum mass M_J is fixed by the condition that sound waves can marginally traverse the (homogenous) matter concentration within its fundamental period of oscillation $(c^2/G\mu)^{1/2}$. When λ_J is its characteristic size, and c_s the speed of sound, we have

$$\lambda_J = c_s(c^2/G\mu)^{1/2},\tag{84}$$

and:

$$c^{-1}M_J = \mu c_s^3(G\mu)^{-3/2} \approx \mu^{-1/2}(kT/Gm)^{3/2},\tag{85}$$

with $m =$ average mass of constituing particles; for photons, $m = 3kT/c^2$.

During the photon era, M_J grows in proportion to R^2. During the nucleon era, $p = \mu_\gamma/3$, whence

$$\left(\frac{c_s}{c}\right)^2 = \frac{dp}{d\mu} = \frac{1}{3}\left(1 + \frac{3}{4}\frac{\mu_N}{\mu_\gamma}\right)^{-1},\tag{86}$$

and M_J is practically constant. When the plasma recombines, photons decouple from matter, and the sound speed drops abruptly from $c(4\mu_\gamma/9\mu_N)^{1/2}$ to $(3kT/m)^{1/2}$. As a result, M_J drops by more than a factor of 10^{10}. Thereafter, M_J decreases slowly due to the fact that matter cools more quickly than radiation, but rises again when matter is reheated by the first (proto-) star generation; compare *Weymann* (1966).

When one compares M_J with the horizon mass $M_H = (4\pi/3)(\mu/c^2)(2ct)^3$, he finds for $c_s = c/\sqrt{3}$:

$$M_J/M_H \approx (l/ct)^3 \approx 1;\tag{87}$$

which means that during lepton and photon era the horizon is both *marginally trapped* and marginally *stable* (against collapse). Thereafter, as discussed above, M_J falls increasingly short of M_H.

From this discussion we learn that there are (at least) *three preferred mass scales:* 1) Masses smaller than $M_1 := \min(M_J) \approx 10^5 M_\odot$ are stabilized by particle pressure. 2) Masses larger than $M_2 \approx 10^{12} M_\odot$ escape severe damping before recombination. 3) Masses larger than $M_3 := \max(M_J) \approx 10^{16} M_\odot$ enjoy uninterrupted growth for all times. These preferred scales are observed in the form of globular clusters, quasars, galaxies, and great clusters respectively.

Again, however, reality is much more complex than this simple classification. And there are *alternative explanations*. For example, *Doroshkevich et al.* (1967) take the view that fluctuations on a galactic scale are produced by thermal instabilities, after a generation of protostars of mass M_1 has reheated the cosmic gas to 10^6 deg, with a consequent increase of M_J to the scale of M_2. Another possible source for fluctuations on galactic, or ultragalactic scales is the matter-antimatter annihilation pressure at the time when the emulsion size freezes in (i.e. right after recombination).

3.3 Initial Power Spectrum

Whereas all authors agree that the universe is unstable to density perturbations in the way described in Section 3.1, one finds rather confusing remarks concerning the initial power spectrum. *Thermal density fluctuations* $\delta\mu/\mu$ are on the order of $1/\sqrt{N}$ for masses composed of N particles, where $N = 10^{68}$ now for $M = 10^{11} M_\odot$. As long as the particle number is conserved, thermal fluctuations are constant. On the other hand, the earlier one starts with thermal fluctuations the larger is the density contrast now. If one had $1/\sqrt{N}$ fluctuations only for all times for which Hagedorn's model yields $N \gtrsim 1$, he would have near infinite density contrast now. The problem, therefore, resides in explaining why there is a finite world time before which fluctuations were smaller than thermal.

At this point there is the common objection that fluctuations are requested for matter domains which extend far *beyond* the *horizon* so that equilibrium theory can no more be justified. True: a Friedmann big bang is spacelike, which implies that spatial homogeneity is a hypothesis rather than a conclusion. However, there is only one simple hypothesis about the initial data of the universe: stochastic homogeneity. The observed homogeneity of the surface of last scattering speaks a strong word in favour of it. For homogeneous initial data, $1/\sqrt{N}$ fluctuations are to be expected as soon as there is thermal equilibrium on scales involving more than one particle.

When is that? In Hagedorn's hadronic gas, the particle number density is almost constant: $n \approx 1$ fm^{-3}. We tentatively estimate the mean free time τ between *hadron collisions* to $\tau = 1/n\sigma v \approx d/v$. From Eqs. (24), (28) we conclude

$$\frac{\tau}{t} \approx \frac{\text{fm } t_h}{ct^2} \approx \left(\frac{10^{-14} \text{ sec}}{t}\right)^2 \tag{88}$$

so that 10^{-14} sec after the big bang an average hadron has performed one collision over another hadron. Collisions with zero rest mass particles occur much more frequently.

How far back in time can we trust our special relativistic quantum gas model? Presumably 10^{-23} sec sets a lower limit on the applicability of *equilibrium* theory based on *strong interactions*. Another lower limit on t is placed by the requirement that the spacetime curvature radius

3*

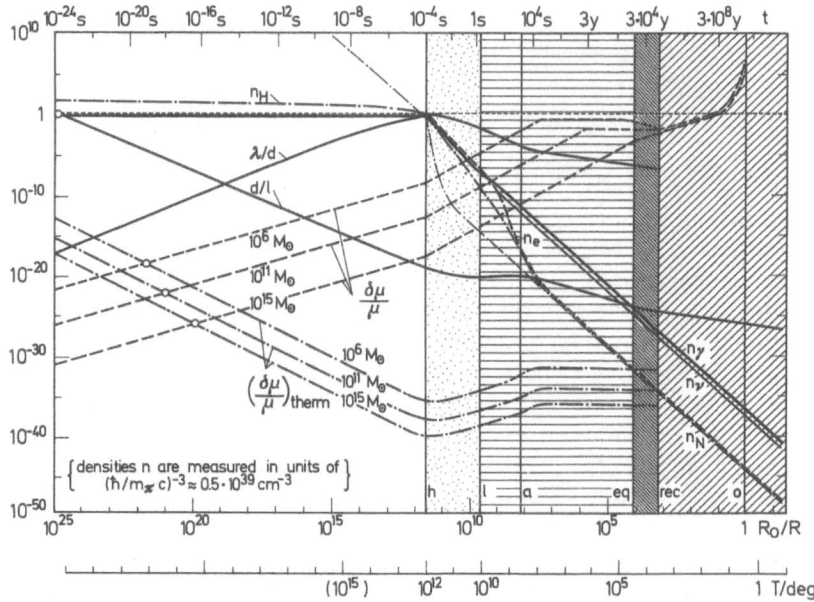

Fig. 2. Number densities of hadrons ($n_{\rm H}$), nucleons (n_N), electrons (n_e), photons (n_γ), and neutrinos (n_ν) in units of $(\hbar/m_\pi c)^{-3} = 0.5 \cdot 10^{39}\ \text{cm}^{-3}$, as functions of inverse world radius R_0/R, photon temperature T, and world time t (lower n_N line applies if baryon-antibaryon separation does not take place). d = average hadron spacing, λ = thermal de Broglie wave length of hadrons, l = trapping length of the universe $\approx$ (Ricci scalar)$^{-1/2}$. $(\delta\mu/\mu)_{\rm therm}$ = thermal density fluctuation, $\delta\mu/\mu$ = estimated density contrast; both on three typical scales

(or trapping length) l be larger than the particle spacing d, because otherwise *gravitational interactions* will dominate strong interactions. $l(t) = 1$ fm holds for $t = 10^{-24}$ sec. These limits are reassuringly far below 10^{-14} sec.

We now suggest the following tentative picture: The first 10^{-23} sec after the big bang are admittedly beyond present theory, but all we need is that *energy* was closely and homogeneously packed in a *non thermalized* way. Thereafter, collisons between hadrons gradually create a thermal spectrum. Due to the strong interaction, this thermalization may only consume fractions of a hadronic collision time τ. We therefore expect to start with a thermal fluctuation spectrum some time between 10^{-23} sec and 10^{-14} sec.

Figure 2 shows both the $1/\sqrt{N}$ density fluctuations, and the present density contrast extrapolated backwards in time for the scales 10^6, 10^{11}, $10^{15}\ M_\odot$. The marked *intersection points* occur at 10^{-19}, 10^{-18}, 10^{-16} sec respectively, well between 10^{-23} sec and 10^{-14} sec. If one insisted on choosing them at equal time, large scale fluctuations were too abundant, or small scale fluctuations too rare. It is amusing to realize that the three intersection times can be characterized by the property that in each case, an average hadron has travelled 10^{-24} times the characteristic size of the fluctuation.

This section has been rather *speculative*. We have endeavoured to stress that a dismission of thermal fluctuations would need a justification as does an adoption. More positively we feel that an understanding of the initial power spectrum is a challenge to the gravitating high energy physicist.

A last word concerning the significance of *Hagedorn's model* for the conclusions of this section. Its role has been threefold: to decrease the number of comoving particles N, to lengthen the time scale via the equation of state, and to reduce the collision rate via non relativistic particle velocities. More conventional equations of state like the one used by *Peebles* (1968c) bend all $\delta\mu/\mu$-curves in Fig. 2 down in such a way that the points of intersection move towards shorter times. Which weakens the belief in their justifiability.

3.4 Present Fine Grained Structure

The title of this section leaves room for a new article which we have neither time nor knowledge to write. It encloses a "*derivation*" of the structure seen in the sky, ranging from the great clusters down to the solar system with its planets, meteors, dust, and solar wind. It encloses the history of stars and star generations. It encloses the formation of the elements, and an explanation of their inhomogeneous fine grained distribution. And it also encloses angular momentum, magnetic fields, and cosmic radiation, both electromagnetic, particle, and gravitational. Instead we want to summarize a few results developed above, and comment on some recent problems and difficulties.

The *cosmic mass spectrum* does not seem to pose problems of principle but problems of detail: Observations suggest that the first units to form had masses $\gtrsim M_1 = 10^5 M_\odot$. This is the smallest unstable mass scale as given by the Jeans criterion. Seed condensations on this scale can only be explained by gravitational growth of fluctuations in the early universe; we suggest that thermal fluctuations do the job. For the galactic scale M_2 one knows three possible sources of initial perturbations − a rather unsatisfactory state of the art. They are: thermal at an early stage (and undamped by photon viscosity), thermal instabilities after reheating, and annihilation forces. We favour the first explanation as it is forced upon us once we want to explain the M_1-scale. Similar remarks apply to the cluster scale M_3 except that reheating can hardly affect it. Mass condensations on scales below M_1 (stars) can only have formed in condensed clouds (in which the Jeans mass is lowered).

The *angular momentum* of *galaxies* has been explained by *Peebles* (1969b) as the result of *rms* tidal interactions between neighbouring protogalaxies at the time when their separation was comparable with their extension. Objections by *Oort* (1969) concern a missing factor 10, which has again been discussed by *Peebles* (1970a). The possibly missing factor 10 may perhaps be explained by *rms* annihilation forces at the interface between matter and antimatter, at the time when the emulsion freezes in. A third and perhaps oldest attempt at an explanation has been recently pursued by *Oort* (1969), *Ozernoi* and *Chernin* (1968), and *Sato et al.* (1970) under the heading "primeval turbulence": Lacking a satisfactory evolutionary explanation, one pushes the difficulty into unknown initial data, destroying their homogeneity. Unfortunately, direct observation cannot rule out such a possibility, compare *Hawking* (1969). However, *Peebles* (1970a) points out that (fine grained) residual turbulence, strong enough to produce galactic angular momenta, would have caused the galaxies to form too soon.

Related to the problem of angular momenta is the problem of *magnectic fields. Harrison* (1970) has advanced an evolutionary explanation based on the (non evolutionary) existence of residual turbulence. His dynamo process is the result of simultaneous rotation and expansion of a 2-fluid system. Observe first that conservation of angular momentum in a turbulent eddy expresses itself in the form $(\mu_m + \mu_\gamma) R^5 \omega = \text{const.}$ where μ_m is the energy density of matter, R the eddy radius, and ω its angular velocity. In the absence of viscous damping, therefore:

$$\omega \sim \begin{Bmatrix} R^{-1} \\ R^{-2} \end{Bmatrix} \quad \text{for} \quad \begin{Bmatrix} \text{radiation} \\ \text{dust} \end{Bmatrix}. \tag{89}$$

During the photon era, if radiation and matter behaved as a single fluid, eddies would spin down according to the first line of (89). There is tight coupling between photons and electrons (due to Thomson scattering), but only millions of times weaker coupling between photons and ions. As a result, the ion gas spins down faster than the photon-electron gas, and differential rotation of the two charged gases produces a magnetic field.

This explanation has two shortcomings: The magnetic fields thus produced are weaker than wanted by a factor 10^8, and they depend on turbulent initial data. One might as well include magnetic fields into the unknown initial data! It appears to me, however, that *baryon-antibaryon annihilation forces* can solve the puzzle: At the beginning of the coalescence period, surface tension between matter and antimatter moves the emulsion at little less than the speed of light, and thereby creates microturbulence. Dissipative energies are minimized when adjacent droplets spin with opposite angular velocity. Adjacent droplets are of opposite baryon number, hence *Harrison*'s dynamo process predicts parallel magnetic seed fields of size $B = 2m_p\omega\,c/e = 2 \cdot 10^{-4}\,\omega$ Gauss sec. Formulae (64), (65) suggest values for ω of order 10^6 sec^{-1} at $T = 3 \cdot 10^8$ deg. Thereafter, conservation of magnetic flux states $B \sim R^{-2}$, and we end up with protogalactic magnetic fields of order 10^{-10} Gauss (it has been used that matter in galaxies is about 10^6 times condensed, compared with the average cosmic density). Gas settling out of the halo may squeeze the protogalactic magnetic field into the disc, thereby increasing its intensity by two orders of magnitude. Another two orders of magnitude can be won for the azimuthal field by differential rotation, and we arrive at the observed 10^{-6} Gauss in our galaxy. — On the other hand, there is a recent suggestion by *Parker* (1970) that all the magnetic fields we are aware of are the result of non-uniform rotation, and cyclonic turbulence in convective, conducting fluid systems.

Let us conclude this article by commenting on *black holes*, and *gravitational radiation*. Black holes are inescapably the final state of matter. As there is little hope for seeing them in the near future (in absorption, by the lens effect, or via infalling matter), observational evidence can only be indirect. *Lynden-Bell* (1969) suggests that almost all quasars have become black holes by now; counting yields roughly one quasar per galaxy! *Wolfe* and *Burbidge* (1970), and *Burbidge* (1970) find that the large mass-to-light ratio of elliptical galaxies, together with the distribution of light, and velocity dispersion of stars speak in favour of central accumulations of black holes. *Kafka* (1970) concludes that *Weber*'s gravitational signals, if true, can only be explained by collisions in a cluster of black holes – a gravar – at the center of our galaxy; cf. *Weber* (1970), or the review by *Kundt* (1970). All problems concerning black holes, and gravitational radiation have been extensively treated by *Ruffini* and *Wheeler* (1970). See also *Tuman* (1970).

Objects falling into a black hole emit part of their energy in form of gravitational radiation. Estimates of the *conversion efficiency* yield a maximum of 20 %, see *Thorne* (1970), compare also *Burke* and *Thorne* (1970). *Hawking* (1970) has shown that the efficiency must be strictly smaller than one. On the other hand, by repeatedly fusing black holes we may approach efficiency one. Baryon number is conserved during gravitational radiation; if a black hole loses energy under collision, the limiting state of an old black hole would have non-vanishing *baryon number* but (almost) vanishing mass.

Gravitational radiation from objects in neighbouring galaxies will be too weak for present detectors to be discerned. What about *gravi-*

tational background radiation? If in the hadron era, cross sections for graviton production were large enough to permit thermal production, gravitational background radiation should have a similar energy distribution as the neutrino background. Considering the difficulties in measuring the (stronger) electromagnetic background (with its large absorption cross section), such an intensity is hopelessly below earthbound sensitivity.

Acknowledgements. I have profited from conversations with Professors and Doctors *R. Hagedorn, H. Leutwyler, M. Rees, E. Schatzmann, H. J. Seifert, J. Stewart*, and *H. D. Zeh.* Most helpful have been guidances to the literature by Doctors *H. Heintzmann, B. Jones, R. Ruffini*, and *D. Sciama.*

Appendix

Hagedorn's hadronic gas. The assumptions of *Hagedorn's* model are listed below Eq. (18). For their evaluation let us recall the derivation of equilibrium thermodynamics from the *grand canonical ensemble:* The grand canonical partition sum $Z(x)$ is defined as

$$Z(x) := \sum_{N=0}^{\infty} Z_N x^N \qquad (A\,1)$$

with

$$Z_N := \sum_{\alpha} e^{-E_{\alpha}/T}, \qquad (A\,2)$$

where we use natural units with $k = 1 = \hbar = c$, and where E_{α} are the N-particle energy eigenvalues, multiple ones listed multiply.

For an *ideal* (quantum) *gas*,

$$E_{\alpha} = \sum_{j} N_{\alpha j} E_j \qquad (A\,3)$$

with one-particle eigenvalues E_j, and occupation numbers $N_{\alpha j} \geqq 0$. The occupation probability w_{α} of E_{α} reads

$$w_{\alpha} = \exp([-E_{\alpha} + gN + K]/T), \qquad (A\,4)$$

where g is the chemical potential or Gibbsian free enthalpy per particle, and where the thermodynamic potential K is fixed through the normalization condition $1 = \sum_{\alpha, N} w_{\alpha}$, or:

$$K = -T \ln Z(e^{g/T}). \qquad (A\,5)$$

For a *one component system,* entropy S is defined as

$$S := -\langle \ln w \rangle \qquad (A\,6)$$

where in the case of a (grand) canonical ensemble:

$$\langle A \rangle := \sum_{\alpha, N} w_\alpha A_\alpha \qquad (A\,7)$$

for any diagonal observable A. Straightforward calculation yields

$$TS = \langle E \rangle - g\langle N \rangle - K. \qquad (A\,8)$$

On the other hand, a reversible variation of the thermodynamic state yields

$$T\,dS = d\langle E \rangle - \delta A - g\,d\langle N \rangle \qquad (A\,9)$$

where δA is the (non integrable) work differential.

The *free energy* F, and *free enthalpy* G of a homogeneous system are defined as

$$\left.\begin{array}{l} F := \langle E \rangle - TS \\ G := \ F \ + pV \end{array}\right\}, \qquad (A\,10)$$

and (A 8) gives the Gibbs-Duhem relation

$$K = F - G = -pV. \qquad (A\,11)$$

The second law of thermodynamics states that (A 9) is in general an inequality, and that consequently, G assumes its minimum for equilibrium. If the total particle number N is not fixed by conservation laws, we therefore have for equilibrium:

$$0 = \partial_N G|_{p,\,T} = g, \qquad (A\,12)$$

i.e. the *chemical potential* vanishes. This applies to the photon, and hadron gas. For them, Eq. (A 5) simplifies to $K = -T \ln Z(1)$.

From *Hagedorn*'s assumptions 1) and 2), the grand canonical partition sum of a *hadronic* gas reads

$$Z = Z(1) = \sum_{N_{ac} \geq 0} \exp\left(-\sum_{a,c} N_{ac} E_{ac}/T\right) \qquad (A\,13)$$

where

$$E_{ac} = (p_a^2 + m_c^2)^{1/2} \qquad (A\,14)$$

is the (relativistic) energy of a particle of species c with 3-momentum p_a, and rest mass m_c. c counts each particle in a spin-isospin-antiparticle multiplet separately. Distinguishing between bosons b and fermions f, and summing geometric series we obtain

$$Z = \prod_{a,b}(1 - e^{-E_{ab}/T})^{-1} \prod_{a,f}(1 + e^{-E_{af}/T}). \qquad (A\,15)$$

When we form the logarithm of Z, products become sums, and the latter are approximated by integrals:

$$\left. \begin{array}{l} \sum_{\alpha} \rightarrow (V/2\pi^2) \int\limits_{0}^{\infty} \mathrm{d}p\, p^2 \\[2ex] \sum_{b,f} \rightarrow \int\limits_{0}^{\infty} \mathrm{d}m [\varrho_b(m) + \varrho_f(m)] \end{array} \right\} , \qquad (A\,16)$$

where ϱ_b, ϱ_f describe the boson and fermion mass spectra respectively. When finally we expand the logarithms of the parentheses into their Taylor series we obtain

$$Z = \exp\left\{ \frac{V}{2\pi^2} \sum_{n=1}^{\infty} \frac{1}{n} \int\limits_{0}^{\infty} \mathrm{d}m \varrho(m;n) \int\limits_{0}^{\infty} \mathrm{d}p\, p^2 \mathrm{e}^{-n\sqrt{p^2+m^2}/T} \right\} \qquad (A\,17)$$

with

$$\varrho(m;n) := \varrho_b(m) - (-1)^n \varrho_f(m). \qquad (A\,18)$$

On the other hand, Z is by definition of the form

$$Z = \int\limits_{0}^{\infty} \mathrm{d}E\, \sigma(E)\, \mathrm{e}^{-E/T}. \qquad (A\,19)$$

Formulae (A 17), (A 19) contain the two unknown functions $\varrho(m;n)$, $\sigma(E)$ whereby with small error, $\varrho(m;n)$ can be replaced by $\varrho(m) := \varrho_b(m) + \varrho_f(m)$, (because the term with $n=1$ dominates in the sum.). $\varrho(m)$ and $\sigma(E)$ are simultaneously tied down by *Hagedorn's* fireball bootstrap listed as assumption 3): Their asymptotic almost equality forces them to take the asymptotic shape

$$\left. \begin{array}{l} \varrho(m) \simeq a\, m^{\alpha}\, \mathrm{e}^{m/T_h} \\[1ex] \sigma(m) \simeq c\, m^{\gamma-1}\, \mathrm{e}^{m/T_0} \end{array} \right\} \quad \text{for} \quad m \rightarrow \infty. \qquad (A\,20)$$

This implies $Z(T) \rightarrow \infty$ for $T \rightarrow T_h$, whence $T_0 = T_h$; and again by comparison of (A 17) and (A 19), (possibly leaving out an alternative):

$$Z \simeq c\, \mathrm{e}^{V} \Gamma(\gamma) \left(\frac{T_h^2}{T_h - T} \right)^{\gamma} \quad \text{for} \quad T \rightarrow T_h \qquad (A\,21)$$

with

$$\alpha = -\frac{5}{2}, \quad \gamma = aV \left(\frac{T_h}{2\pi} \right)^{3/2}. \qquad (A\,22)$$

The partition sum (A 21) describes *Hagedorn's* hadronic gas. It contains the *three unknown parameters* T_h, γ, and c. In order to estimate their values, *Hagedorn* suggested a second model which is obtained from the above one by treating the particles inside a Fermi cell V_h as distinguishable. (Actually, we now present a modified version; *Hagedorn* has given up his original suggestion long since.) This assumption is justified by observing that the above model yields average occupation numbers $\langle N_{ac} \rangle$ in V_h which are small compared to one. It achieves that instead of (A 17) we obtain (via the polynomial sum formula) for $V = V_h$:

$$Z_h = \left\{ 1 - \frac{V_h}{2\pi^2} \int\limits_0^\infty dm \varrho(m) \int\limits_0^\infty dp\, p^2 e^{-\sqrt{p^2+m^2}/T} \right\}^{-1}$$

$$\approx \{ 1 - 2V_h T^3/\pi^2 \}^{-1} \quad \text{for} \quad T \lesssim m_\pi \tag{A 23}$$

$$= T_h^3/(T_h^3 - T^3)$$

with

$$T_h := (\pi^2/2V_h)^{1/3} . \tag{A 24}$$

Here use has been made of the known lower end of the hadronic spectrum: $3\pi, 4K, \dots$. When V_h is chosen as the natural volume of strong interactions (Fermi cell): $V_h = m_\pi^{-3}$, we find from (A 24)

$$T_h = 1.7\, m_\pi = 2 \cdot 10^8 \text{ eV} , \tag{A 25}$$

and comparison of (A 23) with (A 21) shows that T_h has the former meaning. Note that quantum theory forbids to choose V_h much smaller, whereas the assumption of distinguishable particles holds best for the smallest possible equilibrium volume. This explains why our estimate of T_h agrees with the experimental fit given below Eq. (18). Finally we learn from a comparison of (A 21) with (A 23)

$$\gamma = 1, \; a = 7V_h^{-1/2} = 10^4 \text{ MeV}^{3/2}, \; c = (3T_h e^{V_h})^{-1} , \tag{A 26}$$

and

$$\ln Z \simeq \frac{V}{V_h} \ln \left(\frac{T_h}{3\Delta T} \right), \; \Delta T := T_h - T. \tag{A 27}$$

Let us now calculate the *thermodynamic functions*. The average particle number $\langle N \rangle$ is found from (A 13), (A 15) as:

$$\langle N \rangle = -T \sum_{a,c} \partial_{E_{ac}} \ln Z$$

$$= \sum_{a,c} (e^{E_{ac}/T} \pm 1)^{-1} \tag{A 28}$$

$$\simeq \sum_{a,c} e^{-E_{ac}/T} = \ln Z ,$$

whence $n := \langle N \rangle / V$ as given in (19). Similarly from (A 13), (A 27):

$$\langle E \rangle = T^2 \partial_T \ln Z \simeq \frac{V}{V_h} \frac{T^2}{\Delta T}, \qquad (A\ 29)$$

and from (A 11), (A 5), (A 8):

$$\left.\begin{array}{l} pV = T \ln Z \\ TS = \langle E \rangle + pV \simeq \langle E \rangle \end{array}\right\}. \qquad (A\ 30)$$

A little *calculation* is needed to prove Eq. (22), i.e. to find asymptotic expressions (as $T \rightarrow T_h$) for the integrals occurring in

$$\frac{\langle |\boldsymbol{p}| \rangle}{T} = \frac{\int\limits_0^\infty dm\varrho(m) \int\limits_0^\infty dp\, p^3 e^{-\sqrt{p^2+m^2}/T}}{\int\limits_0^\infty dm\varrho(m) \int\limits_0^\infty dp\, p^2 e^{-\sqrt{p^2+m^2}/T}} \qquad (A\ 31)$$

with $\varrho(m)$ given by Eq. (18), or equivalently by (A 20) for m larger than some cutoff mass m_h.

Our last effort is devoted to a determination of the *relative baryon number fluctuations*. Writing the partition sum (A 15) as

$$Z = \prod_{a,\,c} (1 + \varepsilon x_{ac})^\varepsilon (1 + \varepsilon \bar{x}_{ac})^\varepsilon \qquad (A\ 32)$$

with $\varepsilon = \pm 1$ for $\left\{\begin{array}{l}\text{fermions}\\\text{bosons}\end{array}\right\}$ respectively,

$x_{ac} := \exp(-E_{ac}/T)$, and with antiparticles denoted separately by a bar, we have for the baryon number B of non-antibaryons (cf. (A 13)):

$$\left.\begin{array}{l} \langle B^j \rangle = Z^{-1} \left(\sum\limits_c B_c \sum\limits_a x_{ac} \partial_{x_{ac}} \right)^j Z, \\[2mm] Z^{-1} x_{ac} \partial_{x_{ac}} Z = x_{ac}(1 + \varepsilon x_{ac})^{-1}, \\[2mm] Z^{-1}(x_{ac} \partial_{x_{ac}})^2 Z = x_{ac}(1 + \varepsilon x_{ac})^{-1} - \varepsilon x_{ac}^2 (1 + \varepsilon x_{ac})^{-2} \end{array}\right\}. \qquad (A\ 33)$$

For symmetry reasons: $\langle B^j \rangle = \langle \bar{B}^j \rangle$. Moreover: $\langle B\bar{B} \rangle = \langle B \rangle \langle \bar{B} \rangle = \langle B \rangle^2$. We ask for the relative fluctuation φ

$$\varphi := \frac{(\langle (B - \bar{B})^2/2 \rangle)^{1/2}}{\langle B \rangle} = \left(\frac{\langle B^2 \rangle - \langle B \rangle^2}{\langle B \rangle^2} \right)^{1/2}, \qquad (A\ 34)$$

and get from (A 34)

$$\langle B^2 \rangle = \langle B \rangle^2 + \langle B \rangle - \varepsilon \sum_c B_c^2 \sum_a \left(\frac{x_{ac}}{1 + \varepsilon x_{ac}} \right)^2, \qquad (A\ 35)$$

whence

$$\varphi = \left(\frac{1}{\langle B \rangle} \left[1 - \varepsilon \frac{\sum_c B_c^2 \sum_a \left(\frac{x_{ac}}{1 + \varepsilon x_{ac}} \right)^2}{\sum_c B_c \sum_a \frac{x_{ac}}{1 + \varepsilon x_{ac}}} \right] \right)^{1/2} \approx \frac{1}{\langle B \rangle^{1/2}} \quad \text{(A 36)}$$

because the ratio of the double sums is smaller than $B_1 x_{11}(1 + \varepsilon x_{11})^{-1} \ll 1$. But $\langle B \rangle$ is approximately equal to $\langle N \rangle/2$, and $\langle N \rangle$ approaches one for decreasing T, $V = V_{\rm h}$. Which shows that the relative baryon number fluctuations in a Fermi cell approach unity towards the end of the hadron era.

References

Amiet, J. P., Zeh, H. D.: Z. Physik **217**, 485 (1968).
Arp, H.: Ap. J. Suppl. **14**, 1 (1966).
Bahcall, J. N.: Phys. Rev. **136** B, 1164 (1964).
Belinsky, V. A., Khalatnikov, I. M.: to appear (1970), quoted in: *Khalatnikov* and *Lifshitz* (1970).
Burbidge, G.: Comments Astroph. and Space Phys. **II**, 144 (1970).
Burke, W. L., Thorne, K. S.: Proceedings of the Relativity Conference in the Midwest (1970), in press.
Cameron, A. G. W.: Comments Astroph. and Space Phys. **II**, 153 (1970).
Chu, W. T., Kim, Y. S., Beam, W. J., Kiwak, N.: Phys. Rev. Letters **24**, 917 (1970).
Dashen, R., Ma, S., Bernstein, H. J.: preprint Princeton Inst. for Adv. Stud. (1970).
Dautcourt, G., Wallis, G.: Fortschr. Physik **16**, 545 (1968).
De Graaf, T.: Lettere al Nuovo Cimento **4**, 638 (1970a).
— Proceedings of Meeting on Astrophysical Aspects of the Weak Interactions, Cortona, Italy, June (1970b).
Dicke, R. H.: Ap. J. **155**, 123 (1969).
Doroshkevich, A. G., Zeldovich, Ya. B., Novikov, I. D.: Sov. Astron. **11**, 233 (1967).
Ehlers, J., Geren, P., Sachs, R. K.: J. Math. Phys. **9**, 1344 (1968).
Hagedorn, R.: Suppl. Nuovo Cimento **3**, 147 (1965).
— Nuovo Cimento **56** A, 1027 (1968).
— Astron. Astrophys. **5**, 184 (1970a).
— Inv. paper at Coll. on High Multipl. Hadr. Interactions, Paris May (1970b).
Harrison, E. R.: Phys. Rev. **167**, 1170 (1968).
— Monthly Not. **147**, 279 (1970).
Hawking, S. W.: Mon. Not. Roy. Astr. Soc. **142**, 129 (1969).
— Commun. Math. Phys. **18**, 301 (1970).
— *Penrose, R.:* Proc. Roy. Soc. A **314**, 529 (1970).
— *Sciama, D. W.:* Comments Astroph. and Space Phys. **1**, 1 (1969).
Houck, J. R., Harwit, M.: Ap. J. **157**, L 45 (1969a).
— — Science **164**, 1271 (1969b).
Huang, K., Weinberg, S.: Phys. Rev. Letters **25**, 895 (1970).
Jones, J., Jones, B.: Nature **227**, 475 (1970).
Kafka, P.: Nature **226**, 436 (1970).
Khalatnikov, I. M., Lifshitz, E. M.: Phys. Rev. Letters **24**, 76 (1970).

Kundt, W.: Springer Tracts in Mod. Phys. **47**, 111 (1968).
— Fachbericht 34. Physikertagung Salzburg 30. 9. 1969.
— Naturwiss. **57**, 6 (1970).
Libby, M. L., Thomas, F. J.: Physics Lett. **30** B, 88, 400 (1969).
Lynden-Bell, D.: Nature **223**, 690.
Mac Callum, M. A. H.: Commun. Math. Phys. **20**, 57 (1971).
McNutt, D. P., Feldman, P. D.: Science **167**, 1277 (1970).
Misner, C. W.: Ap. J. **151**, 431 (1968).
— Phys. Rev. **186**, 1319 (1969a).
— Phys. Rev. **186**, 1328 (1969b).
Muehlner, D., Weiss, R.: Phys. Rev. Letters **24**, 742 (1970).
Novikov, I. D., Zeldovic, Ya. B.: Ann. Rev. Astron. Ap. **5**, 627 (1967).
Omnès, R.: Phys. Rev. Letters **23**, 38 (1969).
— On the Origin of Matter and Galaxies, preprint 91-Orsay, Lab. de Phys. Théor. et
    Hautes Énergies (1970).
Oort, J. H.: Nature **224**, 1158 (1969).
— Astron. Astrophys. **7**, 381 (1970).
Ozernoi, L. M., Chernin, A. D.: Astron. Zh. **45**, 1137 (1968). Engl. transl.: Soviet Phys.-
    A.J. **12**, 901 (1969).
Parker, E. N.: Comments Astroph. and Space Phys. **II**, 127 (1970).
Partridge, R. B.: American Scientist **57**, 37 (1969).
Peebles, P, J. E.: Ap. J. **153**, 1 (1968a).
— Ap. J. **153**, 13 (1968b).
— Nature **220**, 237 (1968c).
— Ap. J. **154**, L 121 (1969a).
— Ap. J. **155**, 393 (1969b).
— Ap. J. **157**, 1075 (1969c).
— Primeval Turbulence? to appear in: Astrophys. and Space Science (1970a).
— Non-thermal primeval fireball? to appear in: Astrophys. and Space Science (1970b).
— Cosmology and Infrared Astron.-Closing the Gap betw. Theory and Practice, to app.
    in: Comments on Astrophys. and Space Phys. (1970c).
— *Dicke, R. H.:* Ap. J. **154**, 891 (1969).
— *Yu, J. T.:* Primeval Adiab. Perturb. in an Expanding Univ., preprint Princeton, New
    York 1970.
Penrose, R.: Battelle summer rencounters Seattle 1967, contribution to, Benjamin 1968.
Petrosian, V., Bahcall, J. N., Salpeter, E. E.: Ap. J. **155**, L 57 (1969).
Rees, M. J.: Proceedings of Enrico Fermi Summer School Course XLVII, General
    Relativity and Cosmology, Varenna 1969, 1970.
— *Sciama, D. W.:* Comments Astrophys. and Space Phys. **1**, 140, 153 (1969).
Reinhardt, M.: Astrophys. Lett. **7**, 101 (1970).
— two preprints, 5568 Daun (1971).
Rindler, W.: Ap. J. **157**, L 147 (1969).
Ruffini, R., Wheeler, J. A.: Relativistic Cosmology and Space Platforms, preprint Princeton,
    submitted as an ESRO publication 1970.
Ryle, M.: Ann. Rev. Astron. A. **6**, 249 (1968).
Sandage, A.: Ap. J. **152**, L 149 (1968).
— Observatory **88**, 91 (1969).
Sato, H., Matsuda, T., Takeda, H.: Progr. Theor. Phys. **43**, No. 4.
Schatzmann, E.: Physics and Astrophysics, CERN lectures 1970.
Schmidt, B. G.: GRG-Journal **1**, 269 (1971).
Sciama, D. W.: Proceedings of Enrico Fermi Summer School Course XLVII, General
    Relativity and Cosmology, Varenna 1969, 1970.

Seifert, H.-J.: GRG-Journal **1**, 247 (1971).
Shivanandan, K., Houck, J. R., Harwit, M. O.: Phys. Rev. Letters **21**, 1460 (1968).
Smith, M. G., Partridge, R. B.: Ap. J. **159**, 737 (1970).
Steigman, G.: Nature **224**, 477 (1969).
Stewart, J. M.: Ph. D. Thesis Cambridge; subsequent articles in preparation; 1969.
Taylor, R. J.: Nature **217**, 433 (1968).
Thorne, K. S.: Proceedings of Enrico Fermi Summer School Course XLVII, General Relativity and Cosmology, Varenna 1969, 1970.
Tuman, V. S.: Observation of Earth Eigen Vibrations possibly excited by Gravity Waves, preprint Stanislaus State College, Turlock, Calif. 95380, (1970).
Wagoner, R. V.: Ap. J. **156**, 795 (1969).
— *Fowler, W. A., Hoyle, F.:* Ap. J. **148**, 3 (1967).
Weber, J.: Phys. Rev. Letters **25**, 180 (1970).
Weymann, R.: Ap. J. **145**, 560 (1966).
Wolfe, A. M.: Ap. J. **159**, L 61 (1969).
— *Burbidge, G. R.:* Ap. J. **161**, 419 (1970).
Yu, J. T., Peebles, P. J. E.: Ap. J. **158**, 103 (1969).
Zeh, H. D.: Nature **225**, 361 (1970).
Zeldovic, Ya. B.: Comments Astroph. and Space Phys. **II**, 12 (1970).
— *Novikov, I. D.:* Relyativistkaja Astrofisica, Moscow; Engl. reprod.: Univ. of Chicago Press, Vol. 1: Stars and Relativity (1970), Vol. 2: The Universe and Relativity (1971).

Professor Dr. *Wolfgang Kundt*
I. Institut für Theoretische Physik der Universität Hamburg
BRD-2000 Hamburg 36, Germany

Silicon Carbide as a Semiconductor

J. FEITKNECHT

Contents

List of Symbols

A	cross sectional area
a	base vector in hexagonal unit cell
B	magnetic induction
c	base vector in hexagonal unit cell, cell height
d	thickness of crystal platelet
E	electric vector
E_A	acceptor ionization energy
E_C	energy of conduction band
E_D	donor ionization energy
E_G	width of energy gap
E_{GX}	width of exciton energy gap
E_i	ionization energy
E_V	energy of valence band
E_X	binding energy of exciton
e	electronic charge of an electron or base of natural logarithm
g_A	multiplicity factor for an acceptor level
g_D	multiplicity factor for a donor level
h	Planck's constant
I	current or light intensity
k	wave vector
k_{CB}	wave vector of conduction band
k	Boltzmann's constant
m_0	free electron mass
$\left.\begin{array}{l} m \\ m_e \end{array}\right\}$	effective mass of electron
m_h	effective mass of hole
N	number of Si- and C-atoms per unit cell
N_A	acceptor concentration
N_B	doping concentration
N_C	density of states in the conduction band
N_D	donor concentration
N_V	density of states in the valence band
n	free electron concentration
p	free hole concentration
q	pseudomomentum vector
R	resistance or pseudo-resistance, also reflectivity
R_H	Hall constant
T	absolute temperature, also transmission
V	voltage
V_B	breakdown voltage
V_D	diffusion voltage
V_H	Hall voltage
α	absorption constant
ε_0	dielectric constant (permittivity of vacuum)
$\varepsilon(\omega = 0)$	static dielectric constant
$\varepsilon(\infty)$	high frequency (optical) dielectric constant
λ	heat conductivity
μ	Hall mobility
μ_n	electron mobility
μ_p	hole mobility

v frequency of incident electromagnetic radiation
ϱ resistivity
θ_i equivalent temperature of phonons
θ angle

1. Introduction

Natural silicon carbide, the mineral Moissanite, has been found in minute quantities only on rare occasions although the compound is chemically and thermally very stable and its constituents abundant. It was isolated for the first time by *Moissan* [1] from the iron meteorite Canyon Diablo, found in Arizona.

In 1892 [2] *Acheson* devised a method of producing SiC on a large scale for industrial purposes such as grinding and cutting. This product became known as carborundum, and the method is still used in essentially unaltered form to satisfy the heavy demand for abrasives. This is by far the largest market but new applications have been found: The refractory nature of SiC is exploited to manufacture cheap heating elements with long lifetimes; thin coatings are applied to uranium grains to produce "coated particles" for the fuel elements of nuclear reactors; composite materials containing SiC and graphite form part of rocket nozzles. As single crystals became available and the growth methods became more sophisticated, new properties were discovered and consequently new applications became possible. It turned out that whiskers (hair crystals) exhibit unusually high mechanical strength, in addition to being refractory, and it was natural to investigate the properties of SiC reinforced materials.

The first commercial electrical devices made of SiC were polycrystalline aggregates of SiC with various binders. They are still widely employed as lightning arrestors to protect electrical equipment from voltage surges, although the physical processes taking place are largely unknown and the production is more of an art than a science.

Crystallographers were first to show scientific interest in SiC for its very pronounced tendency to crystallize into many different modifications, called polytypes. The first single crystals studied were the result of mishaps in the Acheson furnace. Occasionally platelets of a few square centimeters and a thickness of up to several millimeters were found in so-called blow-holes. Their color ranges from black through blue to different hues of green. Some of them are either very lightly colored or almost perfectly transparent. The crystals are very hard and can only be scratched by diamond and boron carbide. They are often nearly perfect hexagons. SiC is not the only substance occuring in different modifications, for instance ZnS is another well known semiconductor exhibiting this behavior, but what makes SiC unique is the strong effect of polytypism on the electronic properties of SiC. The most striking example of polytype dependence is the bandgap which varies from 2.3 eV for cubic SiC to 3.3 eV for the wurtzite structure. Differences as large as this are typical for differences among distinct semiconductors and were not expected to arise from a change of the configuration of second nearest and farther neighbors while the nearest neighbor bond was unaffected. The question of which long-range forces are causing the existence of so many different modifications and how they are influencing the electronic properties is still unanswered. But the physicists were given a model substance to investigate the effect of purely geometrical changes, on an atomic scale, on the bandstructure, i.e. on semiconductor properties, as opposed to the complex changes taking place when going from one semiconductor to another one. In the latter case the substitution of a chemically different constituent leads to different interatomic forces, thus altering the interatomic distances and quite often the crystallographic structure as well. Many data have been collected during the last decade but a general theory of the geometrical effect on bandstructure is still lacking.

You may consider SiC to be both the oldest and youngest semiconductor. When *Lossew* [3] discovered the electroluminescence of semiconductors in 1907, he was experimenting with a SiC crystal. On the other hand it was only a few years ago that such electroluminescent SiC diodes were indeed marketed. The first systematic analysis of the semiconducting properties was started by *Busch* [4] in 1945 with crystals selected from a large quantity of Acheson material. Due to the tremendous difficulties in growing SiC single crystals of semiconductor quality progress has been disappointingly slow. In 1947 [5] and again in 1955 [6] attempts were made to grow crystals under laboratory conditions to achieve higher purity. The latter process, devised by *Lely*, provided most of the crystals analyzed since then and enabled various electronic devices such as thermistors, electroluminescent diodes, rectifiers, particle detectors, ultraviolet detectors, field effect transistors to be realized.

Only a handful of laboratories have been engaged in SiC research for any length of time. This explains the sparse body of knowledge in comparison to such semiconductors as silicon, germanium, or even gallium arsenide. Heading the group is unquestionably, not only by size and quality, but also in perseverance the Philips Research Laboratory in Eindhoven. Its scope is very broad, embracing such diverse fields as electronic devices, fiber reinforced materials, growth phenomena, and electronic band structure. All other laboratories specialize in one or two areas, not counting crystal growth which is a necessity for every serious undertaking. The Russian groups, mainly in Leningrad and Kiev,

4*

deal with luminescence in a general way from investigating impurity centers to applying the knowledge gained to produce luminescent diodes and display panels. Westinghouse has two groups, one performing basic research on optical properties and another one developing electronic devices for unusual requirements such as high temperature, high radiation fluxes, and high power levels. General Electric's Lamp Division, Norton Research Corporation, and the General Electric Company (Great Britain) put the emphasis on luminescent diodes. Carborundum and Norton, the two big SiC producers in the United States and in Canada, have a big stake in developing better grinding materials and are therefore interested in the growth of heavily doped SiC. Some laboratories in Japan and mainland China have published in the last few years many crystallographic articles. The American groups are encouraged and have been granted research contracts by the U.S. Air Force Research Laboratory, Cambridge.

This review will be restricted to SiC as a semiconductor. After a short summary of its physical and chemical properties a discussion of the crystallographic aspects will follow since many pecularities of SiC as a semiconductor are intimately related to the multitude of polymorphic forms in which SiC occurs. A survey of the growth of single crystals rounds off the picture. The emphasis is laid on the consequences of polytypism on solid state physics, the technology required to produce solid state devices, and their applications.

2. Some Mechanical and Thermodynamic Properties

One of the first observations made on SiC was its high hardness which places it between diamond (10) and topaz (8) as number 9 on Mohs' scale. It also has a high resistance to wear. Little is known about the dependence of these two properties on crystallographic orientation and impurity content [7].

Under certain conditions ribbonlike crystals [8] or whiskers [9] of high mechanical strength can be grown. The whisker cross section may be circular, triangular, or hexagonal (with rounded corners) depending on the growth temperature. Their diameters range from 1 to 30 μm and they may be as long as 15 cm. The most perfect whiskers show a tensile strength of 3200 kg/mm^2 and a modulus of elasticity of 68000 kg/mm^2 for cubic and 69000 kg/mm^2 for hexagonal silicon carbide.

SiC is thermally quite stable. It does not melt at all between atmospheric pressure and 35 atm [10]. At high temperatures it sublimes and dissociates into carbon and a silicon-rich vapor. The carbon residue left behind is in pseudomorphosis with the original crystal. Mass spectrometric analyses of the vapor revealed mainly Si, SiC_2, Si_2C, Si_2 [11]. More recent growth studies performed in various laboratories [12, 13] and thermodynamic calculations [14] seem to suggest that between 2300 and 2600° C a silicon rich liquid, possibly only in a very thin layer on top of the growing crystals, may exist.

No experiments have been performed yet to ascertain the stoichiometry of the compound SiC. Certain findings have led to speculations that SiC grown in a silicon-rich environment may contain excess silicon [15].

Silicides and carbides are abundant compounds but silicon carbide is the only compound between elements of the fourth group of the periodic table. The chemical bond between silicon and carbon atoms is covalent with a slight ionic contribution of about 10% (electron deficiency at the silicon site). The partly ionic bond manifests itself in the pronounced reststrahl absorption band at 12.6 μm which could be well fitted by *Spitzer et al.* [16] to a theoretical model. The covalency is demonstrated by the nearly perfect tetrahedral arrangement of the atoms with a reduced c/a ratio of 1.634, close to the theoretical value of $\sqrt{8/3} = 1.633$. For many order-of-magnitude calculations a suitable average of the silicon and diamond data gives a good estimate for SiC (e.g. thermal conductivity, energy gap).

The silicon carbide surface is covered by a natural layer of SiO_2 which can be dissolved in hydrofluoric acid. But SiC itself is very resistant to chemical attack and not etched by any acids or combination of acids. In contrast, melts of alkaline hydroxides, at temperatures below 1000° C, prove to be effective etchants. Often oxidizing agents such as Na_2O_2 or $Na_2CO_3-KNO_3$ are added. At elevated temperatures gaseous etches can be used such as fluorine, chlorine, or hydrogen. An up-to-date review on etching SiC is given by *Jennings* [17] in the Proceedings of the 2nd International Conference on SiC. Important etchants for device fabrication will be discussed later in the proper context (Section 7.2.3).

3. Crystallographic Properties

Due to the close correlation between crystallographic and semiconducting properties more time is devoted to crystallography than to the chemical and mechanical aspects of SiC.

Silicon carbide caught the interest of crystallographers early because of its pronounced tendency to crystallize in a multitude of different modifications, called polytypes by *Baumhauer* [18]. An excellent review on polytypism written by *Verma* and *Krishna* [19] contains a wealth of information about SiC because this material is one of the standard examples of polytypism (beside ZnS, CdI and others). Polytypism may be defined as the ability of a substance to crystallize into a number of different modifications in all of which two dimensions of the unit cell are the same and the third one is a variable integral multiple of a common length*. For SiC the base vectors of the hexagonal unit cell are $a_1 = a_2 = 3.078$ Å while the third base vector c is an integral multiple of 2.518 Å. The SiC modifications belong to the space groups $F\,\bar{4}3\,m$ (cubic), $P\,6_3\,mc$ (hexagonal), $R\,3\,m$ (rhombohedral), and $P\,3\,m1$ (trigonal) all of which are often described by a hexagonal coordinate system to make comparison simpler. About 150 polytypes of SiC are known and there seems to be no upper limit to the number of possible modifications. The smallest unit cell has a c-axis of 5.04 Å (wurtzite structure), the largest one measures 1500 Å. The more common polytypes contain 8 to 30 atoms per unit cell and have cell heights between 10 and 40 Å.

Table 1. *Nomenclature of the SiC polytypes*

Ramsdell	Zhdanov	Jagodzinski	historical notation
3 C	(∞)	c	SiC IV or β-SiC
24 R	(35)	hcchcccc	
8 H	(44)	hccc	
21 R	(34)	hcchccc	
15 R	(32)	hcchc	SiC I
33 R	(3332)	hcchcchcchc	
6 H	(33)	hcc	SiC II
4 H	(22)	hc	SiC III
2 H	(11)	h	

Each polytype consists of identical layers of close packed Si – C tetrahedra. The difference between them lies in the stacking order of the basic building blocks. Each new layer can be added to the existing crystal in one of two ways, leading to cubic or hexagonal closest packing of the tetrahedra (Fig. 1). Several symbols are in use to specify the modifications. A full description of the structure is provided by *Jagodzinski's* [20] and *Zhdanov's* [21] notations whereas the often used *Ramsdell* symbol [22] is shorter but incomplete insofar as only the number of layers within a unit cell and the symmetry type are given. *H* stands for hexagonal and *R* for rhombohedral (e.g. 6 *H* or 15 *R*). *Jagodzinski* lists the sequence of layers in hexagonal or cubic environment (e.g. *hcc* or *hchcc*), *Zhdanov* counts the number of consecutive layers in cubic arrangement (e.g. 33 or 23). Fig. 2 shows the models and Table 1 lists the more common

---

* Polytypism cannot be regarded as a special one-dimensional polymorphism since different polytypes appear to exist at the same temperature and pressure whereas in polymorphic modifications each one has its own range of thermodynamic stability.

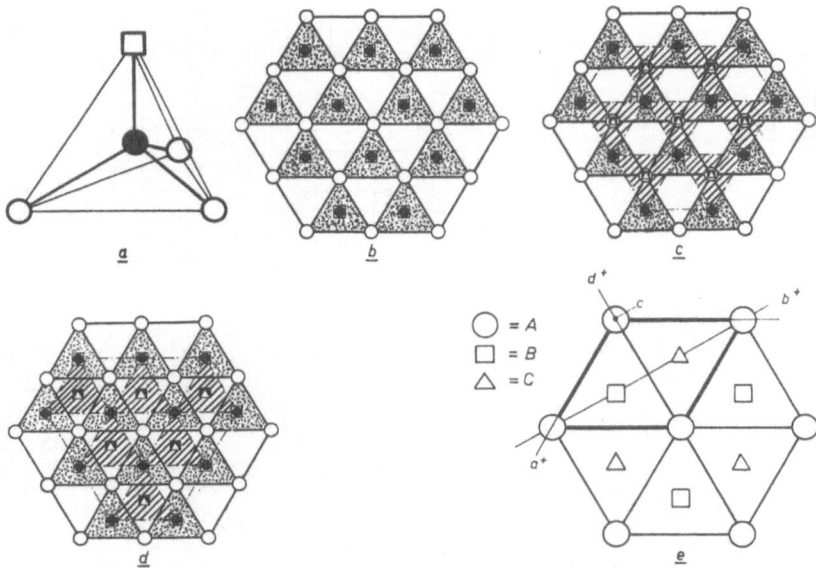

Fig. 1 a–l. Stacking of silicon and carbon atoms in SiC: a) Elementary Si-C tetrahedron. b) Projection of one tetrahedral layer. c) Projection of two adjacent tetrahedral layers of the wurtzite configuration. d) Projection of two adjacent tetrahedral layers of the zinc-blende configuration. e) Projection of the different positions of the atoms. The axes-system and the base of the unit cell are indicated. – Position of atoms in the {11$\bar{2}$0} plane: f) Relative position of Si and C atoms. g) Position of Si atoms in 3 C (wurtzite). h) Position of Si atoms in 2 H (zincblende). i) Position of Si atoms in 15 R. k) Position of Si atoms in 6 H. l) Position of atoms in 4 H. Open symbols = Si, full symbols = C-atoms.

polytypes in different notations (including the old historical ones) in order to facilitate the identification and comparison. For our purpose the *Ramsdell* symbol will suffice.

The determination of the structure of SiC polytypes requires a Weissenberg or a precession camera. Precession camera pictures are especially informative and easy to interpret since they image the reflecting planes in reciprocal space. Thus they can readily be used to construct *Brillouin* and *Jones* zones. For larger unit cells the resolution must be very high. If the sole purpose of the X-ray analysis is the identification of polytypes, Laue pictures are sufficient, once a catalog of the more common polytypes has been assembled.

In all modifications any Si or C atom has the same first and second coordination. Every atom has four first-neighbors of the other kind and twelve second-neighbors of the same kind. Among the polytypes with a larger number of layers per unit cell, two or more may have struc-

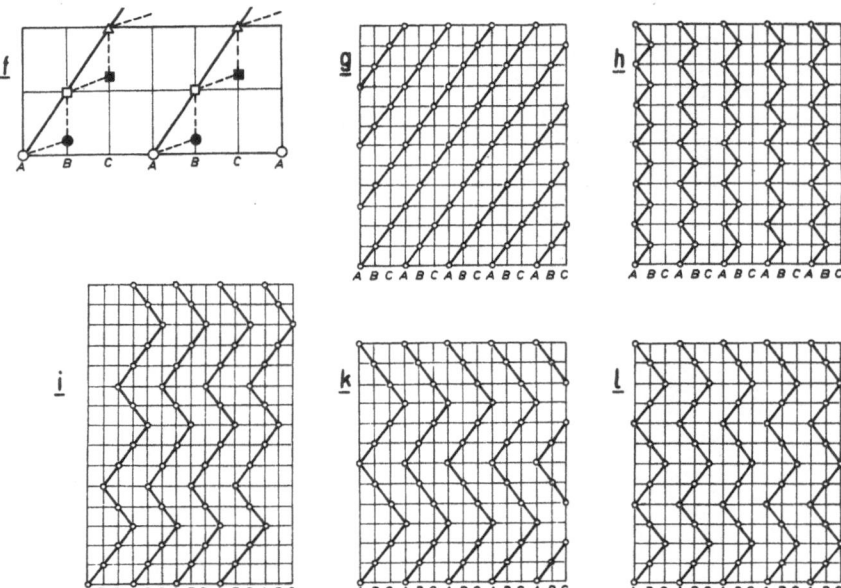

Fig. 1 f–l

tures in which most of the atoms have identical environment up to very distant coordinations since the difference between two polytypes is often due to a few periodic stacking faults. Thus the differences in internal energy of polytypes are negligible and one might assume that disordered structures would prevail. It is true that frequently streaks, caused by disorder along the c-axis, have been seen on X-ray pictures. *Golightly* and *Beaudin* [23] have observed disordered layers of up to 0.4 mm thickness in their samples. However, the bulk of the crystals consists of one or more well ordered polytypes with sometimes very thin disordered layers in between. The occurrence of sandwiches containing several polytypes in epitaxial relation is referred to as coalescence.

Many theories have been put forward to account for the unexpected long-range order because energetically it is not favored over complete randomness in the stacking sequence. The first explanation based on definite principles of crystal growth was given by *Frank* [24]. The concept of his theory relies on the screw-dislocation mechanism of crystal growth and states that the crystal gets its memory from the spiral growth around screw-dislocations. Different Burgers vectors create different polytypes. His ideas were tested by various workers who established in many cases a direct relation between the step height of the growth spiral and the height of the unit cell. *Jagodzinski* [25] on the other hand de-

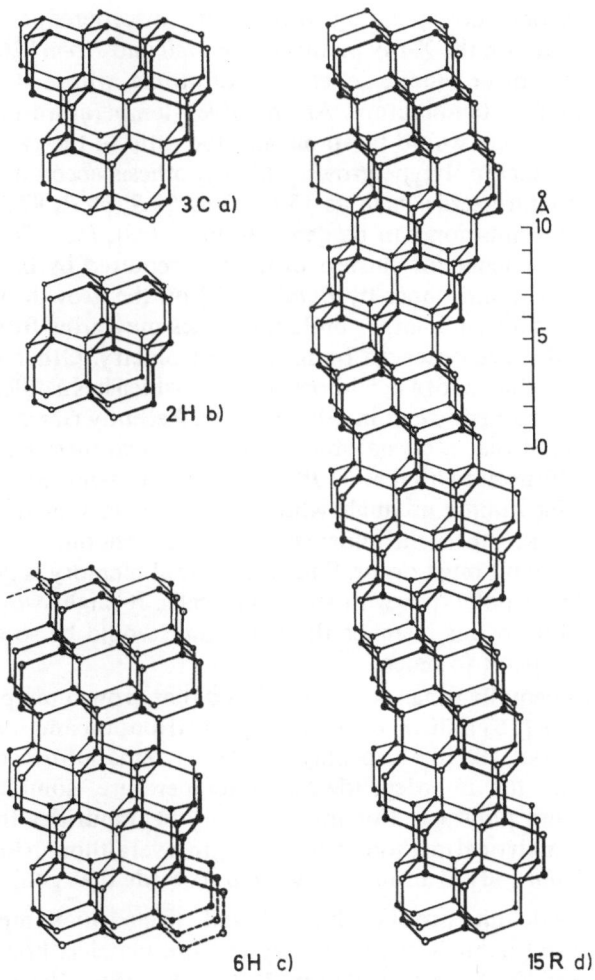

Fig. 2a–d. Models of some SiC polytypes: a) Cubic SiC $= 3\,C$. b) Zincblende structure $= 2\,H$.
c) $6\,H$. d) $15\,R$

veloped a theory based on thermodynamic considerations although at
first sight a formation of an ordered rather than a completely disordered
structure, in one dimension, seems anomalous. In essence he calculates
the total entropy as the sum of the configurational and the vibrational
entropy. The first one increases with rising disorder while the second one
decreases. The sum may therefore exhibit two maxima when plotted
against degree of disorder, the first giving rise to an ordered structure,

the second one corresponding to the partly disordered structures (producing streaks on the X-ray picture). *Ramsdell* and *Kohn* [26] postulated the presence of polymers, i.e. clusters of atoms, whose stability would be governed by temperature. At any given temperature no more than two SiC polymers would be stable and the ratio of the two would then determine which polytype grows. This hypothesis accounts for the well known structure series such as 15 *R*, 33 *R*, 51 *R*, 69 *R*, 87 *R*, etc. which in *Zhdanov*'s notation can be generalized to (33)$_n$ (32). There is no experimental evidence for clusters of the size required by this speculative theory but a recent paper by *Bulakh* [27] on the growth of CdS could be a hint of how such larger building blocks may grow from molecular species of the size observed by mass spectrometry. Other theories emphasize the influence of a neighboring crystal which has a different orientation, a very frequent event in a Lely cavity. Recently *Gomes de Mesquita* [28] observed that a crystal may provide its own memory in a similar manner. He noticed that a 33 *R* specimen was twinned with the two *c*-axes inclined under an angle whose tangent equals the *a/c* ratio. Thus one of the twins stores the information for the other one where the periodic stacking fault must occur. Such an internal memory is geometrically impossible for polytypes with small unit cells. It might work for higher polytypes but in this instance the twin angle would be so small that it would be difficult to detect.

Experimentally it has been established that growth temperature [29] and pressure [15] influence the polytype distribution and it is very likely that impurities [8, 15] also play a role. Systematic investigations are hindered by the interdependence of temperature, solubility of trace elements, and polytype. For many important impurities there are still no proven analytical methods below the ppm level although the pioneering work of *Kuin* [30] is an important step towards that goal.

There is definite proof of the profound effects which impurities have on the crystal habit. Usually the crystals are platelets bounded by the basal planes (0001) and (000$\bar{1}$) but *Knippenberg* and *Verspui* [8] found that adding certain elements (e.g. lanthanum) to the SiC charge leads to the formation of ribbonlike crystals or whiskers. *Shaffer* [15] noticed that heavily boron doped crystals tend to grow as fairly well developed polyhedra.

4. The Growth of Single Crystals

The unique properties of SiC entail very special growth techniques. Well known methods for growing and purifying other single crystals such as Czochralsky growth and zone refining are not feasible because

SiC sublimes at 2830° C even under pressures of up to 35 atm [10]. Three basic processes can be distinguished:
1. growth by sublimation.
2. pyrolytic decomposition of silicon and carbon containing gases,
3. growth from the melt.

4.1 Sublimation Techniques

So far the most successful methods are closely related to *Lely*'s [6] sublimation technique. SiC, or silicon in the presence of graphite, is heated to such a temperature that sublimation occurs or that a high partial pressure of silicon is maintained in the growth chamber. At the colder parts of the system SiC crystals are formed, usually as hexagonal platelets. In the original *Lely* furnace a hollow cylinder of sintered polycrystalline SiC is heated to 2500° C (Fig. 3). The material at the hot outer wall sublimes and diffuses through the porous wall into the cavity where it crystallizes. If the axial temperature profile is sufficiently flat, the growing crystals are platelets whose faces are parallel to the top and bottom of the cylinder. This configuration is energetically favored because it allows the crystals to get rid of the heat of formation by radiating it to the colder ends of the chamber. Even in a small cavity of 40 mm inside diameter it is possible to grow crystals of 10 mm diameter and up to 1 mm thickness (Fig. 4). In general thinner crystals are more perfect than thicker ones which often exhibit a coalescence of several polytypes. The weak points of this method are the difficulty of controlling both the nucleation and the degree of supersaturation.

Kroko [31] has utilized the coloration of SiC when nitrogen is incorporated (cf. 6.3.4 Dichroism) to monitor the growth and to gain insight into its mechanism. At regular time intervals he admitted a burst of nitrogen to the growth vessel. This yielded crystals with dark green bands, comparable to growth rings in a tree trunk, but with equal spacings. From his experiment he concluded that the lateral growth rate is constant with time and further that the majority of the crystals nucleate at the same time.

The highest degree of refinement of the Lely method was achieved by *Kapteyns* and *Knippenberg* [32]. The central part of this furnace is a crucible of pyrolytic graphite which can be produced with extreme purity and has the added advantage of conducting heat very anisotropically. The chosen design leads to a very flat temperature profile in axial direction since the thermal conductivity along the walls is extremely high. On the other hand it is low perpendicular to them (radially). By heating the crucible in an R.F. field its outside temperature can be held

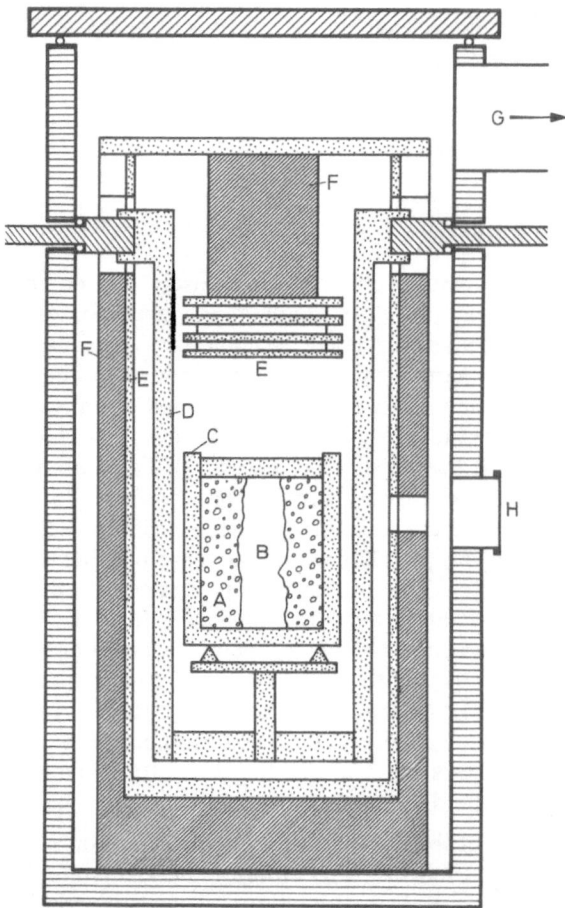

Fig. 3. Cross sectional view of Lely furnace (schematic): A sintered SiC mass, B growth cavity, C graphite crucible, D graphite heating element, resistively heated, E radiation shields made from pyrolytic graphite disks, F graphite felt for heat insulation, G port to vacuum pumps, H viewing port for temperature sensing element (pyrometer)

at 1600° C while the inside reaches the required 2500° C. The relatively low outside temperature permits the use of a water cooled quartz bell jar not much wider than the crucible. This in turn allows the R.F. coil to be placed outside the quartz cylinder and to eliminate all contamination from hot metal parts. Thus good vacuum and high temperature technology can be coupled together to produce the purest crystals made so far.

As mentioned before, the addition of lanthanum to a *Lely* furnace changes the growth pattern radically [8]. Instead of hexagonal platelets ribbonlike crystals grow in the cavity to fill it completely. Typical dimensions are 20 μm thickness, 0.5 mm width, 5 cm length. The growth direction is parallel to [10.0] or [31.0] in hexagonal coordinates. Coalescence of polytypes has been observed exactly as for the usual hexagonal platelets and the epitaxial relation is unchanged. The ribbons seem to be normal crystals exhibiting essentially one-dimensional instead of two-dimensional growth.

Fig. 4. Fragment of the sintered SiC body with crystals grown into the cavity. The largest crystal measures 6 mm on the edge. In the wall the channels are easily seen through which SiC diffuses into the cavity during the growth experiment

Whiskers in the proper sense grow at lower temperatures (1000 to 1800° C) when iron or other metal particles are present [9]. Their growth is dominated by the vapor-liquid-solid (VLS) mechanism advanced by *Wagner* and *Ellis* [33] to explain filamentary growth. Typical, but not always observed, is a little ball or globule at the tip of the whiskers grown by the VLS mechanism. This globule is often rich in the impurity causing the whisker growth. The dominant control parameters are temperature and silicon vapor pressure. The growth direction is [111] for cubic and [11.0] or [00.1] for hexagonal whiskers. Their cross section is circular for the low temperature runs, and polygonal for the higher temperatures.

4.2 Pyrolytic Decomposition

The pyrolytic decomposition of gases containing silicon and carbon does not yield crystals large enough for physical investigations or device work. The method is important, however, because it allows the synthesis of extremely pure feed material for the Lely process if such decomposable gases are chosen which can be purified by multi-stage distillation. Pyrolytic decomposition is also the only known way of producing SiC of the wurtzite structure (2 *H*) [34]. If the gases are decomposed in the presence of a suitably prepared single crystal, epitaxy may take place [35]. For its possible technological impact this aspect will be dealt with later in the context of device fabrication techniques (Section 7.2.2).

4.3 Melt Growth

The growth of SiC from a silicon melt was characterized by *Tanenbaum* [36]: "Those who have had personal experience with growing crystals can appreciate how distressing it can be to attempt to grow a crystal of semiconductor quality while the solvent is evaporating at a great rate and the crucible is liable to disappear at any moment."

A typical arrangement consists of a graphite crucible containing molten silicon. At the colder part of the crucible small and thin crystals are formed (maximum diameter some tenths of a centimeter, thickness $< 100 \,\mu$m). The main problem is the low solubility of carbon in silicon and its small change with temperature. Therefore molten alloys (e.g. Si-Cr, Si-Co) have been tried [37]. Unfortunately the solubility of the metals in SiC is prohibitively high for most applications. A newer variation of this method is liquid epitaxy [38] where large single crystals are exposed to a carbon-saturated silicon melt. If the temperature distribution and the geometrical arrangement of the seed crystals is properly chosen, epitaxial SiC layers of device quality are deposited.

4.4 Purity

Common to all growth processes is the difficulty of growing crystals pure enough for making semiconductor devices. Sources of contamination are the starting material (SiC or Si and C), the containers, and the growth atmosphere.

As mentioned previously very pure SiC can be produced by the pyrolytic decomposition of distillable compounds (e.g. silanes and hydrocarbons), extremely pure silicon is readily available and even very good graphite can be bought.

The crucible material contaminates the growing crystal unless it is high purity graphite itself for all other materials are attacked either because of the high temperature or react with silicon or carbon to release impurities into the system. Pyrolytic graphite has an advantage over normal graphite because of its higher density and lower specific surface area. Therefore degassing of the crucible is a lesser problem and contamination due to adsorbed impurities such as nitrogen is significantly lower.

Considering purity only, ultra high vacuum would seem to be the best growth atmosphere but it is recognized that in a Lely type system better crystals are grown under a pressure of about 1 atm. Extremely pure noble gases are required and the available purity of helium or argon is the limiting factor in the best systems. The nitrogen content of such inert gases leads to n-type SiC with 10^{14} to 10^{15} nitrogen atoms per cm^3 [13].

5. Trace Analyses in SiC

Unless special precautions are taken, SiC contains nitrogen, boron and aluminium. Other important impurities may be beryllium and elements of the iron group.

Without analytical tools capable of detecting the major impurities in the ppb to ppm range (10^{-9} to 10^{-6} impurities) any real progress in the preparation of semiconductor crystals is impossible. In this range three methods have been investigated and compared by *Kuin* [30]. They are mass spectrometry, emission spectroscopy of chemically pre-concentrated (enriched) samples, and thermal neutron activation analysis. The former two suffer from the contamination problem so that activation analysis which is not affected by contamination after irradiation is preferred where feasible. The detection limits of mass spectrometry, emission spectroscopy, and thermal neutron activation are compared in Table 2.

Beside the activation with thermal neutrons irradiation with fast neutrons, gamma rays, and charged particles can also be used. The analysis with charged particles poses additional problems and is much more difficult because charged particles interact very strongly with the matrix. This leads to a pronounced dependence of particle energy, and thus of reaction cross section, upon penetration into the crystal. The charged particles commonly used are protons (^1H), deuterons (^2H), helions (^3He), and to a lesser extent tritons (^3H) and alphas (^4He). *Kuin* [30] has checked the usefulness of proton activation analysis for the determination of boron and nitrogen in SiC since the three principal methods do not yield any results for those two pre-eminent impurities.

Table 2. *Limits of detection in SiC with M.S., E.S. and T.N.A.A., concentrations in p.p.b.* $(1 : 10^9)$ *by weight*

Element	M.S.	E.S.	T.N.A.A.	Element	M.S.	E.S.	T.N.A.A.
H	—	—	—	Ru	300	1000	0.2
He	—	—	—	Rh	100	1000	—
Li	—	1000	—	Pd	400	1000	0.02
Be	0.2	10	—	Ag	900	—	4
B	4	20000	—	Cd	1400	1000	0.2
N	—[c]	—	—	In	200	200	0.001
O	—	—	—	Sn	400	100	1
F	40	—	—	Sb	200	—	0.003[a]
Ne	100	—	—	Te	400	10000	3
Na	100	4000	0.02	I	100	—	0.02[a]
Mg	1200	1000	—	Xe	600	—	3
Al	100	500	—	Cs	800	1000	0.01
P	100	—	1000[a]	Ba	200	20000	0.02
S	400	—	200[a]	La	200	2000	0.01
Cl	200	—	10[a]	Ce	200	20000	2
Ar	800	—	0.2	Pr	200	10000	0.5
K	200	—	1	Nd	700	20000	45
Ca	1000	600	200	Sm	800	20000	0.02
Sc	200	1000	0.1	Eu	400	10000	0.002
Ti	400	300	—	Gd	800	4000	0.2
V	100	3000	—	Tb	200	2000	0.2
Cr	1300	2000	2	Dy	900	10000	0.2
Mn	100	100	0.001	Ho	200	1000	0.01
Fe	2800[b]	2000	200	Er	700	400	0.2
Co	200	1000	2	Tm	200	400	0.5
Ni	800	400	75	Yb	900	40	0.1
Cu	200	100	0.01	Lu	300	400	0.05
Zn	500	4000	0.3	Hf	800	1000	0.5
Ga	900	100	0.01	Ta	4500	20000	0.2
Ge	900	—	0.2	W	900	20000	0.005
As	200	—	0.1[a]	Re	500	100000	0.01[a]
Se	800	—	0.4	Os	800	2000	1
Br	400	0.05[a]	0.05[a]	Ir	500	30000	0.02[a]
Kr	200	—	1	Pt	1000	200	0.2
Sr	1000	4000	0.1	Au	300	200	0.001
Y	600	300	0.1	Hg	1200	200	0.1
Zr	300	1000	5	Tl	500	4000	20
Nb	200	10000	—	Pb	700	100	200
Mo	600	1000	0.1	Bi	400	400	20
Rb	200	—	20	Th	—	6000	0.2
				U	5000	12000	—

Due to the energy dependence of the reaction cross sections the ratio B : N can in principle, be determined if the sample is exposed to 5 MeV protons first and then to 14 MeV protons. At 5 MeV practically only boron is activated whereas at 14 MeV both impurities are transformed at about equal rates into ^{11}C. The results achieved in practice are not yet satisfactory and other methods are being investigated. These are primarily activation with other charged particles which lead to different radio-isotopes for the two impurities.

Radically different methods to determine impurity concentrations include electron spin resonance, nuclear magnetic resonance, and the measurements of electronic properties related to impurities such as Hall analysis or photoelectronic experiments. All of them are not absolute methods and are less specific since they depend on the interaction of the impurities with the host lattice. The most widely used successful technique is the Hall analysis to be described later (Section 6.2.2). In the following the results of the resonance methods will be summarized. Electron spin resonance studies were performed by *van Wieringen* [39], *Woodbury* and *Ludwig* [40], *Hardeman* and *Gerritsen* [41] and *Veinger* [42] and co-workers, to name a few. Although important insights have been gained, paramagnetic and nuclear magnetic resonance techniques have not been developed to an accepted analytical tool for determining impurity concentrations. Most experiments have been performed on rather heavily doped crystals and have had other objectives.

Detailed analyses have been restricted to nitrogen and boron doped SiC. Both impurities substitute for carbon, the electron wavefunction at the N nucleus having mainly *s*-character and the hole at the B site predominantly *p*-character. Since the nitrogen and the boron ions fit almost perfectly on a carbon site, one would expect a similar substitution of aluminum and phosphorus onto silicon sites, but experiments could not confirm this possibility and it is not clear what causes the lack of observability. One important experimental finding is the occurrence of a number of sets of hyperfine lines, explained by the crystallographic inequivalency of the carbon sites. This can easily be visualized by looking at the model of a certain polytype (Fig. 5). In 6 *H* for instance all carbon atoms have a nearest silicon neighbor at a distance 3 *d*. Along the *c*-axis the next nearest silicon neighbor is at 5 *d*, 9 *d*, or 11 *d* depending upon

---

[a] Special chemical treatment needed.
[b] Fe in coincidence with Si_2^+.
[c] N-determination impossible because coincidence with Si^{2+}.
— = not detectable with this method.
M.S. = mass spectrometric analysis.
E.S. = emission spectroscopic analysis.
T.N.A.A. = thermal neutron activation analysis.

the type of site. Thus 6 *H* has three equally abundant but distinct sites which manifest themselves in certain semiconducting and optical properties discussed later.

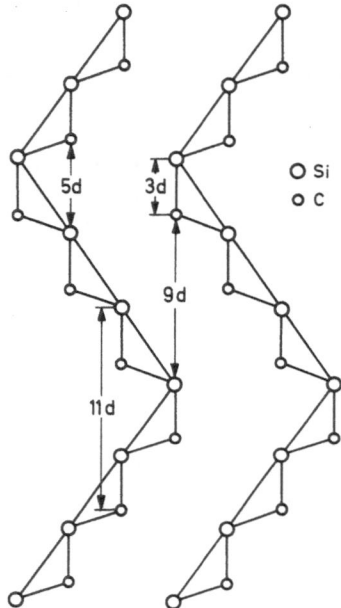

Fig. 5. Diagram showing the three inequivalent sites in 6 *H* (viewed in a {11$\bar{2}$0} plane)

6. Semiconductor Properties

The progress which has taken place during the past decade is caused by two facts. First, refined methods of growing single crystals, the availability of purer materials and of better analytical tools led to much purer and more perfect crystals. Second, it was realized that certain properties are not properties of SiC as such, but the properties of a certain polytype of it. Therefore older published data are incomplete and of dubious value unless a polytype determination was made. Such an analysis is an essential part of most experiments if the results are to be meaningfully interpreted. At the same time it may serve as a check to confirm that one is really dealing with a true single crystal and not a mixture of polytypes in epitaxial coalescence.

    In the following paragraphs an attempt is made to seperate phenomena which depend weakly on polytypism from those which are dominated by it. In addition to polytype effects the common phenomena due to doping, i.e. to impurities, also occur and will be discussed. Therefore it is felt that problems tackled mainly by one group of researchers should be presented as an entity even at the expense of disrupting the aesthetically pleasing classification according to dependence on poly-typism.

6.1 The Large Zone Concept

Most SiC polytypes have large unit cells and hence small Brillouin zones. A comparison of polytypes is facilitated by the use of large complete zones [43a] such that their volumes in k-space are the same for all poly-types. For a polytype with N silicon and N carbon atoms per unit cell (i.e. an NH or NR polytype in the Ramsdell notation) the proper large zone can be reduced to N Brillouin zones. (For the construction of the large zone the reader is referred to the book by *Jones* [43a].) The axial dimension is found to be the same constant $N\pi/c$ for all polytypes, since c (the unit cell length along the symmetry axis) is proportional to the number of close-packed layers in the stacking sequence (i.e. N). For comparison with other polytypes it is convenient to regard cubic SiC as rhombohedral, i.e. polytype $3R$ instead of $3C$.

The large zone for higher polytypes is cut by many planes of energy discontinuity (Fig. 6) which are closely related to X-ray reflections [43b]. The extensive X-ray work can be useful to estimate the importance of these energy discontinuities since the X-ray intensities are in general a good measure of the size of them. Fig. 6 shows half of a large-zone mirror-plane with the triangular steps or cut-outs in the zone boundary of the rhombohedral polytypes on one hand, and the smooth zone boundary on the other hand, characteristic and identical for all hexagonal polytypes. Beside the heavy lines indicating the zone boundary, there are weak oblique lines and broken horizontal lines. The network of oblique lines corresponds to observable X-ray reflections whereas the broken horizontal lines correspond to X-ray reflections which are too weak to be observed. If, for calculating structure factors, one uses a single scattering factor for all silicon atoms and another one for all carbon atoms and considers all interplanar spacings equal, the structure factors corresponding to the broken lines of energy discontinuity vanish (in hexagonal polytypes they vanish identically for certain planes due to the hexagonal symmetry). Thus the discontinuities are very small because the true structure differs only slightly from the ideal tetrahedral structure [44]. There is no such limitation for the oblique lines.

The large zone of any polytype exhibits a central region free of planes with large energy discontinuities, top and bottom planes $(000 \pm N)$ with equally strong discontinuities, and a network of oblique planes charac-

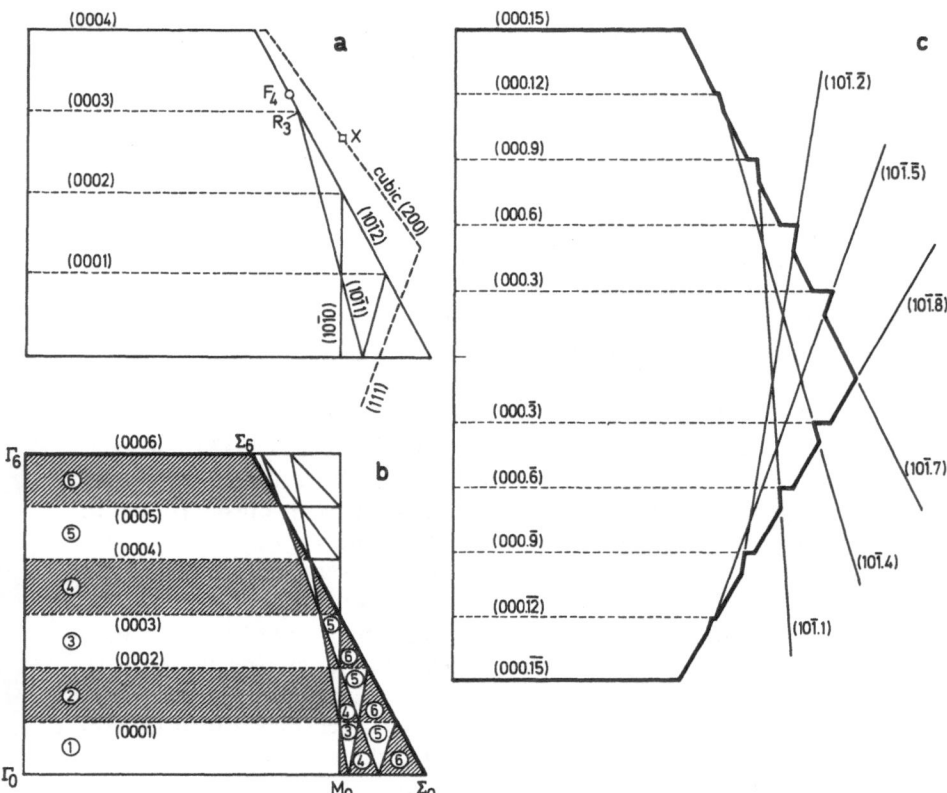

Fig. 6a–c. Comparison of the large (Jones) zones of some SiC polytypes. The figures show the mirror planes containing the conduction band minima. a) Comparison of mirror planes of the 4 H SiC large zone and the cubic Brillouin zone. Boundaries and energy discontinuities are indicated by Miller indices for cubic and by Bravais-Miller indices for 4 H SiC. A cubic conduction band minimum is at point X. The three lines marked (0003), (10$\bar{1}$1), and (10$\bar{1}$2) intersect at R_3. A position F_4, near R_3, is thought to be the location of a 4 H conduction band minimum. (Only one quadrant is shown.) b) A quadrant of a mirror plane in the large zone of 6 H. Application of Jones rule for the reduction of large zones shows that $\Gamma_0 - \Gamma_6 - \Sigma_6 - \Sigma_0$ is a quadrant of a complete large zone. The reduction to six Brillouin zones is indicated by the circled numbers. c) The 15 R large zone is a complete two-dimensional zone reducible to five Brillouin zones (only half the zone is shown in the figure)

teristic of the polytype. The only part of k-space showing pronounced differences between polytypes is therefore the large zone boundary and its neighborhood, excluding the top and bottom boundaries. This led *Patrick* [45] to the classification of properties according to their wave vector dependence:

1. zone boundary wave vector → large polytype differences,
2. axial or zero wave vector → small polytype differences,
3. weighted average over k → usually small differences, unless weighting emphasizes one particular set of wave vectors.

We shall refer to this classification while discussing the semiconductor properties of SiC. It shall suffice for the time being to point out the radically different behavior of n- and p-type SiC with regard to polytypism. This is a consequence of the above k-vector dependence since experiments have shown that the maximum of the valence band is at $k = 0$ and the minima of the conduction band are at or close to the zone boundary.

6.2 Hall Analysis

The most extensively used technique to determine semiconducting properties of SiC is the Hall analysis which yields carrier concentration and Hall mobility as functions of the temperature. Due to the shape of the crystals, usually thin platelets, the *van der Pauw* [46] method has had wide acceptance. At least four ohmic contacts on one surface are required which should be small in comparison to all other dimensions. A current is passed between two adjacent contacts while the potential difference between the other two contacts is measured. By cyclic permutation two independent pseudo-resistances R_1 and R_2 are obtained as voltage/current ratios. It has been shown by van der Pauw that between R_1 and R_2, the resistivity ϱ, and the thickness d of the platelet the following relation holds:

$$\varrho = [\pi d/(2 \ln 2)] (R_1 + R_2) f(R_1/R_2). \tag{1}$$

The function $f(R_1/R_2)$ has been given in graphical form. Since the ratio R_1/R_2 is a function of geometry only, a simple check is provided to exclude inhomogeneous samples by observing the temperature dependence of the ratio R_1/R_2.

The Hall constant R_H and the Hall mobility μ are determined, using the same contacts, by passing a current I_3 through two nonadjacent contacts. The change in voltage is measured across the other two contacts when a magnetic field is applied perpendicular to the sample. The appropriate formulas are

$$R_H = V_H d/(B I_3), \tag{2}$$

$$\mu = R_H/\varrho = V_H (B I_3 \varrho/d)^{-1} \tag{3}$$

with the Hall voltage V_H at the magnetic induction B when a current I_3 is passed through two opposite contacts.

To reduce the error introduced by the use of contacts of finite size, four incisions may be made into the sample giving it the shape of a four-leaf clover.

6.2.1 Mobility

An extensive analysis of mobility data of *n*- and *p*-type crystals was undertaken by *van Daal* [47]. He concluded that the temperature dependence of the mobility in hexagonal SiC above 400° K was due to the scattering of the charge carriers by polar optical modes as well as by non-polar modes. In the latter case the ratio between the optical and acoustical deformation potentials amounts to about 2.5. The contribution to the total scattering effect from polar acoustical modes is very weak and will only be perceptible in pure samples at quite low temperatures. By taking into account the effect of ionized impurity scattering the mobility of charge carriers can be explained throughout the whole temperature range from 78 to 1300° K. He found a great similarity between the hole mobility in cubic and in hexagonal SiC. The electron mobility determined by lattice vibrations turned out to be significantly higher in cubic than in hexagonal SiC. Since effective masses and deformation potentials are not known, van Daal and others have used them as adjustable parameters in fitting the mobility data. This leads to a certain ambiguity and more recent data suggest that polar modes are only important at high temperatures. *Patrick* [48] concludes from a comparison with silicon that in relatively pure *n*-type samples a combination of acoustic and intervalley scattering limits the mobility between 300 and 800° K. At lower temperatures ionized and neutral impurity scattering seem important, at higher temperatures polar-optical scattering. It is impossible to make a unique fit to the experimental data because there are too many free parameters even if only the "principal" phonons (with wave vector in the large zone) are taken.

The experiments of *Barrett* and *Campbell* [49] have shown that electron mobility is polytype dependent. This could be exploited to determine the scattering mechanism by comparing polytype mobility ratios. In a temperature range dominated by polar scattering, for instance, this ratio would give immediately the mass ratio. Since in the temperature range investigated intervalley and acoustic scattering limit the mobility, not even the mobility ratio provides new clues as effective masses and coupling constants combine to determine this ratio. If the experimental range were wide enough to include two distinct regions dominated by different scattering mechanisms, a change of the mobility ratio could be found. No such change was observed up to 800° K.

The high electron mobility of cubic SiC already noted by van Daal is due to selection rules which severely restrict intervalley scattering [50]. Only longitudinal acoustic (LA) phonons are permitted and these have such a high energy (79.4 meV) that they are unimportant at room temperature.

Due to the crystal habit of SiC no measurements have been made of the anisotropy of the Hall constant. Crystallographically good samples seldom exceed a thickness of $100 \, \mu$m. It has been frequently observed (e.g. *Bosch* [51]) that less perfect platelets contain internal barriers which simulate ratios of the resistivity parallel and perpendicular to the c-axis of 100 to 1000.

6.2.2 Conductivity and Carrier Concentration

Electrical conductivity is more difficult to compare with theory. Due to the high ionization energies of donors and acceptors (typically 0.1 to 0.5 eV) the temperature dependence is dominated by carrier concentration, e.g. the crystal purity, since intrinsic SiC has not yet been grown. The second factor entering conductivity is mobility which changes much less rapidly with temperature than carrier concentration (a power law against an exponential dependence) but depends, as we have seen, also on purity, crystallographic perfection, and for n-type SiC on crystal symmetry.

The electron (or hole) concentration in n- (or p-)type SiC can be described in general quite accurately by a model containing a single donor (or acceptor) level compensated by acceptors (or donors). This model was discussed by *de Boer* and *van Geel* [52] and is governed by the equations:

$$\frac{n(n + N_A)}{N_D - N_A - n} = N_C/g_D \exp(-E_D/kT), \qquad (4)$$

$$\frac{p(p + N_D)}{N_A - N_D - p} = N_V/g_A \exp(-E_A/kT). \qquad (5)$$

A plot of the logarithm of the carrier concentration vs. the inverse of the absolute temperature yields in general three distinct regions. (In order to be definite, we restrict ourselves in the following to n-type SiC.) At high temperatures the carrier concentration saturates due to the exhaustion of the donor level, yielding $n = N_D - N_A$. At low temperatures an exponential rise is observed, the slope of which is $(-E_D/k)$, characteristic of compensated semiconductors. Under proper conditions, i.e. when $N_D \gg N_A$ and N_C', where $N_C' = N_C/g_D \exp(-E_D/kT)$, an intermediate range appears with a logarithmic slope of $(-E_D/2k)$ which is well known from other semiconductors where compensation is generally unimportant. From such an analysis important data can be gained: the energetic position of the donor level in the forbidden gap, the concentration of the excess donors $(N_D - N_A)$, and under favorable conditions independently the donor concentration N_A. Since nitrogen is the dominant impurity

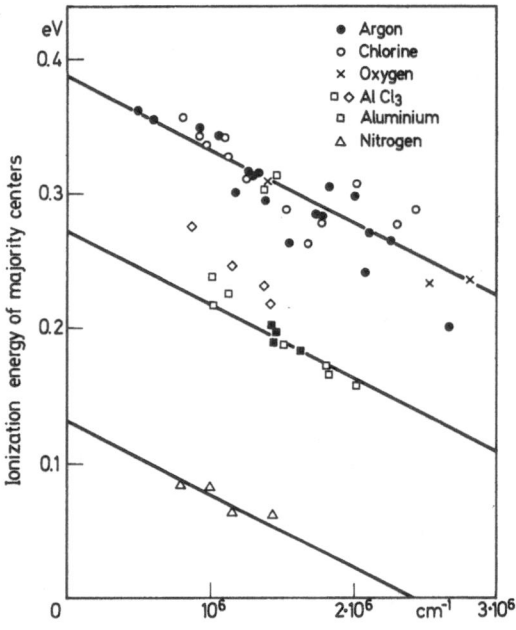

Fig. 7. Experimental data collected by *van Daal et al.* [55] showing the dependence of the ionization energy of majority centers on the reciprocal mean distance of the minority centers. The two upper lines refer to *p*-type samples, the lower line to *n*-type samples. The *p*-type dopant of the top line could not be positively identified, it could be boron. The dopant responsible for the second line is aluminum, the one for the lowest line nitrogen

in *n*-type SiC and it is so far impossible to determine its concentration radiochemically at the level in question (below 10^{18} per cm³), Hall analysis is the most widely used method to check nitrogen doping.

Departures from the above model have been observed around liquid nitrogen temperature [53] and were interpreted as resulting from a second unidentified shallow donor. Purer samples fail to show the saturation typical for extrinsic semiconductors at high temperature (before intrinsic conduction sets in). Instead a second steep rise appears [54] which is probably associated with a deep donor obscured in less pure samples. Thus at least three different donor levels have to be considered to explain the temperature dependence of the electron concentration. Due to the wide energy gap of SiC and the high ionization energies of impurities in SiC the quasi-Fermi level can move through a broad energy range and high temperatures are required to achieve this. The initial plateau ascribed to a shallow donor is observed at liquid nitrogen temperature, the saturation of the nitrogen level starts above 500° K, and the final steep rise around 1000° K. The extension of the measuring range is hampered by the very high resistance of the crystals below 77° K and the instabilities above 1500° K. At 2000° K SiC begins to dissociate, i.e. silicon evaporates.

The hole concentration in p-type SiC was carefully analyzed by *van Daal et al.* [55] and, in contrast to nitrogen doped crystals, its temperature dependence could be adequately described by Eq. (5), i.e. one compensated acceptor was dominant. Comparing samples of different purity and degree of compensation N_A/N_D it was found that the ionization energy of the majority centers was a function of the minority center concentration (Fig. 7) in accord with the theory of *Debye* and *Conwell* [56]:

$$E_A = E_{A0} - 2.66 \, e^2 \varepsilon_0^{-1} N_D^{1/3}. \tag{6}$$

6.3 Properties Related to Zone Boundary Wave Vectors

In this chapter we are going to discuss properties which show a pronounced dependence on polytypism since the zone boundary is determined by the complex network of energy discontinuities characteristic of the polytype. The dominant feature is the variation of the indirect bandgap with polytype due to the position of the conduction band minimum at or close to the zone boundary.

6.3.1 Bandedge Absorption and Exciton Luminescence

Research performed at the Westinghouse Research Laboratories during the past decade has elucidated many aspects of polytypism and its influence on bandstructure. The two basic experiments which have brought forward a great wealth of information are low temperature absorption measurements at the bandedge and low temperature excitation of luminescence by ultraviolet radiation. A summary of the exciton luminescence data in comparison to other semiconductors has been compiled by *Dean* [57].

The absorption spectra of all SiC polytypes are characteristic of indirect transitions which create excitons (electron-hole pairs). In indirect transitions the conservation of momentum requires the emission or absorption of one or more phonons. A simple theory of indirect absorption yields the formula [58]

$$\alpha = \sum_i A_i [h\nu - E_G - k\theta_i(k_{CB})]^2 + B_i [h\nu - E_G + k\theta_i(k_{CB})]^2. \tag{7}$$

The first term is due to phonon emission and the second one to phonon absorption. When the square root of the absorption coefficient is plotted versus photon energy, fairly distinct kinks close to the bandgap are observed which are the result of phonon emission (phonon absorption occurs only at higher temperatures when the states are populated). Resolution is not sufficient to identify all the phonons observed in the luminescence experiment but there are no inconsistencies. The phonons prominent in absorption are called "principal" phonons and are thought to be at the extended zone edge. They vary surprisingly little from polytype to polytype (with the exception of $2H$) if one considers the complex network of energy discontinuities close to where the conduction band minima are supposed to be (Table 3). Therefore the absorption curves can be shifted along the energy axis and, when multiplied by an appropriate

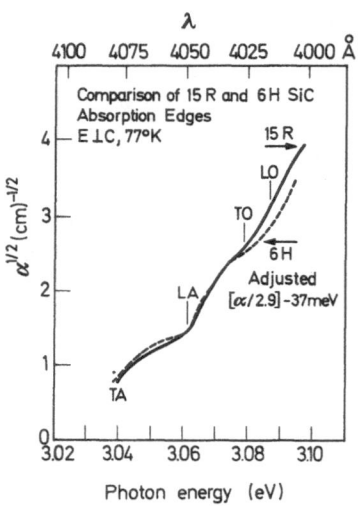

Fig. 8. Comparison of the absorption edges of $15R$ (solid line) and $6H$ (broken line) at 77° K. The abcissa and the ordinate refer only to the $15R$ curve. The $6H$ absorption coefficients have been divided by 2.9, and the $6H$ curve has been translated along the energy axis by 37 meV to bring it to near coincidence with the $15R$ curve

factor, be brought to near coincidence (cf. Fig. 8). This allows one to compare the absorption strengths which vary by a factor of ten among polytypes. Absorption curves for the electric vector parallel to the c-axis $(E \parallel c)$ have also been obtained but the structure is not as well resolved as for $E \perp c$. This latter arrangement is experimentally much easier to realize because SiC has the tendency to crystallize in thin platelets with the c-axis perpendicular to the large face.

The temperature dependence of the exciton energy gap E_{GX} was obtained from absorption measurements by observing, at low temperatures, the break in the curve which indicates the beginning of the phonon emission and, at higher temperatures, the break which indicates the onset of phonon absorption.

The phonon energies are much more accurately determined from luminescence spectra where one gets sharp peaks in contrast to the mere kinks observed in absorption. The combination of the absorption with the luminscence data gives the values of the exciton energy gaps E_{GX} plotted in Fig. 9. The independent variable in this plot, called hexagonality, is an empirical parameter with no theoretical justification for class 1 (zone boundary wave vector) properties. It is defined as the percentage of hexagonally stacked layers per unit cell and can most readily be obtained from *Jagodzinski*'s *hc* notation (cf. 3. Crystallographic Properties).

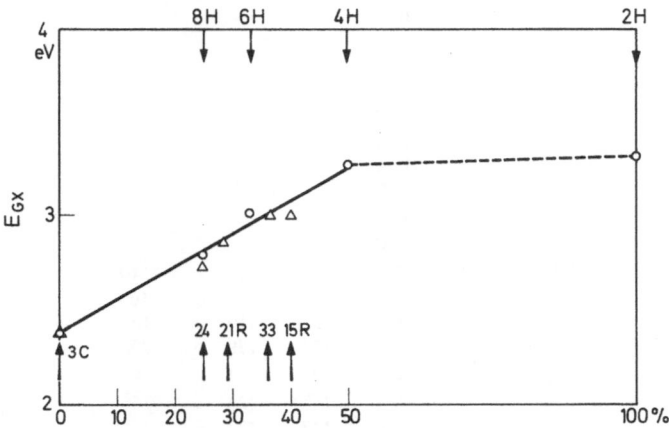

Fig. 9. Diagram showing the experimentally observed linear relation between exciton energy gap and percent hexagonal stacking up to 50% hexagonality

The photoluminescence spectra of even relatively simple polytypes are very complex and have not yet been unravelled completely. *Choyke, Hamilton*, and *Patrick* [59] succeeded in explaining the lines due to exciton complexes, but a more sophisticated understanding of the energy levels of donors and acceptors responsible for donor-acceptor pair spectra is needed to interpret this part of the spectrum. In this review the analysis of the spectrum due to exciton complexes will be illustrated, taking one of the more common polytypes, 15 R, as an example. The results from other polytypes are summarized in Table 3.

Table 3. *Phonon energies in eight polytypes (in meV)*

	TA$_1$	TA$_2$	LA	TO$_1$	TO$_2$	LO
3 C	46.3		79.5	94.4		102.8
24 R	46.0		77.3	94.4		104.5
21 R	46.5	53	77.5	94.5		104
15 R	46.3	51.9	78.2	94.6	95.7	103.7
33 R	46.3	52.3	77.5	94.7	95.7	103.7
6 H	46.3	53.5	77.0	94.7	95.6	104.2
4 H	46.7	$\begin{cases}51.4\\53.4\end{cases}$	$\begin{cases}76.9\\78.8\end{cases}$	95.0		$\begin{cases}104.0\\104.3\end{cases}$
2 H	52.5	61.5		91.2	100.3	103.4

Exciton energy gap, exciton binding energies, and estimated ionization energy of nitrogen donors

	E_{GX} (eV)	E_{4x} (meV)	E_{3x} (meV)	$E_{3x} - E_{4x}$ (meV)	E_i (meV)
3 C	2.390	10			
24 R	2.728				
8 H	2.80				
21 R	2.853	10.0			
		11.2			
		13.4			
		32.7			
		34.2			
		40.0			
15 R	2.986	7	125	118	140
		9	148	139	160
		19	158	139	160
		20	199	179	200
33 R	3.003	—	171	171	190
		9.2	143	134	150
		9.6	143	133	150
		11.2	142	131	150
		15.8	180	164	180
		21	180	159	180
		22.5	180	158	180
		23.4	216	193	220
		23.9	216	192	220
		25.7	220	194	220
		33.2	—	197	230
6 H	3.023	16	162	146	170
		31	203	172	200
		33	237	204	230
4 H	3.265	7			
		20			
2 H	3.330	10			

In 1958 *Lampert* [60] postulated "effective-mass-particle complexes" analogous to either molecular or polyelectronic complexes. Table 4 lists such complexes and their atomic analog, using *Lamperts*'s as well as *Patrick's* shorter symbols.

Table 4. *Effective-mass-particle complexes*

Particle or complex	*Lampert's* notation	*Patrick's* notation	Atomic analog
electron	—	e	electron e^-
hole	+	h	positron e^+
exciton	+ −	x	positronium $e^+ e^-$
donor ion	$\oplus$	①	H^+
un-ionized donor	$\oplus -$	②	H
exciton bound to donor ion	$\oplus + -$	③	H_2^+ if $m_h \gg m_e$
electron bound to un-ionized donor	$\oplus - -$		H^-
hole-exciton complex	+ + −		$\begin{cases} H_2^+ \text{ if } m_h \gg m_e \\ e^+ e^+ e^- \text{ if } m_h = m_e \\ H^- \text{ if } m_h \ll m_e \end{cases}$
exciton-exciton complex	+ + − −		$\begin{cases} H_2 \text{ if } m_h \gg m_e \\ \quad \text{ or } m_h \ll m_e \\ e^+ e^+ e^- e^- \text{ if } m_h = m_e \end{cases}$
exciton bound to un-ionized donor	$\oplus + - -$	④	$\begin{cases} H_2 \text{ if } m_h \gg m_e \\ H^- e^+ \text{ if } m_h = m_e \end{cases}$

Luminescence due to the decay of four-particle complexes was first observed by *Haynes* [61] in silicon which is also a semiconductor with indirect, phonon-assisted transitions. In materials with direct transitions such complexes have also been found (e.g. in CdS by *Thomas* and *Hopfield* [62]).

The atomic analogs are typical, but not the only ones possible, since it is known that muons (μ-mesons) can replace electrons in atoms. Many of the above mentioned atomic complexes have been theoretically treated and estimates about binding energies and mean separations have been computed.

The SiC samples of *Choyke et al.* [59] were n-type with nitrogen as main impurity and other compensating impurities in various concentrations. Ultraviolet radiation creates excitons which can be captured by neutral nitrogen to form a four-particle complex or by a singly ionized nitrogen donor to form a three-particle complex. The complexes decay by electron-hole annihilation, emitting photons and, in most cases, one or more phonons. Fig. 10 shows a possible level diagram of the complexes

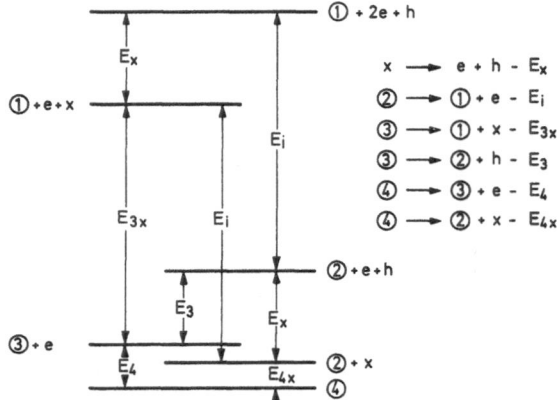

Fig. 10. Energy-level diagram and decay reactions of effective-mass-particle complexes. The levels represent possible relative energies of various combinations of the particles which form a four-particle complex. The complex itself is the lowest level, being the most strongly bound combination. The top level represents the completely separated particles

observed in SiC and lists the six decay reactions. The six equations are not all independent. E_{4x} and E_{3x} are directly measured in the experiment and permit to calculate a lower limit and a "probable" value of the ionization energy of nitrogen in silicon carbide polytypes according to the equation:

$$E_i = E_{3x} - E_4 + E_4 > E_{3x} - E_{4x} \tag{8}$$

provided both the four- and the three-particle spectrum are recorded.

The four-particle spectrum results from the decay of the exciton bound to the un-ionized donor into a free exciton and a neutral donor, the three-particle spectrum from the decay of the exciton bound to a donor ion into a free exciton and a singly ionized nitrogen donor. Electron spin resonance experiments have shown [40] that nitrogen substitutes in SiC on C-sites which are not all equivalent. Since there are $N/2$ inequivalent sites in a hexagonal polytype NH and $N/3$ inequivalent sites in a rhombohedral NR, we expect as many different lines in the photoluminescence.

The decay of the complexes may take place without the emission of a phonon although the hole has $k = 0$ and the electron $k = k_{CB}$ since the exciton is localized. This recoilless recombination entails extremely sharp lines (halfwidth of about 0.2 meV). In other transitions crystal momentum is preserved by the emission of one or more phonons. Thus each phonon-free line is accompanied by a series of lines due to the emission of different phonons. Table 5 lists the number of possible exciton lines resulting from these multiplicative effects of crystal symmetry.

Table 5. *Multiplicity of phonon lines*

Polytype	Atoms of C and Si in the unit cell	Number of phonon branches	Inequivalent sites in the unit cell	Bound exciton lines	
				no-phonon	one-phonon
3 C	2	6	1	1	6
8 H	16	48	4	4	192
24 R	16	48	8	8	384
21 R	14	42	7	7	294
15 R	10	30	5	5	150
33 R	22	66	11	11	726
6 H	12	36	3	3	108
4 H	8	24	2	2	48
2 H	4	12	1	1	12
N R	$2N/3$	$2N$	$N/3$	$N/3$	$2N^2/3$
N H	$2N$	$6N$	$N/2$	$N/2$	$3N^2$

In the four-particle spectrum of 15 R only four out of the five possible no-phonon lines were observed, labeled P_0, Q_0, R_0, and S_0, displaced by 7, 9, 19, and 20 meV from E_{GX}. The fifth line may be missing because E_{4x} is too small to bind an exciton at 6° K or too large to produce an observable intensity. (It was noted by *Patrick et al.* [63] that the relative capture cross sections of the inequivalent sites decreases with increasing E_{4x}.) Each of the four no-phonon lines is followed by a series of lines due to the phonon emission. Since the binding energies E_{4x} vary significantly from site to site, it is possible to identify the lines belonging to each series by observing the temperature dependence of their intensities. The relative intensities are $I(P) > I(Q) \gg I(R) > I(S)$. Measuring the displacement of the other P-lines from P_0 gives the energies of 18 phonons* with much greater accuracy than the absorption data mentioned earlier. The combination of the two measurements yields a value for the exciton energy gap $E_{GX} = 2.986$ eV at 6° K which is smaller than the usual energy gap E_G by the still unknown exciton binding energy E_X.

At 20° K the P- and Q-series disappear because of the small binding energies E_{4x}. On the other hand duplicate R- and S-lines appear which are interpreted as thermally excited states of the four-particle complex. They are displaced by 4.8 meV and attributed to the presence of a second valence band (Γ_5) split off by spin-orbit interaction. No sign of a third valence band (Γ_1) due to the crystal field splitting could be observed. However, it is thought to be present in analogy to ZnO and CdS which

---

* It is quite possible that the other 12 theoretically possible phonon lines could not be found because not all multiple peaks could be resolved. Within the experimental error the phonon energies are the same for all the series P, Q, R, and S.

have been analyzed by *Hopfield* [64]. The data suggest a minimum value for the crystal field splitting of 30 meV but it could be much larger.

The three particle spectra (due to excitons bound to ionized donors) can still be observed at higher temperatures since the binding energy E_{3x} is about an order of magnitude higher than E_{4x}. Thus the four-particle spectrum is already quenched when the three-particle spectrum still has high intensity. The higher binding energy E_{3x} also leads to a stronger interaction with the lattice and as a consequence to multi-phonon emission. The intensity of the three-particle spectrum is higher (in comparison to the four particle spectrum) in more strongly compensated crystals because, for a given donor concentration, the fraction of ionized donors is the larger the more compensating acceptors are present (see 6.2 Hall Data).

Again only four of the possible five series of lines could be observed, called A, B, C, and D. Each series has a narrow, phonon-free line, and a number of broader lines displaced towards lower energies. The lower the energy, the broader the peaks get due to the overlapping of the four series and due to the appearence of multiphonon bands. The phonon-free lines yield binding energies E_{3x} of 125, 148, 158, and 199 meV (compared to E_{4x} between 7 and 20 meV). Again the lines of the series A and B, the ones with the lower binding energies, are an order of magnitude more intense than those of the series C and D. Resolution is good enough in the series A and B to obtain the phonon energies. Except for some fine structure the spectra are the same and do not differ from the ones of polytype 6 H. Fine structure of the series (A, B, and D consist of doublets) and excited states observed at higher temperature (48° K) are interpreted as Jahn-Teller and valley-orbit splitting respectively and are found in other polytypes of SiC, too.

Taking the difference between E_{3x} and E_{4x} results in a minimum ionization energy of the nitrogen donor according to Eq. (8). An estimate of E_4 is given by *Hamilton et al.* [65]. The value E_i calculated by inserting this estimated E_4 is the "probable" ionization energy of the donor.

Table 3 summarizes the results collected by the Westinghouse group on the polytypes 2H, 4H, 6H, 33R, 15R, 21R, and 3C and comparable data on 24R published by *Zanmarchi* [66]. All Westinghouse data are the result of a detailed analysis such as the one described above, whereas *Zanmarchi* deduced his phonon energies from a least square fit to absorption data taken over a broad temperature range.

No three-particle spectra could be obtained in 2H, 21R, and 3C. The three-particle spectrum observed in 4H is thought not to originate from nitrogen complexes. The reason for the absence of three-particle complexes may be the vanishingly small binding energy E_3, i.e. in certain polytypes the excitons are not bound to the nitrogen ions.

6.3.2 Broad Band Luminescence

Photoluminescence in SiC is excited by ultraviolet radiation as is exciton luminescence, whereas electroluminescence is due to the recombination of carriers injected through a contact. The latter therefore requires a $p-n$ junction. The technology of making junctions will be described below (Section 7.2.1) since the phenomenon of injection luminescence can be treated without a detailed knowledge of the structure of the junction. Photoluminescence and electroluminescence have in common the property that minority carriers, either created by illumination or by injection, recombine and as a consequence emit visible light. In electroluminescence two cases have to be distinguished, emission under forward bias which has a spectral dependence similar to that of photoluminescence and emission under reverse bias. The latter is ascribed to impact ionization and occurs at wavelengths corresponding to the energy gap. It will not be dealt here since little effort has been devoted to its understanding in SiC.

The investigations center on the influence of dopants and polytypes on the wavelength of the emitted radiation and on the recombination mechanism itself. In the case of electroluminescence the injection mechanism has to be studied too. In contrast to the line spectrum observed in exciton luminescence, the radiation considered here is in general emitted in a rather broad band*. This is not restricted to SiC and can be explained by the interaction of the impurity centers and, at room temperature, by thermal line broadening. The broad spectrum resulting from transitions with photon energies considerably smaller than the bandgap is due to donor-acceptor pair spectra which cannot be resolved at high doping levels. Another band, giving rise to blue luminescence, is only observed in heavily doped samples whereas more lightly doped crystals yield line spectra in the same spectral range, thus proving the interaction of like centers with each other.

The details of the recombination and the levels involved have not yet been worked out although many different techniques were employed to throw light on it. Most frequently electroluminescent and photoluminescent spectra and their variation with temperature have been compared. Also the influence of the kind of dopants and of the doping concentrations on the spectra has been studied. Additional experiments include Hall measurements [68], absorption [70], photoconductivity, decay analysis of luminescence and photoconductivity [71], quenching by light and temperature [78], thermoluminescence [72], etc.

The main scientific result of all these investigations is the observation of a pronounced polytype effect consisting of a one to one correlation

---

* *Brander* [38] and *Gorban* [67] have observed line spectra in aluminum doped crystals.

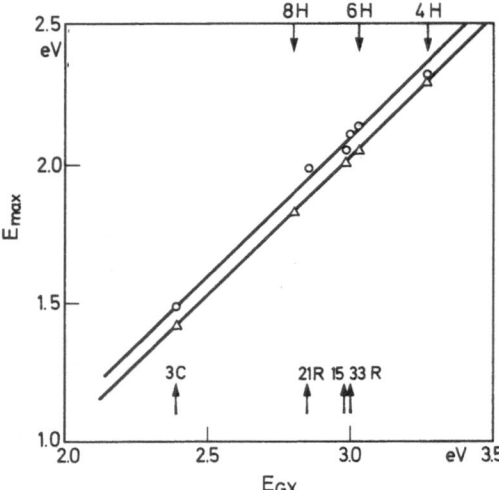

Fig. 11. Dependence of the emission peak on exciton energy gap in photoluminescence (○) and electroluminescence (△)

between the bandgap and the energy of the emission maximum, shown in Fig. 11 [73]. The explanation for this striking behavior is that the centers involved are more or less rigidly coupled to the allowed bands. The nature of the levels responsible for the radiative transitions, however, is still undetermined, for the ionization energies of the dopants are too low to account for the photon energies.

It has been speculated that the radiative transitions occur between donors and excited states of the acceptors [74], that the nitrogen can be incorporated in SiC to form a deeper level [75], or that still another kind of impurity is involved [38].

Beside polytype effects also impurity effects play an important role since a change in the spectral distribution of the emitted light can be promoted by varying the dopants. Unfortunately only on very rare occasions a "pure" spectrum is observed. The recorded bands result usually from the superposition of "pure" spectra related to acceptor-nitrogen pair luminescence, to transitions from the conduction band to the acceptor level, to transitions from the nitrogen donor level to the valence band, and to contamination. In this latter case luminescence can be due to an unknown coactivator or the presence of unwanted acceptors such as aluminum. The intensity of the different bands depends upon doping and temperature.

Table 6 lists the maxima typical for acceptors in 6 H SiC. The difference in acceptor ionization energy (deduced from Hall experiments) does not

Table 6. *Peak of the emission band in 6 H electroluminescent diodes at room temperature due to different impurities*

Impurity	Peak emission (eV)	Observer
Be	2.1	*Violin et al.* [130]
B	2.1	*Brander* and *Sutton* [38]
Al	2.5–2.65	*Brander* and *Sutton* [38]
Ga	2.4–2.6	*Kholuyanov* [75]
Cr	2.65	*Bukke et al.* [72b]
N	2.75	*Brander* and *Sutton* [38]

Table 7. *Ionization energies in SiC polytypes (in eV)*

Impurity	6 H		3 C	
	Energy	Method	Energy	Method
Be	0.38	Hall [130]		
B	0.39	Hall [69]		
Al	0.28 0.39 0.49	EL [76]	0.23	EL [77]
	0.27	Hall [55, 80]		
Ga	0.35	EL [75]		
	0.23	PL [74b]		
In	0.12	PL [74b]		
N	0.17 0.20 0.23	XL [65]	0.1	{ EL [77] XL [59]
	0.10	Hall [80]		

Methods: Hall = Hall analysis,
EL = electroluminescence,
PL = photoluminescence,
XL = exciton luminescence.

The discrepancies between the ionization energies determined by Hall analysis and those determined from optical experiments have not been explained satisfactorily. Hall measurements generally do not show the splitting of the impurity levels, observed in exciton luminescence, since the resolution of the method is not sufficient.

account for the wavelength change of the emission peak since $E_{Al} = 0.27$ eV and $E_B = 0.39$ eV (Table 7) whereas the emission shifts from 2.5 to 2.1 eV. Thus different recombination centers or different states of excitation of the same centers have to be assumed as mentioned before. *Gorban' et al.*

[76] ascribe the luminescence at 2.5 eV in SiC : N, Al (6 *H*) to donor-acceptor pair spectra. They are supported by *Zanmarchi*'s findings [77] in cubic SiC. On the other hand, in boron doped crystals the transition responsible for the light emission is still in doubt since the energy difference between the boron level and the nitrogen level is about 0.5 eV larger than that of the observed radiation. (The ionization energies are known from Hall experiments, and in the case of nitrogen supported by exciton luminescence.) Several authors made suggestions to account for the missing 0.5 eV. *Kholuyanov* [75] has speculated that the replacement of silicon by nitrogen (which usually occupies the carbon site) might give rise to a deeper nitrogen level. More complex is the level diagram proposed by *Addamiano et al.* [74a] who have postulated a transition from a donor (e.g. chlorine or oxygen), separated from the conduction band by 0.4 to 0.5 eV, to an excited state of the boron center. They were led to this donor energy by *Choyke*'s and *Patrick*'s paper [78] who had established the existence of an impurity level at this energetic position in heavily doped crystals. Recently several researchers [54, 79] have found evidence for such a deep donor also in Hall experiments but its chemical nature could not be determined. Thus it seems likely that a donor of unknown origin at about 0.5 eV is responsible for the luminescence in boron doped SiC.

The shift of the emission peak with temperature is usually smaller than differences among individual crystals. If it occurs, it can often be traced to the superposition of two or more bands with different temperature dependence which simulate such a wavelength shift.

Boron doping has yielded bright luminescence up to room temperature and higher. Other dopants may give room temperature luminescence under suitable preparation conditions but thermal quenching has been observed down to temperatures as low as 120° K.

6.3.3 Electron Effective Masses

There are no generally accepted values of electron masses. The first estimate of the electron mass in 6 *H* dates back to *Lely* and *Kröger* [80]. They concluded from Hall data that $m = 0.6\,m_0$ which is in disagreement with the newer findings of *Ellis* and *Moss* [81] who interpret their Faraday experiments to give $m(6\,H) = 0.25\,m_0$ and $m(15\,\mathrm{R}) = 0.28\,m_0$. Owing to erroneous assumptions about the position of the allowed bands this value cannot be trusted. No other, direct determinations, such as cyclotron resonance experiments, have been made.

Rather indirect methods to estimate the relation among polytypes were used by *Choyke* and *Patrick* [59]. The strength of the optical absorption at the bandedge is given, among others, by the multiplicity of the conduction band minima, thought to be the most important factor in SiC, the matrix elements for photon and phonon transitions and the densities of states. The latter are proportional to $m^{3/2}$ and allow a rough estimate to be made. The exciton luminescence data provide a check as to the general trend, since the reaction energies listed on Fig. 9 are related to the mass, if the effective mass approximation is trusted. Though it is known that the effective mass approximation is not a good approximation in SiC, it should not fail predicting the general trend. The mass ratio $m(6H)/m(15R)$ estimated from the optical measurements [82] lies between 1.27 and 2.28. The mobility ratio $\mu(15R)/\mu(6H)$ of about 1.7 [49] is consistent with the lower estimate if one assumes that the coupling constants of the two polytypes are the same. The ambiguity of the interpretation of the Hall data does not warrant to dig deeper, especially as long as the component parallel to the c-axis of the conductivity tensor is not known.

Recently *Lomakina* [83] has determined

$$m(6H) = 1.2 \, m_0 \,,$$

$$m(15R) = 1.0 \, m_0 \,,$$

$$m(4H) = 0.9 \, m_0 \,.$$

There seems to be no question that the electron effective masses are polytype dependent but there is some doubt about the validity of the values reported in the literature.

6.3.4 Dichroism

It is possible to identify certain polytypes of heavily doped n-type SiC by the color: $3C$ has a bright egg-yolk yellow, $6H$ appears flask green, $15R$ is green with a brownish tint, to name the more common polytypes. Except for the cubic modification the colors are not due to bandgap transitions. *Biedermann* [84] has determined the absorption coefficient of several polytypes ($4H$, $6H$, $8H$, and $15R$) as a function of wave number for light polarized parallel and perpendicular to the c-axis. He found a number of fairly broad absorption bands which are especially pronounced for light polarized parallel to the c-axis. More recently *Dubrovsky* and *Radovanova* [85] repeated his experiment and extended the measurements to include the temperature dependence of the absorption (77° K to 1100° K). When going to higher temperatures, the absorption peak of $6H$ is shifting to lower energies and at the same time the height decreases and the peak broadens. Starting at 640° K the absorption strength increases again. This is interpreted (cf. Fig. 12) as the result of transitions

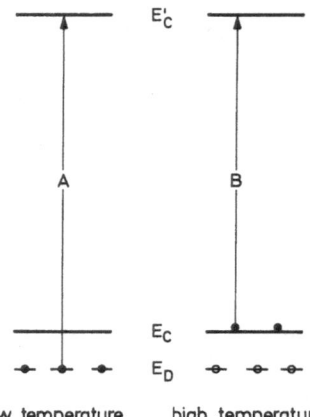

low temperature   high temperature

Fig. 12. Energy-level diagram explaining the shift of the impurity absorption peak with temperature. Transition A is dominant at low temperatures. Its importance decreases with increasing temperature due to the exhaustion of the donor level E_D. Transition B gains importance as the conduction band E_C is thermally populated. E'_C is a higher empty conduction band

(A) from the donor levels E_D to a higher, empty conduction band E'_C. As the temperature is increased, this contribution decreases because the concentration of neutral donors decreases due to thermal ionization. At its place transitions (B) from the lowest conduction band E_C, which is getting thermally populated, to the same empty band are gaining importance. The high absorption strength is evidence for direct transitions, i.e. it is thought, that the higher bands are at the edge of the Jones zone just like the minima of the lowest conduction band (cf. 6.6 Band Structure). This interpretation is in complete agreement with *Biedermann*, but it should be kept in mind that it is not the only possible explanation, although it is the most plausible one. One of the difficulties of it arises from unpublished data on other polytypes [86] which provide evidence of a shift to higher energies with increasing temperature. To explain this, still another empty band E''_C, favored by selection rules, would be required or one would have to assume that the empty band E'_C shifts considerably with temperature relative to the lowest conduction band E_C. At the present time little information exists about higher bands and it would seem reasonable to accept the above interpretation as a working hypothesis and use the absorption data to gain information about higher lying bands at the zone edge.

    Blank and *Potter* [87] use the absorption strength at 6000 Å to determine the relative doping concentrations of different $6H$ samples. In order to do so, they measure the transmission at 5000 and 6000 Å and take the ratio. If one assumes that the denominator in the transmission formula

$$T = I/I_0 = \frac{(1-R)^2 e^{-\alpha d}}{1 - R^2 e^{-2\alpha d}} \qquad (9)$$

does not differ much from unity, the transmission ratio is given by

$$\frac{T(6000)}{T(5000)} = \exp\{(\alpha_{5000} - \alpha_{6000})\, d\}\,.$$

Since the transmission at 5000 Å, close to the bandedge, does not depend on doping, the transmission at 6000 Å, however, is related to the nitrogen concentration through the absorption coefficient α, the logarithm of the transmission ratio is thus a measure of the doping concentration:

$$-\frac{1}{d}\ln\frac{T(6000)}{T(5000)} = \alpha_{6000} - \alpha_{5000} \sim N_D\,.$$

6.4 Properties Related to Axial or Zero Wave Vector

Wave vectors at the origin of the large zone and those along the zone axis are very similar in all polytypes since the energy discontinuities are so small that in the usual approximation they disappear. Thus we expect only minor polytype differences and no impurity effects.

6.4.1 Direct Energy Gap

Very little information has been gathered about the direct energy gaps at the zone center. *Choyke* and *Patrick* [88] made an attempt to determine the width of the direct gap in $6H$ but the experiment was not conclusive. The absorption coefficient was measured very carefully on samples ranging in thickness down to 1.8 μm. Successive fitting procedures allowed three indirect gaps to be identified at 3.0, 3.7, and 4.1 eV. The fourth gap at 4.6 eV also appears to be indirect, but it is difficult to distinguish between direct and indirect transitions at this energy. As shown previously (Eq. (7)) the simple theory of indirect transitions yields a linear relation between the square root of the absorption coefficient and the energy difference $(h\nu - E_{Gi})$. The absorption coefficient of direct transitions is proportional to $(h\nu - E_{Gi})^{1/2}$ for allowed transitions and $(h\nu - E_{Gi})^{3/2}$ for forbidden transitions. Due to the experimental uncertainties which are larger at higher photon energies and due to the fitting procedure which consists of successive subtractions, it cannot unequivocally be distinguished between the different kinds of energy dependence. However, an indirect gap at 4.6 eV would not explain the reported structure in the reflectivity [89].

The published band calculations [90] (see Section 6.6) use the first indirect gap as fitting parameter and indicate that the direct gap at $k = 0$ does not vary much from polytype to polytype. This would be expected from *Patrick*'s classification. There is also experimental evidence for such a behavior: ZnS, another polytypic substance, is a direct semiconductor and its energy gap at the zone center does not vary by more than 3 % [91].

6.4.2 Hole Masses and Mobilities

No polytype comparisons of hole masses and mobilities have been made either in SiC or in ZnS. Only small differences are expected except perhaps for the cubic modification. Cubic SiC is the only polytype whose valence band has a higher symmetry and hence a higher degeneracy which may lead to different hole masses and mobilities. For other polytypes the exciton luminescence experiments (Section 6.3.1) have suggested that spin-orbit splitting (4.8 meV) and crystal field splitting (> 30 meV) lift the degeneracy at $k = 0$ and are the same for all polytypes.

6.4.3 Infrared Absorption

The first data on phonons in SiC were derived from the strong residual ray absorption analyzed by *Spitzer et al.* [16] who in order to fit the reflectivity curve, had to use both a strong and a weak oscillator. Their results show little difference between $6H$ and cubic SiC. Newer measurements with higher resolution were performed by *Ellis* and *Moss* [81] who found additional weak absorption lines in $6H$ and $15R$ which they could not assign.

These data and (at that time) preliminary Raman experiments led *Patrick* [92] to consider the possibility that all SiC polytypes have a common phonon spectrum in the axial direction of the large zone. This allowed the three reported weak lines to be identified as fundamental lattice modes. The extension of this work culminated in the discovery that in polytypic substances the phonon dispersion curve in the axial direction can be derived from Raman scattering experiments (see 6.4.4).

In contrast to the three weak absorption lines and the Raman active modes there is the two-phonon infrared absorption [93] which is indicative of the values at the large zone boundary rather than of phonons on the symmetry axis. It is due to the combination bands of vibrational modes of the lattice and does not show a polytype dependence within the limits of the experimental errors (1 meV).

Finally there are the phonon energies obtained from exciton luminescence experiments which are also known not to represent points on the symmetry axis, because they are, as discussed previously, linked to the conduction band minima which lie off axis.

Table 3 summarizes the phonon energies observed in different polytypes and experiments, and classifies them according to the large zone concept wherever applicable.

6.4.4 Raman Scattering

An elegant confirmation of the large zone concept was provided by the Raman scattering experiments of *Feldman et al.* [94]. Specimens of various polytypes were cut into rectangular parallelepipeds a few millimeters on the side and a fraction of a millimeter thick. The spectra were excited with a 400 mW argon laser. The observation of the Raman spectrum in different arrangements, with different propagation and polarization directions, was used to determine the symmetry types of the vibrational modes. The phonons active in first-order scattering have representations A_1, E_1, or E_2 in hexagonal polytypes and A_1 or E in rhombohedral [92]. In both cases they fall into two distinct groups:

1. The strong modes at $k = 0$ (strong in infrared absorption) have almost the same frequencies for all polytypes. They are longitudinal or transverse, and have energies which depend slightly on the angle θ between propagation direction and c-axis.

2. The weak modes at $k \neq 0$ are classified according to their atomic motion as axial (parallel to the c-axis) or planar (perpendicular to the c-axis), the latter ones being double*. The weak mode frequencies are characteristic of the polytype and independent of the angle θ. These modes can be used to construct dispersion curves equivalent to those commonly derived from neutron scattering experiments (Fig. 13).

The experimental uncertainties of the Raman frequencies were less than $2 \, \text{cm}^{-1}$, thus the doublets observed in the spectra give a measure of the energy discontinuities (2 to 8 meV) within the large zone at $q = x q_{max}$. Here q_{max} is common to all polytypes and is given by $N\pi/c$ for a polytype containing N double layers per unit cell, x is the special value corresponding to $q = 0$ in the Brillouin zone [92].

Polytypes $4H$, $6H$, $15R$, and $21R$ were analyzed and the great majority of the predicted lines was found. Plotting the Raman frequencies of all polytypes vs. x yields the combined dispersion curve mentioned earlier (Fig. 13). The higher the polytypes the more closely spaced are the values of x, accessible to Raman experiments, since they have many modes and their large zones have many internal energy discontinuities. On the other hand the data of *Feldman et al.* and incomplete data on

---

* The more common notation of longitudinal and transverse is somewhat inaccurate for hexagonal and rhombohedral polytypes. This is discussed in Ref. [92] in detail.

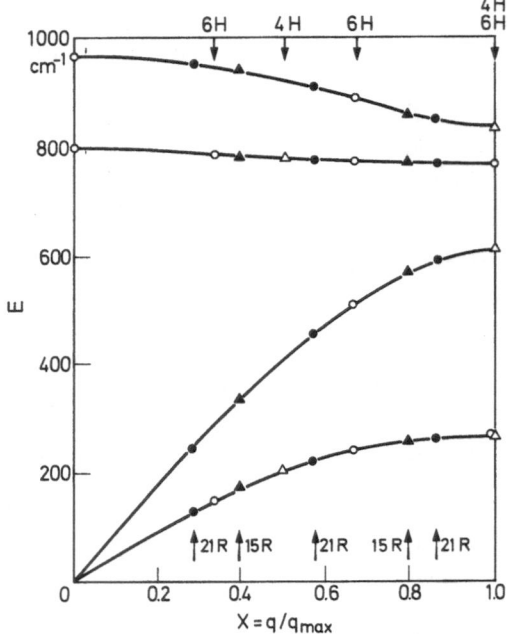

Fig. 13. Combined dispersion curves using data of four polytypes. For each polytype the Raman accessible values of $x = q/q_{max}$ are marked at the top of the figure

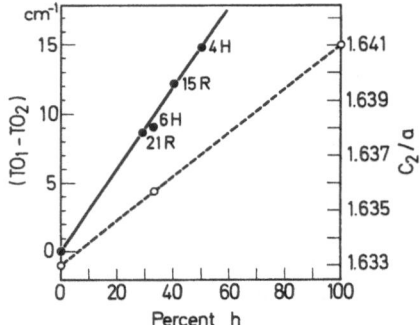

Fig. 14. Two measures of polytype anisotropy in SiC plotted against "percent h", i.e. hexagonality. The closed circles represent the anisotropy of the transverse optical phonon modes, from Raman experiments. The open circles are reduced c/a axial ratios for polytypes $6H$ and $2H$, from X-ray measurements

polytype $69R$ [95] seem to indicate that both the line intensity and the doublet splitting decrease as their number increases.

Two of the three strong modes (called LO and TO_2) have an angular variation of energy whereas the second transverse mode TO_1 is independent of θ. It was shown that the TO_1 mode is also independent of polytype while the TO_2 mode depends on polytype. For the longitudinal mode the observed variation was comparable to the experimental uncertainty so that no systematic dependence on polytype could be established. The angular dependence of the TO_2 mode may be taken as a measure of crystal anisotropy. Fig. 14 shows that the energy difference $TO_1 - TO_2$ and the c/a axial ratio are both linear functions of the percentage hexagonal packing or hexagonality defined earlier.

Thus the empirical "percent h rule" seems to account for the small polytype differences of the class 2 properties. These differences are not predicted by the wave vector classification but due to the observed tetrahedral distortion (c/a axial ratio) [44]. The "percent h rule" is not restricted to SiC and ZnS polytypes but holds also for certain metallic alloys with structures analogous to the polytypes.

6.5 Properties Related to a Weighted Average over k

6.5.1 Dielectric Function

The dielectric function involves all wave vectors but it is strongly weighted for photon energies near direct transition thresholds. In general the dielectric function will be similar for all polytypes except for the thresholds where transitions occur at a wave vector on the zone boundary. Then a strong polytype effect may be observed. The dielectric constants have only been determined for $6H$: $\varepsilon(\omega = 0) = 10.2$ [96], and $\varepsilon(\infty) = 6.7$ [97], but are expected to be approximately equal for all polytypes.

The dielectric function determines the reflectivity and birefringence. The experimental data are sparse but indicate again that the general appearance is the same for all polytypes except for energies which are connected to direct transitions at the zone boundary. Thus the reflectivity spectra are in general quite similar [89], but may show distinctive structure as reported by *Choyke et al.* [98]. The better known ZnS polytypes exhibit the same behavior, i.e. a general similarity with some distinct differences [99]. Birefringence has only been measured in ZnS and, because *Brafman* and *Steinberger* [91] used approximately bandgap light, it is very likely that zone center transitions are dominant, thus giving the birefringence a class 2 appearance.

6.5.2 Thermal Properties

The thermal properties should have little polytype dependence except for a slight anisotropy which may vary with hexagonality. *Bosch* [100] reported an anomalous heat conductivity below 30° K which can be fitted by $\lambda = 1.1 \times 10^3 T^2 W/(\text{cm} \times \text{degree})$ and by $\lambda = 23 \times T^{-0.65} W/(\text{cm} \times \text{degree})$ between 110 and 200° K. Other researchers found widely differing values which might be attributed to polytype variations. But *Slack* [101] has shown that this large spread of the thermal conductivity values is caused by the variation in the impurity content of the samples. He found that representative specimens of pure SiC agree within a factor of 2 with theoretical predictions below 300° K and that the values are close to the geometrical mean of pure silicon and diamond, and thus very high.

Taylor and *Jones* [102] measured the thermal expansion of hexagonal and cubic SiC between 77 and 1470° K. The lattice parameters are nearly constant up to 400° K, then they change almost linearly with temperature. The c/a ratio exhibits an inflection around 800 to 900° K. The expansion coefficients are all about the same up to 400° K, then they diverge. $(1/c)$ (dc/dT) levels at 4.68 and $(1/a)$ (da/dT) at 4.20×10^{-6}. The coefficient of cubic SiC is intermediate. Around 800° K the situation is reversed in so far as now the expansion along the a-axis is increasing again at about the same rate as the coefficient of the cubic modification. A comparison with the *Mie-Grüneisen* theory [103] also indicates a sharp change in the mode of lattice vibrations at 400° K. This may be explained by the high density of phonon states associated with the flat planar acoustic branch in the axial direction. The phonon dispersion curves (Fig. 13) indicate a density of states peak at 265 cm^{-1} (380° K). Therefore at this temperature there is a preferred population of modes propagating along the c-axis, leading to a preferred expansion direction. This situation persists until the higher frequency branches are excited, notably the flat optical phonon branches. A similar anomaly is expected in all hexagonal and rhombohedral polytypes because they all have the same high density of low-frequency axial modes. To what extent the axial acoustic branch is responsible for the change around 800° K (560 cm^{-1}) is unknown.

6.6 Bandstructure Calculations

The first calculations of the bandstructure were made by *Kobayasi* [104] and *Bassani* and *Yoshimine* [105] for cubic SiC. They show a triply degenerate valence band maximum at the center of the Brillouin zone (Γ point) and three conduction band minima at the zone boundary in $\langle 100 \rangle$ directions (at the point X) (Fig. 6a). The large size of the unit cells of the more common polytypes has prevented the computation of the bandstructure of other polytypes for many years. Only recently *Herman et al.* [106] and *Junginger* and *van Haeringen* [90] have taken advantage of the greater knowledge of SiC polytypes and the refined computer facilities. They used a modified OPW method (orthogonal plane wave)

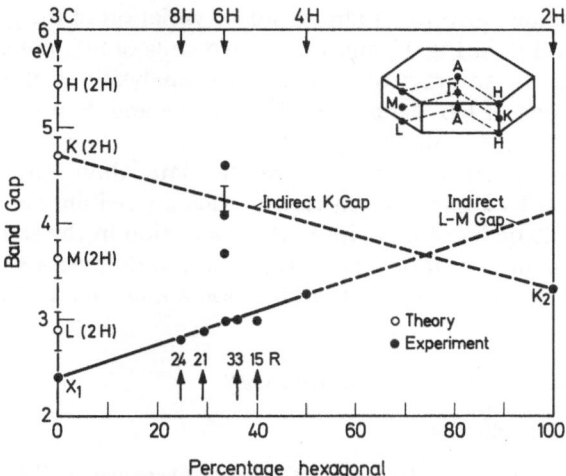

Fig. 15. Variation of indirect band gap with hexagonality in various SiC polytypes. The experimental points are based on the work of the Westinghouse group (Ref. [59] and [88]). The theoretical points at $3C$ denote energy levels calculated at those positions in reciprocal space which correspond to the H, K, M, and L positions for the $2H$ reduced zone

to compute the bandstructure of $2H$, $4H$, and $6H$, inserting the experimentally determined indirect bandgaps as fitting parameters. Work in progress [107] on improved ultraviolet reflection spectra should provide an independent check of the theoretical work.

The results suggest that the direct energy gap of all polytypes is closely the same and that the large differences of the optical (indirect) gaps are due to a delicate balance which causes the conduction bands to move up and down. In line with this is ZnS which shows a much smaller variation of the width of the forbidden zone (a few percent [91]) than SiC, because it is a direct semiconductor.

Herman et al. [106] have speculated that the striking linear relation between bandgap and hexagonality may have a deeper reason. They notice that the lowest conduction band minima of $2H$ lie at different positions in the reduced zone than the minima of the other polytypes. For $2H$ *Patrick et al.* [108] have proved that the conduction band minima are at the point K (Fig. 15). For the other polytypes *Herman et al.* [106] have concluded that they lie along a line joining points L and M. Thus *Herman* suggests that the K-gap and the L-M-gap are linearly related to the hexagonality, the former one decreasing and the latter one increasing with increasing hexagonality. Such a correlation between hexagonality and bandgap is plausible since the hexagonality is reflected in the axial ratio c/a which in turn is a measure of the lattice distortion.

This may ultimately lead to an almost linear variation of the gaps. This model is illustrated in Fig. 15 and the known gaps of SiC polytypes are marked. Within the accuracy of the theoretical analysis one of the higher conduction bands of $6H$ determined by *Choyke* and *Patrick* [88] can be identified with the indirect gap at K.

It would be worthwhile to analyze the data about the impurity absorption (cf. 6.3.4 Dichroism) in view of such a possibility and to plot the higher bands deduced from impurity absorption in the same figure. If the proposed interpretation is correct, the puzzle pieces should fall into place and provide additional information about such higher bands.

7. Silicon Carbide Device Technology

7.1 Historical Introduction

Lossew [3] discovered in 1923 injection luminescence in SiC crystals. A review article by *Destriau* and *Ivey* [109] gives the historical background with many references. A satisfactory explanation of the phenomenon was offered much later by *Lehovec et al.* [110] who suggested that the light emission was due to recombination of minority carriers injected across a $p - n$ junction. The difficulty of growing single crystals and of making electronic devices from them hampered progress for the next three decades.

In 1957 *Patrick* [111] published a model for an actual $p - n$ junction consisting of a three layer structure analogous to a $p - i - n$ junction. Instead of an intrinsic middle layer, a high resistivity region due to compensation was assumed. Even today such a high resistivity middle zone of less than $5\,\mu$m thickness is frequently invoked to explain the unusual logarithmic slope of $e/2\,k\,T$ of the current-voltage characteristics and the anomalous behavior of the differential capacitance of SiC diodes [133].

Consequently *Patrick* and *Choyke* [78] analyzed many diodes made from aluminum doped crystals having an *n*-type skin. Under forward bias visible light is emitted uniformly over the whole junction area, while only isolated "blue spots" are observed under high reversed bias. These spots are often microscopically small and are thought to arise from local breakdowns. *Patrick* and *Choyke* concluded from a comparison of electroluminescence, photoluminescence, and current-voltage characteristics at different temperatures that the formation of an impurity band between 0.35 and 0.5 eV below the conduction band is significant and that the decrease of the luminescent efficiency with increasing current density is due to the large cross section for the radiationless recombination of electrons in this band.

Another comprehensive study of SiC diodes is the thesis of *Greebe* [112]. He had a Lely furnace at his disposition and grew crystals in an atmosphere of pure argon which yielded *p*-type crystals. At the end of the growth experiment he changed the atmosphere to nitrogen which doped the outer layer as *n*-type. His analysis again indicated a highly compensated middle region between the bulk *p*-type and the *n*-type skin. The current-voltage curves are typical of many SiC diodes: a leakage current ("foot") at voltages below 1.5 V, then an exponential rise with a logarithmic slope of $e/(2kT)$, and finally the much more gradual rise which was ascribed by *Greebe* to the series resistance of the diode.

These papers and others triggered the research on SiC devices which in turn required new technologies. Although we are far from a standard method of producing devices, there are some well established techniques which are described below.

7.2 Device Fabrication

7.2.1 Junction Formation

Two types of junctions are discussed here, distinguished by the fabrication method, namely diffused and grown junctions. The diffused junctions are the result of diffusing an excess of minority dopants into a bulk crystal exactly as in the well known silicon technology. In SiC the difficulties are much greater because all known impurities diffuse very slowly at temperatures below 1800° C and yield very shallow junctions. Above this temperature care has to be taken to prevent sublimation of the bulk crystal. Typical for this method is the procedure developed by *Blank* [113] where *n*-type single crystals are heated in a graphite crystal holder to 1800 to 2300° C and exposed to aluminum containing vapors. In order to avoid the decomposition of the crystals, SiC powder is added to control the partial pressure of the constituents. Beside aluminum, both boron [114] and beryllium [115] have been used to produce *p*-layers on *n*-type bulk material. The reverse, *n*-type diffusion into *p*-type crystals has also been attempted [116], but has met less success because nitrogen diffusion is much slower. The published diffusion constants show wide scatter and should be regarded as guidelines only. One reason for this scatter is the rather limited temperature range accessible to experiment, a second one is the lack of knowledge concerning the diffusion mechanisms, and last but not least is the difficulty of ascertaining the boundary conditions. It seems to be a matter of great importance whether we deal with solid-solid or vapor-solid diffusion [117]. It is also known that diffusion is sensitive to crystal orientation, often being fastest perpendicular to the c-axis, slower on the carbon side, and slowest on the silicon side.

Older than the diffusion technique is the *in situ* preparation of junctions mentioned above in context with *Greebe*'s thesis. These so called grown junctions may consist of a p-type core with an n-type skin, when during the final stage of the growth run nitrogen is added to the system, or of an n-type core embedded in a p-type skin, when an acceptor is admitted at the end of the experiment. The main advantage of this method over straight diffusion is the ability of growing deep junctions. Since the growth is carried out at high temperatures (typically 2500° C), inter-diffusion at the original interface is not negligible. At these temperatures the impurities are fully ionized and an additional diffusion mechanism has to be considered. *Pell* [118] proposed that the driving force for the ionized impurities is the built-in electric field at the junction which leads to an almost perfectly compensated high resistivity region.

It is difficult to assess the importance of ion drift in a particular growth run since a delicate balance governs the mechanism. At high temperatures and low doping concentrations the free carriers should swamp the crystal and short-circuit the field. Thus ion drift can only be operative for as long as the concentration of ionized impurities is larger than the intrinsic concentration n_i. Some unpublished results [119] suggest that ion drift is indeed small in very pure crystals.

Knippenberg and *Verspui* [8] improved the technique of the grown junctions. They observed that aluminum enhances crystal growth and that aluminum doped layers could be grown on nitrogen doped crystals at temperatures as low as 1800° C. This reduces interdiffusion greatly and leads to sharper junctions. This low temperature growth does not differ much from an epitaxial process with the distinction of

a) the substrate crystals are grown *in situ* and

b) the source material is not a gas mixture but a hot body of sintered SiC.

7.2.2 Epitaxy

If we expose a "seed" crystal to a stream of atoms in such a way that the arriving atoms arrange themselves relative to the seed so as to form a single crystal, we call this epitaxy. Important parameters controlling epitaxy are the temperature of the seed or substrate crystal and the rate of incidence of the mobile atoms. In gaseous or vapor phase epitaxy the seed is exposed to a vapor stream, in liquid epitaxy it is submerged in a liquid which provides the material to build up the layer. Epitaxy is a very attractive way of producing junctions, especially in SiC, because

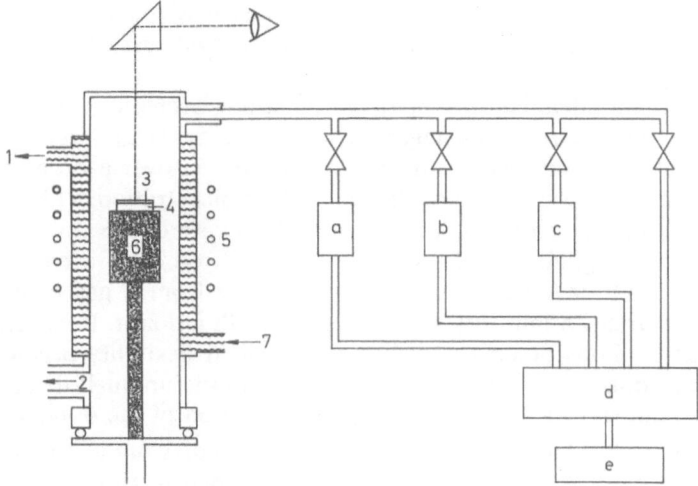

Fig. 16. Schematic drawing of an epitaxial system for growing SiC epitaxially by the vapor-phase method. *1* cooling water outlet, *2* exhaust, *3* epitaxial layer, *4* SiC substrate crystal (seed), *5* rf-coil, *6* rf-susceptor and support for substrate crystal, *7* cooling water inlet, a, b, c vessels containing silane, hydrocarbon, and dopants respectively, d Pd-cell to produce ultrapure hydrogen, e hydrogen container

other methods have severe drawbacks due to the high processing temperatures and the accompanying effects which were described above.

The first investigations were centered on vapor phase systems and many combinations of carbon and silicon rich compounds in varying ratios were tried [35, 120]. Today a typical system (Fig. 16) employs methylchlorosilanes (e.g. CH_3Cl_2SiH, CH_3Cl_3Si) and hydrocarbons (e.g. hexane) and hydrogen as carrier gas. The partial pressures, i.e. the ratio C : Si, are easily controlled by the temperature of the storage containers. In order to dope the epitaxial layer, a third vessel containing the dopant in suitable form (e.g. B_2H_6 or $Al(C_2H_5)_3$ for *p*-type layers and PH_3 or N_2 for *n*-type doping) is provided. The mixture of the gases is directed towards the seed crystal which is resting on a susceptor, heated by an R.F. coil outside of the all-quartz system. Quartz is a requisite for both the high temperatures and the high purity. A vacuum pump completes the equipment since purging with the purest hydrogen available is usually not sufficient. Typical settings are: molar ratios of hydrogen : silane : hexane = 1000 : 2 : 1, a substrate temperature of 1700 to 1800° C, and growth rates of 0.1 to 1 μm/min. Already *Spielmann* [35] noted that for a given substrate temperature the growth rate determines the quality of the deposit. At high rates polycrystalline layers are formed, at intermediate rates cubic SiC with many twins, and at the lowest rates the polytype of the substrate is reproduced. Since SiC is etched by hydrogen at epitaxial temperatures the net growth may be minute for good layers.

The main advantages of epitaxial junctions over diffused or grown ones, beside the more manageable temperatures, is the fact that the profile of the doping concentration can be varied over a broad range, from a step to a very gradual change of the concentration. This may be accomplished by closely controlling the flow rate of the dopant since diffusion is negligible under the actual conditions. It is even possible to grow multi-layered structures by switching from a gas containing an acceptor as an impurity to another one with a donor. In contrast to this the methods described earlier have limited flexibility because the concentration profile is essentially a diffusion profile, uniquely determined by the temperature and the surface concentration of the dopants.

Brander and *Sutton* [38] developed liquid epitaxy of SiC to the point that device quality layers could be grown. In their method SiC crystals are immersed as seeds into liquid silicon which is in turn kept in a graphite crucible. The correct choice of geometry and temperature leads to a transport of carbon from the crucible walls to the seed crystals and to epitaxial growth. Doping can be achieved by adding dopants to the silicon melt. However, only abrupt profiles will result unless special steps are taken or the crystal is subjected to an annealing cycle which causes diffusion to take place at the original interface. The occurrence of a compensated intermediate layer has been observed which may be caused by the dissolution of the original surface and the subsequent re-incorporation of the impurities in the early growth stage.

7.2.3 Ancillary Techniques

The techniques for mechanically shaping SiC crystals, such as lapping, abrasive and ultrasonic cutting, are all standard, but diamond and boron carbide powders must be used since these are the only materials harder than SiC.

Etching techniques are extensively used in device fabrication. SiC is a structurally polar crystal with a carbon and a silicon face. Etching rates for the same etch are usually much faster on the carbon face than on the silicon face.

Molten salt etches have been described in detail by *Faust* [121] and *Gabor* and *Jennings* [122]. Sodium salts such as NaF, NaOH, Na_2O, and their mixtures with oxidizing agents such as Na_2O_2, KNO_3 etc. were studied. The very high etch rates of 1 to 2 μm/sec and the lack of suitable etch masks make the molten salts unfit for most device work.

An exception is their application in quickly removing damaged material or exposing structural defects.

Gaseous etching with chlorine or chlorine-oxygen [123] mixtures between 850 and 1050°C yields the smooth surfaces desirable for epitaxial and device work in general. By adjusting the temperature or the oxygen content the etch rate can be varied over two orders of magnitude. The optimum composition was found [123] to be: 68 parts Ar, 26 parts Cl_2, and 6 parts O_2. At a total flow rate of 175 cm^3/min and at 900°C an etch rate of 0.25 ± 0.02 μm/min results on the C-face.

Another gaseous etch is conveniently used in epitaxial systems. Hydrogen attacks SiC above 1500°C. At 1700°C and at a flow rate of 3000 cm^3/min typical etch rates of 0.25 μm/min and very smooth surfaces could be observed [124].

Oxidation of SiC is also a very useful tool. Silicon dioxide can be grown on SiC with steam oxidation techniques similar to those employed for silicon. Oxidation is carried out at 900 to 1200°C in a carrier gas saturated with water vapor. It provides an easy method of determining the C- and Si-faces of the SiC crystal due to the different attack on the two faces. Oxidation can also be used to delineate the junction since the rates on p- and n-type regions of a single crystal are different. After beveling a crystal containing a $p - n$ junction, it is exposed to water vapor or dry oxygen. The different growth rates of the oxide lead to thin film interference colors which permit to determine the junction depth, provided the doping concentrations are high enough to influence the oxidation rate.

In hydrofluoric acid only the thin natural oxide film on top of SiC crystals is removed. If the immersed SiC platelet is connected to a voltage source, electrolytic etching occurs [125]. This is specific for p-type material and has been used to etch mesa structures [126] and to determine junction depth [123].

Of great technological importance are masking techniques. *Campbell* and *Berman* [123] have devised a method, called the self-masked diffusion-etching process, to fabricate planar devices. The essential steps are as follows (Fig. 17):

1. Aluminum diffusion to form a p-type skin into an n-type platelet which was lapped to a prescribed thickness.

2. Removal of material by precision lapping from the C-face of the platelet until the n-type layer reaches a prescribed thickness (the diffused junction on the Si-face constitutes the lower gate of a field effect transistor).

3. Growth of silicon dioxide over the entire C-face surface of the platelet by the steam oxidation technique.

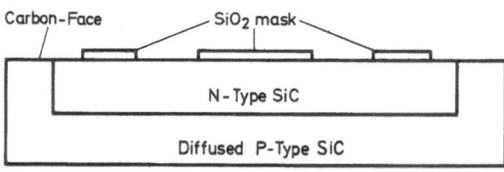

a) Section view, oxide mask after photoresist etch

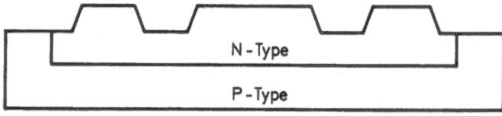

b) Following Cl_2-O_2 etch and removal of oxide

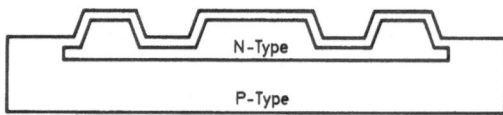

c) Following second diffusion of P-Type impurity

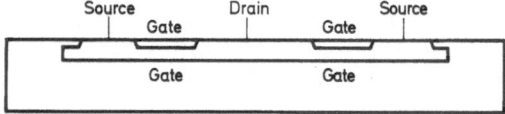

d) "Self-Mask" removed leaving finished structure

Fig. 17. Self-masked diffusion and etching technique developed by the Westinghouse group to make SiC integrated circuits. Details are given in the text

4. Photoresist etching of the oxide layer to expose areas of *n*-type surface for chlorine etching (Fig. 17a).

5. Chlorine etching of the SiC through the openings of the SiO_2 mask and removal of the oxide in a buffered HF solution (Fig. 17b).

6. Second aluminum diffusion (Fig. 17c).

7. Removal of the SiC self-mask to obtain a planar structure (Fig. 17d).

8. Applying evaporated Au-Ta contacts to source, gate and drain regions using the photoresist aluminum rejection mask technique.

This self-masked diffusion-etching process is the key technology which provides precise control of the device structure. No other masking material than SiC itself can withstand the high processing temperatures and the aggressive etches required for a SiC device technology.

Many metals and alloys have been tested as contact materials. Most work has been confined to ohmic contacts needed not only in device work but also for determining the electrical properties of single crystals. The requirements are stringent since the material should not only make an ohmic contact but should also adhere well, have good mechanical properties, and should withstand high temperature in order to fully exploit the capabilities of SiC. For n-type crystals a gold-tantalum alloy (3 to 5% tantalum), nickel and tungsten fulfill the requirements as long as the conductivity is not too low. For p-type SiC the contacts are generally less acceptable. However, relatively heavily doped crystals can be satisfactorily contacted with gold-tantalum-aluminum alloys or nickel. Most contacts are fused contacts but evaporated or sputtered metal electrodes have also successfully been made. Nickel can be applied by electroless plating.

Little has been published on the nature of metal/SiC contacts. *Mead* [127] and *Hagen* [128] investigated Schottky barrier contacts. Both investigators cleaved n-type and p-type crystals (polytypes $6H$ and $15R$) in a stream of evaporating metal under ultra high vacuum conditions. Hagen showed that the barrier height is essentially independent of metal work function and polytype. However, it should be noticed that the two polytypes investigated have very similar band gaps and one would not expect to find significant differences. The result indicates that the Fermi level at the metal/semiconductor interface is fixed with respect to the bandedges. This is generally attributed to surface states. *Mead* [127] observed that the fixation of the Fermi level occurs predominantly on covalent semiconductors and is absent for the more ionic semiconductors, which are characterized by a linear relation between metal work function and barrier height.

Experimental methods to determine the barrier height are capacitance measurement, photoresponse measurement, and fitting current-voltage characteristics [129]. The last one is the least successful since in many cases the characteristics deviate considerably from the theoretical diode equation. The capacitance data depend to a certain extent on the model chosen and are therefore not quite as reliable as the photoresponse measurements.

7.3 Applications

The first commercial electronic SiC devices were thermistors made from $6H$ single crystals. Electroluminescent diodes followed since the requirements with regard to purity are not as severe as for rectifier diodes. Almost from the beginning of the SiC device development two main lines evolved, luminescent diodes on the one hand and high voltage rectifiers on the other one.

7.3.1 Electroluminescent Diodes

The main reason for making electroluminescent SiC devices is not their refractory nature but the unique feature that they can emit light covering the whole visible spectrum from red to blue. No other semiconductor achieves this and there is no other blue-emitting semiconductor with a comparable efficiency.

As mentioned in Section 6.3.2 the different colors can be obtained either by doping one and the same polytype with different acceptors or, alternatively, by doping different polytypes with one and the same acceptor. For the lack of good single crystals of the rarer polytypes in large quantities the first approach is favored.

Fabrication techniques do not appear to be crucial since different techniques do not lead to a radically different performance of the diodes. Usually diffusion [113, 130] or liquid epitaxy [38] are used. Most laboratories start with nitrogen doped crystals. The p-type material usually contains boron but sometimes aluminum is added. The practical problem is to achieve high efficiencies and high brightness at the same time. As current density is increased, the light output does not increase proportionally, a phenomenon not unique to SiC but here it occurs at lower levels than in other semiconductors. The reason for this behavior is not known. It could be caused by low injection efficiency, internal barriers, large series resistance, decreasing carrier lifetime with increasing injection, and so on.

The General Electric diodes emit light perpendicular to the junction through the thin p-type layer on top [113]. The Norton diodes emit parallel to the junction from a very narrow and parallel region [12]. The group at the Electrical Engineering Institute in Leningrad has made alphanumeric displays using single crystals of a few millimeters on the side. *Brander* [131] has reported a solid state "magic eye".

The current voltage characteristics of luminescent devices behave in a similar manner to the diodes described by *Greebe* [112]. Over a wide current range (up to seven decades) they follow an exponential law of the form $\exp{(eV/2kT)}$. Some exceptional diodes exhibit an exponential rise of the form $\exp{(eV/kT)}$ at higher current densities indicating that one carrier injection is becoming dominant [131]. At still higher currents saturation occurs.

The light output is proportional to the junction current at low current densities. At higher current densities it becomes sublinear, often obeying a square root law. The break in the logarithmic brightness curve is related to the fabrication techniques. Solution grown junctions (liquid epitaxy) tend to be linear over a wider current range than junctions made by vapor phase epitaxy [132] or diffusion [133].

The efficiency of luminescent diodes varies with temperature and injection level. Therefore a meaningful comparison between diodes made in different laboratories is difficult. The highest external efficiencies reported at room temperature are 1 % [134] for diodes emitting yellow light (presumably boron doped SiC). It should be mentioned, however, that this is true for low current densities only because at higher injection levels the sublinear brightness/current behavior reduces the efficiency. The temperature dependence can be explained by the presence or absence of additional impurities and is therefore related to the preparation of the junction. *Todkill* and *Brander* [132], for instance, found a steadily decreasing efficiency in solution grown devices and a maximum slightly below 200° K for diodes made by vapor phase epitaxy. This behavior is due to two competing processes. In general the efficiency decreases with increasing temperature because the probability of non-radiative recombination increases faster than that of radiative transitions. If the carriers recombine non-radiatively through traps, the higher temperature can reduce the recombination probability by emptying the traps. A limited range of increasing efficiency with increasing temperature has also been observed by *Potter* [133] and *Violin* and *Kholuyanov* [114] for diffused junctions. The latter two authors reported a monotonic decrease of the efficiency for alloyed junctions.

Detailed investigations about response times of luminescent devices are sparse. *Todkill* and *Brander* [132] noted fast response times ($<0.75\,\mu$s) for solution grown junctions and much slower ones (up to $750\,\mu$s) for epitaxial devices. The slow response times do not represent carrier lifetimes but are caused by severe trapping effects. This follows from an analysis of the response time as a function of temperature and current density. Tungsten is suspected to be the trapping center.

A rather unique device was developed by *Brander* [131]. It is the solid state analog to the magic eye of the electron tube industry. It consists of a p-type crystal and an n-type epitaxial layer (Fig. 18). Three terminals are provided, a large area ohmic contact to the bulk and two small ohmic contacts far apart on the opposite side. Since the conductivity of the epitaxial layer is chosen to be low, a voltage gradient is maintained between the upper contacts across the n-layer. One contact is biased at a fixed voltage above the knee of the forward characteristic of the $p - n$ junction, while the signal to be indicated is applied to the second contact. If the potential at any point along the strip is in excess of the forward voltage drop of the diode, then current will flow through that region and light will be emitted. If the voltage falls below the knee voltage, there will be insufficient current flow in that region to cause luminescence. Since both current and brightness vary exponentially with voltage, a sharp cut off between light and dark regions can be obtained.

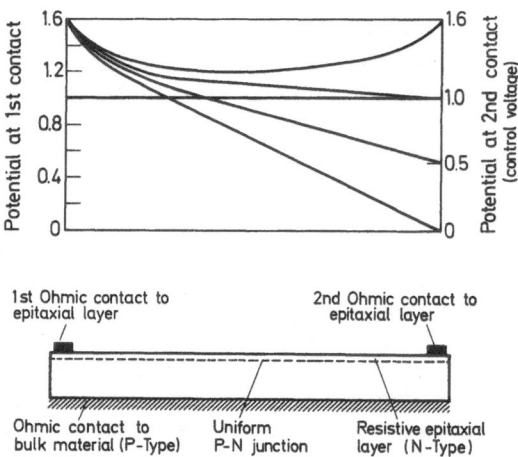

Fig. 18. Solid state "magic eye". Schematic cross section along the length of the device and the voltage distribution under various conditions of bias

7.3.2 Rectifier Diodes

Rectifier diodes are more difficult to produce than luminescent diodes because much purer crystals are needed in order to achieve high blocking voltages. The $p-i-n$ structure has prevailed in the silicon power technology for its superior performance [135]. By selecting judiciously the width of the intrinsic region a compromise between good forward and blocking characteristics can be achieved. Thus a similar approach for SiC seems justified, but truly intrinsic SiC has not yet been made. The closest one can get, are SiC diodes with a high resistivity region, accomplished by compensation, adjacent to more heavily doped p- and n-type SiC. This happens almost naturally when grown junctions are produced.

Campbell and *Berman* [123] have found that, starting with p-type SiC, time and rate of the introduction of nitrogen into the growth chamber determine to a large extent the electrical characteristics of the junction. The width of the high resistivity zone and the degree of compensation control the forward voltage drop and the blocking capability. The devices made by this method can be operated at temperatures of 500° C. The forward voltage drop decreases with increasing temperature but even at 500° C it is always larger than two volts. Rectifiers capable of up to 50 A peak have been prepared and specially processed units have exhibited a reverse capability of 600 PIV [123]. The reverse characteristics show generally a "soft" breakover rather than the sharp onset of avalanche breakdown known from several silicon diode types. This is attributed to the imperfection of the junction which leads locally to very high internal fields where breakdown occurs. Upon increasing the voltage new weak areas reach the critical field. Thus the avalanche breakdown is smeared out over a voltage range of considerable width. Table 8 lists typical diodes and their properties at 30 and 500° C.

Table 8. *Electrical properties of typical SiC rectifiers*

Unit type	Forward voltage (V) half wave av.		Forward current (A) half wave av.		Reverse voltage (V) peak		Reverse current (mA) peak	
	30° C	500° C	30° C	500° C	30° C	500° C	30° C	500° C
Power rectifier	4.2	2.8	1.5	1.5	200	150	0.01	0.40
Low current instrument device	5.0	2.5	0.5	0.5	500	300	0.001	0.008
High voltage (stacked unit)	31	20	0.01	0.01	2000	1500	0.001	0.005
Low voltage blocking diode	5.1	4.3	15.0	15.0	120	100	0.8	1.6

The encapsulation design is as important to successful high temperature operation as the SiC itself. It should fulfill among others the following requirements: protect the electrodes against oxidation, allow for the different thermal expansion coefficients of the components, and dissipate effectively the internally generated heat. Units have been made with thermal impedances as low as 0.25 degrees/Watt above 200° C. Such encapsulations permit operation at $100 \, A/cm^2$ continuously and at $3000 \, A/cm^2$ intermittently.

Measurements on diodes with junctions grown at lower temperatures [126] have verified the predicted high blocking capability. The experimental results even surpass the theoretical estimates by *Sze* and *Gibbons* (cf. Fig. 19) [136].

Another approach to high voltage devices is the epitaxial deposition. Again its main advantage over other methods of producing junctions is its flexibility and the well defined experimental conditions. The epitaxial growth of a homogeneous and uniform film over a large area is still a problem. Since diffusion is negligible, an optimal structure can be realized. As will be shown below, it is advantageous to start with a fairly heavily doped *n*-type substrate. Successively a high resistivity (no intentional doping) and a *p*-type layer are deposited by epitaxy. The width and the doping concentrations of these layers can be optimized with regard to the desired performance. In general heavily doped contact regions are desirable to ensure good injection. The high resistivity zone should not be wider than twice the ambipolar diffusion length [135].

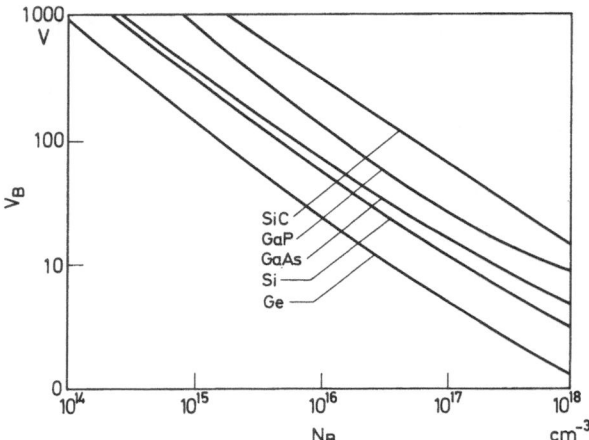

Fig. 19. Breakdown voltage V_B of various semiconductors as a function of background doping concentration N_B in abrupt junctions (after *Sze* and *Gibbons* [136]). The SiC line is estimated from the dependence of the breakdown voltage on energy gap which was derived for the other semiconductors

An important design consideration is the choice of the best structure. The doping of the compensated layer determines the breakdown voltage and its width the forward voltage drop as long as the heavily doped contact regions have a negligible resistance. In an epitaxial diode the resistance of the top layer can usually be neglected due to its thinness. Thus the problem reduces to choosing a suitable substrate. Conductivity is determined by the free (mobile) carrier concentration times the mobility. From the data sheet (Table 9) is seen that $\mu_n/\mu_p \approx 10$ to 15 at room temperature. Since the carrier concentration is an exponential function of the ionization energies and the donor ionization energy is about 3 times smaller than the acceptor ionization energies, the carrier concentration is one to two orders of magnitude higher in an *n*-type crystal than in a *p*-type substrate. If we take compensation into account the situation is even more favorable for the same degree of compensation. This is illustrated in Table 10 where the resistance of different substrates of 1 mm^2 area and a thickness of $200\,\mu\text{m}$ is shown. It is seen that with higher doping concentrations the disparity of *n*- vs. *p*-type crystals is lowered. Too high a doping is of no practical use however, since the mobilities decrease due to charge carrier scattering at impurities and the carrier concentration grows only as the square root of the doping concentration (a consequence of equations 4 and 5). It should also be kept in mind that experimentally it is relatively easy to reduce the acceptor concentration to 10^{17} cm^{-3}, but quite difficult to achieve that with

Table 9. *Semiconductor properties of SiC*

Property	Symbol	6 H	15 R	3 C	Unit
Exciton energy gap at 4° K [59]	E_{GX}	3.023	2.986	2.390	eV
Exciton energy gap at 300° K [59]		2.975	2.95	2.36	eV
Energy gap at 300° K [29]	E_G	2.86	2.9	2.2	eV
Temperature shift of energy gap	dE_G/dT	−3.3 [137]	—	−5.8 [138]	10^{-4} eV/°
Electron mobility at 300° K	μ_n	600 [47]	500 [49]	750 [139]	cm²/Vs
at 600° K		75 [47]	90 [49]	130 [139]	cm²/Vs
Hole mobility at 300° K	μ_p	55 [47]	—	20 [47]	cm²/Vs
at 600° K		8 [47]	—	7 [47]	cm²/Vs
Electron effective mass	m_e	1.2 [83]	1.0 [83]	0.4 [140]	m_0

Table 10. *The effect of compensation on the resistance of a crystal having a thickness of 200 µm and a cross sectional area of 1 mm²*

		Uncompensated		Compensated by 10^{17} impurities/cm³	
n-type	$N_D\,(\text{cm}^{-3})$	$n\,(\text{cm}^{-3})$	$R\,(\Omega)$	$n\,(\text{cm}^{-3})$	$R\,(\Omega)$
	10^{18}	3.6×10^{17}	0.12	3.0×10^{17}	0.14
	10^{20}	4.5×10^{18}	0.009	4.0×10^{18}	0.010
p-type	$N_A\,(\text{cm}^{-3})$	$p\,(\text{cm}^{-3})$	$R\,(\Omega)$	$p\,(\text{cm}^{-3})$	$R\,(\Omega)$
	10^{18}	8×10^{15}	52	5×10^{14}	830
	10^{20}	8×10^{16}	5.2	4×10^{16}	10

Mobility values of $\mu_n = 300\ \text{cm}^2/\text{Vs}$ and $\mu_p = 30\ \text{cm}^2/\text{Vs}$ were assumed. At doping levels of $10^{20}\ \text{cm}^{-3}$ actual mobilities may be lower due to impurity scattering. The donor ionization energy was taken as 0.1 eV (typical for nitrogen doping) and the acceptor ionization energy as 0.3 eV (typical for aluminum, the most common acceptor).

nitrogen since all the known acceptors are metallic and nitrogen is present even in high purity helium or argon (cf. 4.4 Purity).

Compensation may also affect the resistance of graded profiles severely. Solving Eq. (5) for a graded junction reveals that to a good approximation the free hole concentration is reduced by the factor N_D/N_C' $\approx N_D/10^{14}$ due to the background level of nitrogen. Under certain conditions this may be an important factor (e.g. acceptor diffusion into heavily doped n-type SiC).

7.3.3 Miscellaneous Experimental Devices

A natural spin-off of the rectifier development constitute $p - n$ junction detectors for photons or nuclear particles. Electron-hole pairs created by ionizing radiation are separated in the electric field of the junction and collected at the contacts to give rise to a charge or a voltage. The basic requirements for detectors are the same as for high voltage recti-fiers, namely high resistivity and purity. High resistivity is needed to maintain a high accelerating field for the electron-hole pairs. But this resistivity should not be achieved by compensation since the high con-centration of impurities gives rise to many levels which can act as trapping or recombination centers, thus effectively reducing the diffusion length of the generated carriers. This in turn leads to a low efficiency of the counter.

A design analysis shows that the surface layer, acting as a window for the radiation, should be very thin in comparison to the penetration depth in order to minimize the energy loss in this low-field region. This energy is wasted as far as the detector is concerned because the generated carriers recombine before being collected. On the other hand the depletion region (the high-field middle zone) should be nearly as wide as the range of the ionizing particles to ensure the largest possible signal. The width of the depletion region is controlled by the preparation technique for the junction and by the application of a reverse voltage on the junction. The bias voltage changes the distribution of holes and electrons which is determined by the field and their diffusion lengths. As the voltage is increased the depletion region is widened and the detec-tor capacitance lowered according to the equation

$$1/C^2 = 2/(\varepsilon e A^2) (1/N_A + 1/N_D) (V_D - V) \tag{10}$$

which is valid for an abrupt junction. The more the width matches the range of the particles, the higher the collection efficiency becomes. This might lead to the conclusion that any junction can be used efficiently as a detector. Beside the square root dependence ($d \sim \sqrt{V}$), it turns out in practice that by increasing the bias voltage, to match both range and depletion width more closely, the noise is increased too. For a given geometry there is an optimal bias voltage which maximizes the signal to noise ratio.

Two techniques of producing $p - n$ detectors have been proved successful, a third one is still in an exploratory stage. The first one [123] consists of diffusing aluminum into a relatively lightly doped n-type crystal ($N_D \approx 10^{16}$ per cm^3), followed by chemically thinning the front layer to reduce the junction depth. The second one [141] is to deposit epitaxially the p-type surface layer on a lightly doped n-type crystal or to go one step further and to deposit on a fairly heavily doped n-crystal an almost intrinsic layer to produce the depletion layer with the proper width, followed by the deposition of the top p-layer*. The epitaxial process makes the chemical thinning superfluous. A third technique, ion implantation, could eventually become important for very shallow junctions. In this method ions are accelerated to typically 100 keV and shot into the crystal lattice. So far junction formation by ion implantation has been observed in Ge, Si, and GaAs [142] but has not been very successful in SiC [143].

SiC diodes have been prepared which are capable of counting alpha particles up to 700° C [123]. About 10 eV are required to produce an electron-hole pair in SiC. These pairs have short lifetimes, of the order of 0.1 μs, which is the reason for reducing the thickness of the top layer to a minimum.

If a film of ^{235}U is placed in front of the detector, it can be used as neutron counter [144]. Thermal neutrons captured in the uranium produce fission. The fission products are ionizing particles and create electron-hole pairs in the solid state detector. *Ferber* and *Hamilton* [145] have studied such counters for flux mapping in a nuclear reactor. They concluded that the detectors could be used over a dynamic flux range of at least 10^4. The only limitations are the diode size and the thickness of the neutron conversion layer.

SiC, with bandgaps of typically 3 eV, is well suited for the detection of ultraviolet radiation. In this case photons are absorbed in the crystal and create electron-hole pairs. A simple model for a photovoltaic SiC detector [146] could relate the wavelength of maximum response to the junction depth and the depletion width of the diode. A more rigorous calculation [147] including such parameters as carrier lifetime, mobility, surface recombination velocity, and their temperature dependence agreed quite well with experimental data on peak response wavelength and its temperature variation.

A very important and perhaps crucial point in the future development of SiC devices is the ability to produce active, three-terminal devices, such as transistors. The short carrier lifetimes make a majority carrier device the logical choice. *Chang et al.* [148] developed the planar technology mentioned before to produce by diffusion and etching techniques a unipolar field effect transistor. Although the device characteristics deteriorate at 500° C, there is still a useful gain (Fig. 20). Epitaxial methods have not yet been successful because most of the fabrication steps require processing of the carbon face and good epitaxial growth has only been achieved on the silicon face.

---

* Annealing at 1800 to 2100° C seems to improve the counting characteristics because it was noted that the junction should neither be too abrupt nor highly graded.

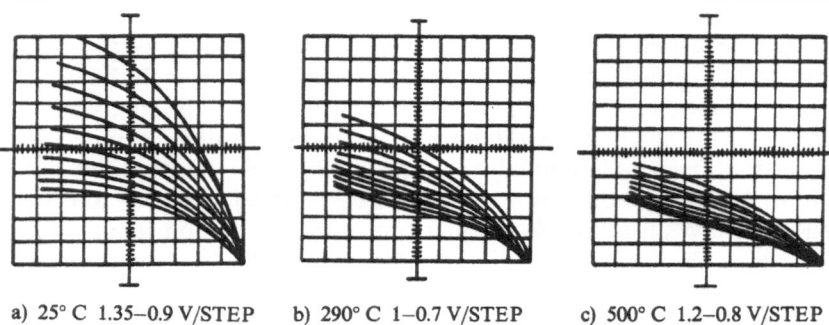

a) 25° C 1.35–0.9 V/STEP   b) 290° C 1–0.7 V/STEP   c) 500° C 1.2–0.8 V/STEP

Fig. 20 a–c. Family of drain characteristics of an experimental SiC field effect transistor at the indicated temperatures and gate voltages (horizontal 5 V/div, vertical 0.5 mA/div). a) 25° C, b) 290° C, c) 500° C

A $p-n$ junction shaped in such a way as to allow the emission of electrons into vacuum has been described first by *Patrick* and *Choyke* [149]. Later the work has been taken up again by *Brander* and *Todkill* [150] who constructed a cold cathode for cathode ray tubes. SiC is a particularly well suited semiconductor for this application since its high thermal stability and its excellent heat conductivity can cope with the heat generation and dissipation of such a device. The high energy gap facilitates electron emission into vacuum without the use of work function lowering coatings. To make the emission of hot electrons efficient, the $p-n$ junction should be as close to the surface as possible, possibly within one diffusion length, and the number of trapping and recombination centers should be low. Ion implantation is the best method for producing extremely shallow junctions but this technique has not yet reached the point where it can be utilized in SiC. Therefore small recesses were cut into the 10 μm thick epitaxial n-layer which ensured a reasonably low spreading resistance. The emission pattern observed on a cathode ray screen enlarges the junction area and shows distinct regions of avalanche breakdown distributed around the circumference of the cut out. Retarding field experiments yield the energy distribution of the electrons in the microplasma. The equivalent electron temperature is a steeply rising function of the bias voltage and reaches a maximum of 6500° K at about 12 V.

7.4 Problems and Promises of SiC

Presently the most serious difficulty for producing SiC devices in large quantities and economically is the lack of knowledge about the growth mechanism and the causes of polytypism. Although we have come far, many questions are still unanswered. The main reason for this is the difficulty of finding and controlling the parameters responsible for polytypic growth. Factors closely related to it are growth temperature, temperature distribution in the growth cavity, pressure, impurities, fluctuations of temperature and of impurity concentrations. It is quite possible that still other factors are important which have not yet been identified. Attempts have been made to keep the above parameters constant but the most successful growth method, the Lely technique, is not suited for precise, meaningful temperature readings and even less for the observation of the growing crystals. The most significant advances may come from a completely new method of growing SiC single crystals. It would not matter whether a multitude of parallel platelets of uniform quality results, or a solid bar of monocrystalline SiC similar to the well known silicon single crystals. The technologically important aspect is the uniformity in crystallographic respect, with regard to the impurity distribution, and thus the semiconducting properties. Even without such a new method significant improvements will result from better materials becoming more and more available. A first step in this direction has been made by *Kapteyns* and *Knippenberg* [32] who are using pyrolytic graphite as crucible material which assures higher purity and better control of the temperature distribution.

The availability of large quantities of identical single crystals of high purity may solve many of the other problems plaguing todays researchers. One such problem is certainly the occurrence of internal barriers manifesting themselves in abnormally high resistances parallel to the c-axis, large voltage drops in certain luminescent diodes not associated with the junction, and unusually high forward voltages in some rectifier diodes. These barriers are thought to result from coalescence on a microscopic scale or to be connected with disordered layers.

From the point of view of the device designer all characteristic properties are acceptable, some even excellent, with the exception of the carrier lifetimes (Table 9). No reliable data from pure crystals have been published. The generally accepted typical value, however, is $0.1 \, \mu s$ or less which is extremely low. It is hoped that better crystals, meaning crystals with no internal barriers, fewer crystalline defects, and higher purity, will show higher lifetimes because there are no theoretical reasons for the low figures.

Some of the difficulties are offset by clear advantages of SiC over other semiconductors and some of the properties which cause problems have their advantages too. *Shockley* [151] put it this way: "SiC suffers from the very same thing that makes it good." Certainly one of the latter is the occurrence of polytypes and its overwhelmingly advantageous influence on physical properties. This is especially true for light emitting and detecting devices. Here polytypism gives great flexibility and thus adds a new dimension since response cannot only be varied by different doping but also by using different polytypes. Another possibility arises in making heterojunctions which should be nearly ideal since the internal stresses are negligible and interfacial states due to lattice mismatch, chemical reactions, or compositional gradients are nonexistent.

Many positive aspects are related to the wide bandgap. Foremost among them are the capability of emitting blue light, high temperature performance, high voltage applications, and great reliability. Some of them have been covered on the preceding pages and will be omitted here.

Sze and *Gibbons* [136] have deduced a relation between semiconductor doping, bandgap, and breakdown voltage. If one inserts values of SiC into their general formulas, SiC leads all the other known semiconductors. *Van Opdorp* and *Vrakking* [126] have shown that the theoretical predictions are underestimates and that the breakdown voltage of SiC exceeds the one of silicon by an order of magnitude. This extraordinarily high breakdown strength compensates to a certain extent for the disappointingly low carrier lifetimes because it enables us to make the middle zone of a $p-i-n$-structure narrow enough to get the carriers across and still maintain high reverse blocking capabilities.

High reliability has been demonstrated in every laboratory. Although the test data are limited, life tests are encouraging indeed. Small rectifier diodes have run successfully 1000 hours at 500° C with no measurable change in electrical characteristics [123]. Electroluminescent diodes have not shown any detectable degradation over 15'000 hours at temperatures ranging from 20 to 400° C and carrying current densities of 5 to 20 A/cm² [131].

With its high temperature and high voltage capabilities, its ability to emit blue light, and its proven reliability, SiC has truly outstanding and attractive properties and will therefore capture its share of the semiconductor device market.

8. Conclusions

At the present state of development a device engineer may be reluctant to choose SiC, primarily because he cannot count on a vast supply of identical crystals as it is common for the well established semiconductors. Still, if he is faced with the problem of designing a rectifier operable at 500° C or a diode emitting blue light, he has no other choice than SiC and he will be surprised how closely most of the typical design parameters match those of silicon. A major reason that the exceptional properties of SiC have not been exploited commercially is the very limited effort put into SiC research as compared to the resources allocated for silicon, germanium, or even gallium arsenide.

We know that compound semiconductors are inherently more difficult to grow because they do not only pose the question of purity, a major project by itself, but also of stoichiometry. If the vapor pressures of the constituents vary widely, as they do in the case of SiC, severe problems may be encountered. Other difficulties treated extensively are the high growth temperature, the ease with which nitrogen, commonly considered to be inert, is incorporated as a donor, and the tendency to crystallize into a multitude of significantly different modifications.

These problems are balanced, however, by superior performance in many applications as well as by the scientific insights to be gained from the study of this material. In SiC the electronic properties are not only influenced by impurities but also by the crystalline structure. The first effect is common to all semiconductors but the second one is masked in a semiconductor series (e.g. II–VI or III–V compounds) by the changes due to the chemical composition.

Each impurity has its characteristic ionization energy which differs from semiconductor to semiconductor. In germanium and silicon it can be computed using the effective mass approximation so that reasonable agreement with experiment is found for shallow levels. The information about SiC is limited to the detailed analysis of nitrogen as a donor, some data on aluminum, and the deductions made from electroluminescent experiments. All these findings indicate that SiC fits the general picture only if each polytype is considered to be a different semiconductor, in clear contrast to the effective mass approximation. But SiC is different in still another respect. The same impurity has several ionization energies in one and the same polytype due to the existence of the inequivalent sites in the large unit cell. Therefore no other semiconductor provides the opportunity to study

a) the effect of slight changes in neighbor arrangements on the properties of an impurity (by comparing the same impurity at inequivalent lattice sites in the same polytype) and

b) the effect of changes in the bandstructure, but not the composition, on the semiconductor properties (by comparing different polytypes).

Up to now it was assumed that semiconductor properties were adequately described by nearest neighbor interactions and long range forces were generally neglected. In SiC it has been demonstrated convincingly that at least the properties related to zone boundary wave vectors (transitions involving the conduction band) are affected by the mutual position of all the atoms in a volume whose height is large in comparison to the nearest neighbor distance. Further it is not unreasonable to speculate that the occurrence of so many well defined polytypes with distinct electronic properties is caused by a unique and complex interdependence of crystal growth and bandstructure. A model, able to account for the observed effects in SiC, should lead to a much deeper understanding of the solid state, since a theory which explains the mutual influences of the detailed atomic configuration and the electronic properties could also provide a new basis for the theoretical treatment of amorphous semiconductors or liquid crystals.

Undoubtedly the strong crosslinks between basic investigations and applied physics which proved fruitful in the past will continue to be a major factor and SiC research will primarily be stimulated by the need for new materials in new applications, but it will by itself promote a deeper understanding of semiconductor physics and the general knowledge of the solid state.

Acknowledgments: Thanks are due to Brown Boveri & Co., Ltd. for the permission to publish this review. I am especially grateful to *A. J. Perry* for his suggestions with regard to content and style, and to my coworkers, *M. Königer* and *G. Gramberg,* for their valuable assistance before and during the preparation of this manuscript. I would also like to thank all my colleagues in the "International Committee on SiC" for the friendly and open discussions and the important advance information, so necessary for an up-to-date review. Last but not least I appreciate the great patience of my dear wife and our children while I was busy writing this article.

References

1. *Moissan, H.:* C. R. Acad. Sci. Paris **140**, 405 (1905).
2. *Acheson, A. G.:* Brit. Pat. 17911 (1892).
3. *Lossew, O. W.:* Telegraphy and Telephony **18**, 61 (1923).
4. *Busch, G.:* Helv. Phys. Acta **19**, 167 (1946).
   — *Labhart, H.:* Helv. Phys. Acta **19**, 463 (1946).
5. *Kendall, J. T., Yeo, D.:* Proc. Int. Congr. Pure Appl. Chem. London **11**, 171 (1947).
6. *Lely, J. A.:* Ber. Deut. Keram. Ges. **32**, 229 (1955).
7. *Shaffer, P. T. B.:* Am. Ceram. Soc. **47**, 466 (1964).
8. *Knippenberg, W. F., Verspui, G.:* Mat. Res. Bull. **4**, S. 45 (1969).
9. — — Mat. Res. Bull. **4**, S. 33 (1969).
10. *Scace, R. I., Slack, G. A.:* Silicon Carbide – A High Temperature Semiconductor, p. 24 (Ed. *J. R. O'Connor* and *J. Smiltens*). Oxford: Pergamon Press 1960 (in the following abbreviated "SiC–1960").
11. *Drowart, J., de Maria, G.:* SiC – 1960, 16 (1960).
12. *Kamath, G. S.:* Mat. Res. Bull. **4**, S. 57 (1969).
13. *Knippenberg, W. F.:* personal communication.
14. *Yudin, B. F., Borisov, B. G.:* Refractories **7–8**, 499 (1967).
15. *Shaffer, P. T. B.:* Mat. Res. Bull. **4**, S. 13 (1969).
16. *Spitzer, W. G., Kleinmann, D., Walsh, D.:* Phys. Rev. **113**, 127 (1959).
17. *Jennings, V. J.:* Mat. Res. Bull. **4**, S. 199 (1969).
18. *Baumhauer, H.:* Z. Krist. **50**, 33 (1912).
    — Z. Krist. **55**, 249, (1915).

19. *Verma, A. R., Krishna, P.:* Polymorphism and Polytypism in Crystals. New York: John Wiley & Sons, Inc. 1966.
20. *Jagodzinski, H.:* Acta Cryst. **2**, 201 (1949).
21. *Zhdanov, G. S.:* Dokl. Akad. Nauk. S.S.S.R. **48**, 39 (1945).
22. *Ramsdell, L. S.:* Am. Min. **32**, 64 (1947).
23. *Golightly, J. P., Beaudin, L. J.:* Mat. Res. Bull. **4**, S. 119 (1969).
24. *Frank, F. C.:* Phil. Mag. **42**, 1014 (1951).
25. *Jagodzinski, H.:* Neues Jahrb. Miner. Monatsh. **3**, 49 (1954).
    — Acta Cryst. **7**, 300 (1954).
26. *Ramsdell, L. S. Kohn, J. A.:* Acta Cryst. **5**, 215 (1952).
27. *Bulakh, B. M.:* J. Cryst. Growth **5**, 243 (1969).
28. *Gomes de Mesquita, A. H.:* J. Cryst. Growth **3**, 747 (1968).
29. *Knippenberg, W. F.:* Philips Res. Rept. **18**, 161 (1963).
30. *Kuin, P. N.:* Mat. Res. Bull. **4**, 273 (1969).
31. *Kroko, L. J.:* J. Electrochem. Soc. **113**, 801 (1966).
32. *Kapteyns, C. J., Knippenberg, W. F.:* J. Cryst. Growth **7**, 20 (1970).
33. *Wagner, R. S., Ellis, W. C.:* Appl. Phys. Letters **4**, 89 (1964).
34. *Merz. K. M., Adamsky, R. F.:* J. Amer. Chem. Soc. **81**, 250 (1959).
    Adamsky, R. F., Merz, K. M.: Z. Krist. **111**, 350 (1959).
35. *Spielmann, W.:* Z. Angew. Phys. **19**, 93 (1965).
    Campbell, R. B., Chu, T. L.: J. Electrochem. Soc. **113**, 825 (1966).
36. *Tanenbaum, M.:* "SiC – 1960", 132 (1960).
37. *Hall, R. N.:* J. Appl. Phys. **29**, 914 (1958).
    Halden, F. A.: "SiC – 1960", 115 (1960).
    Rosengreen, A.: Mat. Res. Bull. **4**, S. 355 (1969).
38. *Brander, R. W., Sutton, R. P.:* Brit. J. Appl. Phys. Ser. 2, Vol. **2**, 309 (1969).
39. *van Wieringen, J. S.:* Semiconductors and Phosphors, p. 367. New York: Interscience Publ. Inc. 1958.
40. *Woodbury, H. H., Ludwig, G. W.:* Phys. Rev. **124**, 1083 (1961).
41. *Hardeman, G. E. G., Gerritsen, G. B.:* Mat. Res. Bull. **4**, S. 261 (1969).
42. *Veinger, A. I.:* Sov. Phys.-Semicond. **3**, 52 (1969).
    — *Shulekin, A. F.:* Sov. Phys.-Semicond. **3**, 113 (1969).
43. a) *Jones, H.:* The Theory of Brillouin Zone and Electronic States in Crystals, Chap. 5. Amsterdam: North-Holland Publ. Comp. 1960.
43. b) *Verma, A. R.:* Crystal Growth and Dislocations, Chapter 7. London: Butterworths Sci. Publ. Ltd. 1953.
    Mitchell, R. S.: Z. Krist. **109**, 1 (1957).
    Ott, H.: Z. Krist. **63**, 1 (1926).
    Smith, F. G.: Am. Mineralogist **40**, 658 (1955).
44. *Gomes de Mesquita, A. H.:* Acta Cryst. **23**, 610 (1967).
45. *Patrick, L.:* Mat. Res. Bull. **4**, S. 129 (1969).
46. *van der Pauw, L. J.:* Philips Res. Rept. **13**, 1 (1958).
47. *van Daal, H. J.:* Philips Res. Rept. Suppl. No. 3 (1965).
48. *Patrick, L.:* J. Appl. Phys. **38**, 50 (1967).
49. *Barrett, D. L. Campbell, R. B.:* J. Appl. Phys. **38**, 53 (1967).
50. *Patrick, L.:* J. Appl. Phys. **37**, 4911 (1966).
51. *Bosch, G.:* J. Phys. Chem. Solids **27**, 795 (1966).
52. *de Boer, J. H., van Geel, W. C.:* Physica **2**, 186 (1935).
53. *Lomakina, G. A.:* Sov. Phys. – Solid State **8**, 1038 (1966).
54. *Moore, M. J.:* J. Electrochem. Soc. **116**, 109 (1969).
55. *van Daal, H. J., Knippenberg, W. F., Wasscher, J. D.:* J. Phys. Chem. Solids **24**, 109 (1963).

56. *Debye, P. P., Conwell, E. M.:* Phys. Rev. **93**, 693 (1954).
    Pearson, G. L., Bardeen, J.: Phys. Rev. **75**, 865 (1949).
57. *Dean, P. J.:* Luminescence of Inorganic Solids, p. 120. (Ed. by *P. Goldberg*). New York: Academic Press 1966.
58. *McLean, T. P.:* Progress in Semiconductors, Vol. 5, p. 55. London: Heywood and Co. Ltd. 1960.
59. *Choyke, W. J., Patrick, L.:* Phys. Rev. **127**, 1868 (1962) (6 H).
    Hamilton, D. R., Choyke, W. J., Patrick, L.: Phys. Rev. **131**, 127 (1963) (6 H).
    Patrick, L., Hamilton, D. R., Choyke, W. J.: Phys. Rev. **132**, 2023 (1963) (15 R).
    Choyke, W. J., Hamilton, D. R., Patrick, L.: Phys. Rev. **133**, A 1163 (1964) (3 C).
    Patrick, L., Choyke, W. J., Hamilton, D. R.: Phys. Rev. **137**, A 1515 (1965) (4 H).
    Hamilton, D. R., Patrick, L., Choyke, W. J.: Phys. Rev. **138**, A 1472 (1965) (21 R).
    Choyke, W. J., Hamilton, D. R. Patrick, L.: Phys. Rev. **139**, A 1262 (1965) (33 R).
    Patrick, L., Hamilton, D. R., Choyke, W. J.: Phys. Rev. **143**, 526 (1966) (2 H).
60. *Lampert, M. A.:* Phys. Rev. Letters **1**, 450 (1958).
61. *Haynes, J. R.:* Phys. Rev. Letters **4**, 361 (1960).
62. *Thomas, D. G., Hopfield, J. J.:* Phys. Rev. **128**, 2135 (1962).
63. *Patrick, L., Hamilton, D. R., Choyke, W. J.:* Phys. Rev. **132**, 2023 (1963).
64. *Hopfield, J. J.:* J. Phys. Chem. Solids **15**, 97 (1960).
65. *Hamilton, D. R., Choyke, W. J., Patrick, L.:* Phys. Rev. **131**, 127 (1963).
66. *Zanmarchi, G.:* Philips Res. Rept. **20**, 253 (1965).
67. *Gorban', I. S., Suleimanov, Yu. M., Shvaidak, Yu. M.:* Sov. Phys.-Semicond. **3**, 101 (1969).
68. — *Mishinova, G. I., Suleimanov, Yu. M.:* Sov. Phys.-Solid State **7**, 2991 (1966).
69. *Violin, E. E., Kholuyanov, G. F.:* Sov. Phys.-Solid State **6**, 1331 (1964).
70. *Gorban', I. S., Rud'ko, S. N.:* Sov. Phys.-Solid State **5**, 995 (1963).
    Tomashpol'skii, F. G., Kholuyanov, G. F.: Sov. Phys.-Solid State **3**, 1960 (1962).
71. *Kholuyanov, G. F.:* Sov. Phys.-Solid State **4**, 2322 (1963).
72. a) *Bukke, E. E., Vinokurov, L. A., Gorban', I. S., Gumenyuk, A. F., Suleimanov, Yu. M., Fok, M. V.:* Sov. Phys.-Semicond. **2**, 163 (1968).
72. b) — — — — — — Sov. Phys.-Semicond. **1**, 1164 (1967).
73. *Addamiano, A.:* J. Electrochem. Soc. **111**, 1294 (1964).
    Kholuyanov, G. F.: Sov. Phys.-Solid State **6**, 2668 (1965).
74. a) *Addamiano, A., Potter, R. M., Ozarow, V.:* J. Electrochem. Soc. **110**, 517 (1963).
74. b) — J. Electrochem. Soc. **113**, 134 (1966).
75. *Kholuyanov, G. F.:* Sov. Phys.-Solid State **7**, 2620 (1966).
76. *Gorban', I. S., Marazuev, Yu. A., Suleimanov, Yu. M.:* Sov. Phys.-Semicond. **1**, 514 (1967).
77. *Zanmarchi, G.:* J. Phys. Chem. Solids **29**, 1727 (1968).
78. *Patrick, L., Choyke, W. J.:* J. Appl. Phys. **30**, 236 (1959).
79. *Knippenberg, W. F.:* Personal communication.
    Feitknecht, J.: unpublished.
80. *Lely, J. A., Kröger, F. A.:* Semiconductors and Phosphor, p. 525. New York: Interscience Publishers, Inc. 1958.
81. *Ellis, B., Moss, T. S.:* Proc. Royal Soc. London **299**, Ser. A, 383 and 393 (1967).
82. *Choyke, W. J., Hamilton, D. R., Patrick, L.:* Phys. Rev. **139**, A 1262 (1965).
83. *Lomakina, G. A.:* Author's Abstract of Dissertation for Candidate's Degree (in Russian), IPAN SSSR, Leningrad, 1967.
84. *Biedermann, E.:* Solid State Comm. **3**, 343 (1965).
85. *Dubrovsky, G. B., Radovanova, E. I.:* Phys. Letters **28** A, 283 (1968).
86. — Personal communication.
87. *Blank, J. M., Potter, R. M.:* 2nd Meeting of the International Committee on SiC, Vienna, 1969 (unpublished).

88. *Choyke, W. J., Patrick, L.:* Phys. Rev. **172**, 769 (1968).
89. *Wheeler, B. E.:* Solid State Comm. **4**, 173 (1966).
90. *Junginger, H. G., van Haeringen, W.:* Phys. Stat. Sol. **37**, 709 (1970).
91. *Brafman, O., Steinberger, I. T.:* Phys. Rev. **143**, 501 (1966).
92. *Patrick, L.:* Phys. Rev. **167**, 809 (1968).
93. — *Choyke, W. J.:* Phys. Rev. **123**, 813 (1961).
94. *Feldman, D. W., Parker, J. H., Jr., Choyke, W. J., Patrick, L.:* Phys. Rev. **170**, 698 (1968) and **173**, 787 (1968).
95. *Brüesch, P., Feitknecht, J.:* unpublished results.
96. *Hofman, D., Lely, J., Volger, J.:* Physica **23**, 236 (1957).
97. *Spitzer, W. G., Kleinman, D. A., Frosch. C. J., Walsh, D. J.:* Phys. Rev. **113**, 133 (1959).
98. *Choyke, W. J., Patrick, L., Cardona, M.:* Bull. Am. Phys. Soc. **11**, 764 (1966).
99. *Baars, J. W.:* II–VI Semiconducting Compounds, p. 631. (*D. G. Thomas*, Ed.). New York: W. A. Benjamin, Inc. 1967.
100. *Bosch, G.:* Philips Res. Rept. **16**, 445 (1961).
101. *Slack, G. A:* J. Appl. Phys. **35**, 3460 (1964).
102. *Taylor, A., Jones, R. M.:* "SiC – 1960", 147 (1960).
103. *Grüneisen, E.:* Handbuch der Physik **10**, 1 (1926).
104. *Kobayasi, S.:* J. Phys. Soc. Japan **13**, 261 (1958).
105. *Bassani, F., Yoshimine, M.:* Phys. Rev. **130**, 20 (1963).
106. *Herman, F., Van Dyke, J. P., Kortum, R. L.:* Mat. Res. Bull. **4**, S. 167 (1969).
107. *Choyke, W. J.:* Personal communication.
108. *Patrick, L.: Hamilton, D. R., Choyke, W. J.:* Phys. Rev. **143**, 526 (1966).
109. *Destriau, G, Ivey, H. F.:* Proc. IRE **43**, 1911 (1955).
110. *Lehovec, K., Accardo, C. A., Jamgochian, E.:* Phys. Rev. **83**, 603 (1951) and **89**, 20 (1953).
111. *Patrick, L.:* J. Appl. Phys. **28**, 765 (1957).
112. *Greebe, C. A. A. J.:* Philips Res. Rept. Suppl. No. 1 (1963).
113. *Blank, J. M.:* Mat. Res. Bull. **4**, S. 179 (1969).
114. *Violin, E. E., Kholuyanov, G. F.:* Sov. Phys.-Solid State **6**, 465 (1964).
115. *Kal'nin, A. A., Pasynkov, V. V., Tairov, Yu. M., Yas'kov, D. A.:* Fiz. Tverd. Tela **8**, 2982 (1966) (Sov. Phys.-Solid State **8**, 2381).
116. *Slack, G. A.:* J. Chem. Phys. **42**, 805 (1965).
     Kroko, L. J., Milnes, A. G.: Solid State Electr. **9**, 1125 (1966).
117. *Knippenberg, W. F.:* Personal communication.
118. *Pell, E. M.:* J. Appl. Phys. **31**, 291 (1960).
119. *Gramberg. G., Königer, M.:* Unpublished results.
120. *Bonnke, M., Fitzer, E.:* Ber. Deut. Keram. Ges. **43**, 180 (1966).
121. *Faust, J. W., Jr.:* SiC – 1960, 403 (1960).
122. *Gabor, T., Jennings, V. J.:* J. Electrochem. Tech. **3**, 31 (1965).
123. *Campbell, R. B., Berman, H. S.:* Mat. Res. Bull. **4**, S. 211 (1969).
124. *Gramberg, G., Königer, M.:* To be published.
125. *Brander, R. W., Boughey, A. L.:* Brit. J. Appl. Phys. **18**, 905 (1967).
126. *van Opdorp, C., Vrakking, J.:* J. Appl. Phys. **40**, 2320 (1969).
127. *Mead, C. A.:* Solid State Electron. **9**, 1023 (1966).
128. *Hagen, S. H.:* J. Appl. Phys. **39**, 1458 (1968).
129. *Goodman, A. M.:* J. Appl. Phys. **34**, 329 (1963).
     Heine, V.: Phys. Rev. **138**A, 1689 (1965).
130. *Violin, E. E., Kal'nin, A. A., Pasynkov, V. V., Tairov, Yu, M., Yas'kov, D. A.:* Mat. Res. Bull. **4**, S. 231 (1969).
131. *Brander, R. W.:* Mat. Res. Bull. **4**, S. 187 (1969).
132. *Todkill, A., Brander, R. W.:* Mat. Res. Bull. **4**, S. 293 (1969).

133. *Potter, R. M.:* Mat. Res. Bull. **4**, S. 223 (1969).
134. *Lorenz, M. R.:* Trans. Met. Soc. AIME **245**, 539 (1969).
135. *Herlet, A., Spenke, E.:* Z. Angew. Phys. **7**, 99/149/195 (1955).
136. *Sze, S. M., Gibbons, G.:* Appl. Phys. Letters **8**, 111 (1966).
137. *Choyke, W. J., Patrick, L.:* SiC – 1960, 306 (1960).
138. *Dalven, R.:* J. Phys. Chem. Solids **26**, 439 (1965).
139. *Nelson, W. E., Rosengreen, A., Bartlett, R. W., Halden, F. A., Marsh, L. E.:* AFCRL-67-0218, Final Rept. Air Force Cambridge Research Laboratories, Bedford, Mass. (1967).
140. *Pickar, P. B., et al.:* Summary Tech. Rept. No. OR 8394, AD-641 198 (1966).
141. *Königer, M.:* Thesis Techn. University, Braunschweig (1970).
142. *McCaldin, J. O.:* Progress in Solid-State Chemistry, Vol. 2. (H. Reiss, Ed.). Oxford: Pergamon Press 1965.
     Mayer, J. W., Marsh, O. J., Mankarious, R., Bower, R.: J. Appl. Phys. **38**, 1975 (1967).
     Mayer, J. W., Marsh, O. J., Shifrin, G. A., Baron, R.: Can. J. Phys. **45**, 4073 (1967).
143. *Matzke, Hj., Königer, M.:* Phys. Stat. Sol. (A) **1**, 469 (1970).
144. *Canepa, P. C., Malinaric, P., Campbell, R. B., Ostroski, J. W.:* IEEE Trans. NS–11, 262 (1964).
145. *Ferber, R. R., Hamilton, G. N.:* Unpublished.
146. *Campbell, R. B., Chang, H. C.:* Solid State Electr. **10**, 953 (1967).
147. — — Internat. Electron Devices Meeting, Paper 9.1, Washington D. C., Oct. 1967 (to be published).
148. *Chang, H. C., Formigoni, N. P., Roberts, J. S.:* Final Rept. to Marshall Space Flight Center (Huntsville, Alabama), Contract NAS 8-11861 (1966).
149. *Patrick, L., Choyke, W. J.:* Phys. Rev. Letters **2**, 48 (1959).
150. *Brander, R. W., Todkill, A.:* Mat. Res. Bull. **4**, S. 303 (1969).
151. *Shockley, W.:* SiC – 1960, XVII, 1960.

Dr. *J. Feitknecht*
Grundweg 2
CH-6460 Altdorf
Switzerland

High Energy Treatment of Atomic Collisions

K. DETTMANN

Contents

Introduction

In the theory of atomic scattering one investigates the cross sections for collisions between two atomic systems, like atoms, ions or electrons. A scattering process can be elastic or inelastic. In an elastic collision no electronic transitions occur and the kinetic energy of the collision partners is the same after the collision and before. Elastic collisions are often described approximately by potential scattering assuming a static potential for the interaction of the colliding systems. This theory is well known and the high energy limit has been investigated by *Glauber* [1].

Inelastic collisions are much more complicated as at least a three body problem is involved. There are two types of inelastic collisions, direct and rearrangement collisions. In a direct collision an electron is excited or ionized, while it is replaced or captured in a rearrangement collision. The simplest example is a proton-hydrogen collision.

There are two theoretical descriptions of inelastic collisions. The common one is the *Lippmann-Schwinger* approach [2], which is well known from potential scattering. Recently *Faddeev* [3] proposed a different description, in which for instance a three body problem is decomposed in a series of two body collisions. Consequently not the potential, as in the *Lippmann-Schwinger* equation, but the *T*-matrix for a two body collision is the fundamental quantity. For a high energy treatment the *Lippmann-Schwinger* approach is the natural one, so that we will not deal with *Faddeev* equations.

The scattering formalism is treated in a number of textbooks [4—9], reproducing the theories of *Lippmann-Schwinger* [2, 10] or *Gell-Mann* and *Goldberger* [11]. There are many applications in atomic physics, which for high energies are based on the Born approximation. We will list here some of those, which are concerned with proton-hydrogen collisions and seem typical to us for the general development. *Bethe*'s review article [12] deals with excitation and ionisation processes in the Born approximation. A particular review paper on Born expansions has been given by *Holt* and *Moiseiwitsch* [13]. *Bates* [14] and recently *Bransden* [15] review the charge exchange collisions. Excitation- and ionisation cross sections have been calculated in the Born approximation by *Bates* and *Griffing* [16]. Many authors [17—25] have investigated the high energy limit for the charge exchange cross section. In the high energy limit of the first Born approximation the internuclear repulsion contributes substantially to the charge exchange cross section. This seems unreasonable (see Chapter IV), and other approximations, like the impulse- and distorted wave Born approximation [26, 27] have been proposed. Some authors [28, 29] suggested that the Born series for rearrangement collisions diverges for all energies. But a simple counter example [30, 31] and an exactly solvable three body problem [32] showed that their arguments are not conclusive. The correct high energy limit is given by the first and second Born approximation and has been calculated by *Drisko* [22]. Here the internuclear repulsion cancels.

In the impact parameter treatment of atomic collisions the nuclei are assumed to be infinitely heavy and to move on a straight path with fixed impact parameter and constant velocity. By this semiclassical theory only the electronic transitions are treated quantummechanically, and the scattering theory is reduced to a one body problem with time dependent interaction. *Moiseiwitsch* [33] and *Mittleman* [34] discussed the validity of the impact parameter treatment and calculated [34] direct- and capture cross sections for proton-hydrogen scattering.

The following article is intended to establish the basic theory for a high energy treatment of atomic collisions. We will always stick to the three body problem illustrated in Fig. 1, though most of the formalism can be generalized without any change to more complicated systems. As far as the general theory is concerned we will not specify the potentials and masses, so that also nuclear scattering is included. To illustrate the theory, proton-hydrogen collisions or potential scattering will serve as an example.

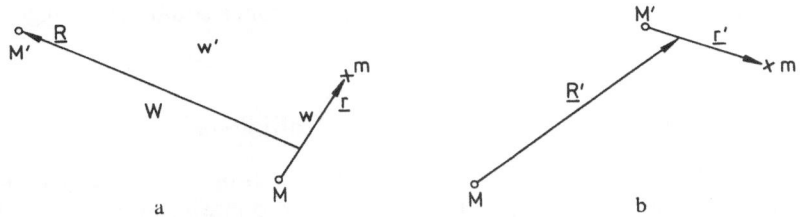

Fig. 1. Three body collision $M' + (M, m)$ with interactions w, w', W, and a direct and b rearrangement coordinates

In Chapter I we set up a wave packet theory for direct and rearrangement collisions. The result is well known from the common formal scattering theory [2, 10, 11]. It provides the prescription how to calculate the various differential cross sections from the T-matrices. The wave packet treatment is not only the proper quantum mechanical method but in Coulomb scattering it is absolutely necessary because here formal stationary scattering theory breaks down. The Coulomb case is treated in Chapter II. This solves an old problem, as Coulomb scattering does not fit in the framework of the ordinary formal scattering theory. Recently the interest in Coulomb scattering revived after *West* [35] had shown that the Coulomb functions are solutions of the homogeneous *Lippmann-Schwinger* equation and the T-matrix vanishes on the energy shell. In Chapter III we formulate the impact parameter treatment and justify it for a one dimensional model. It turns out that the impact parameter theory is valid for $\dfrac{m}{M, M'} \ll 1$ (Fig. 1) and $|\varepsilon_{i,f}|/E_i \ll 1$, where $\varepsilon_{i,f}$ are the binding energies of the bound state subsystem before and after the collision with total energy E_i. In this sense the impact parameter theory is a high energy treatment. In Chapter IV, we calculate in an asymptotic expansion for high energies E_i the leading term of the direct and re-arrangement cross section. This is possible in a Born series approach and it turns out that high energy here means $\dfrac{E_i}{|\varepsilon_i|}\dfrac{m}{M} \gg 1$. For atomic collisions this requires energies being three orders of magnitude larger than in the impact parameter theory. As an example we calculate for proton-hydrogen collisions the total inelastic and charge exchange cross section, and the inelastic angular dispersion for fast protons in a hydrogen gas. In Chapter V we set up the *Schwinger* variational principle [2] for the T-matrix of direct- and rearrangement collisions. With a simple ansatz for the wave function a renormalized second Born approximation results. This is exact in the high energy limit and seems to be useful even for low energies, as the results for potential scattering from a square well suggest.

I. Wave Packet Theory of Inelastic Collisions

In scattering theory one investigates the collision between two systems, which are thought to be initially in well defined quantum states and well separated such that their interaction is vanishingly small. The simplest case is potential scattering where the systems are two masses M_1 and M_2 interacting by a potential dependent on the relative coordinate of M_1 and M_2. In this case only elastic collisions occur, i.e. the kinetic energy of M_1 and M_2 is the same after the collision and before. Inelastic scattering is possible for collisions with at least three particles. The simplest situation is illustrated in Fig. 1. Particle M' collides with the bound state system (M, m). The interactions are the potentials $w(r)$, $w'(r')$, $W(r - r')$ depending on the relative coordinates of (M, m), (M', m) and (M, M') respectively. In atomic physics this corresponds to a proton-hydrogen collision. The inelastic collisions for this or any more complicated system can be divided up in direct and rearrangement collisions. In the former ones m is either excited into a higher bound state or ionized into a free state, while in the latter ones m is picked up by M'.

The quantity to be calculated is the differential cross section for any of these collisions. For the sake of simplicity we will set up the scattering theory for the three body system of Fig. 1. The generalisation to more involved systems is straightforward.

In the centre of mass system the wave function depends on the relative coordinates r, R (Fig. 1) and t, and the *Schroedinger* equation is

$$H\Psi(r, R, t) = i\hbar\dot\Psi(r, R, t) \tag{1.1}$$

with the Hamiltonian

$$H = -\frac{\hbar^2}{2v}\,\partial^2_{\mathbf{R}} - \underbrace{\frac{\hbar^2}{2\mu}\,\partial^2_{\mathbf{r}} + w}_{H_i} + \underbrace{w' + W}_{V_i}$$

$$= -\frac{\hbar^2}{2v'}\,\partial^2_{\mathbf{R}'} - \underbrace{\frac{\hbar^2}{2\mu'}\,\partial^2_{\mathbf{r}'} + w'}_{H_f} + \underbrace{w + W}_{V_f}.$$

$$(1.2)$$

The second representation of H in (1.2) follows from the first by the coordinate transformation

$$\begin{aligned} r' &= -\mathbf{R} + \zeta r \\ \mathbf{R}' &= \xi \mathbf{R} + \eta r . \end{aligned} \quad ; \quad \xi = \frac{M'}{M'+m}, \quad \zeta = \frac{M}{M+m}, \quad \eta = 1 - \zeta\xi \quad (1.3)$$

We have introduced in (1.2) the reduced masses

$$\begin{aligned} v &= \frac{M'(M+m)}{M'+M+m} & v' &= \frac{M(M'+m)}{M+M'+m} \\ \mu &= \frac{mM}{m+M} & \mu' &= \frac{mM'}{m+M'} . \end{aligned}$$

$$(1.4)$$

The scattering state $\Psi(t)$ follows from (1) with the initial condition $\Psi(t \to -\infty) = \phi_i(t)$. Here $\phi_i(t)$ is the incoming wave packet describing the collision – partners approaching each other from large distance. The collision takes place at $t = 0$, when we choose a wave packet of the form

$$\begin{aligned} \phi_i(t) &= \int d\mathbf{K}\, A(\mathbf{K})\, e^{-\frac{i}{\hbar} E_i(\mathbf{K})t}\, \varphi_i(\mathbf{K}) \\ \phi_i(\mathbf{K}) &= e^{i\mathbf{K}\mathbf{R}}\varphi_i(r); \quad H_i\phi_i(\mathbf{K}) = E_i(\mathbf{K})\,\phi_i(\mathbf{K}); \\ E_i(\mathbf{K}) &= \frac{\hbar^2 K^2}{2v} + \varepsilon_i \end{aligned} \quad (1.5)$$

with real $A(\mathbf{K})$. The initial normalized bound state of (m, M) is $\varphi_i(r)$ with binding energy ε_i. The representation (1.5) of $\phi_i(t)$ is in terms of eigenstates of H_i, the initial Hamiltonian defined in (1.2). Therefore

$$H_i\phi_i(t) = i\hbar\dot{\phi}_i(t) \quad (1.6)$$

i.e. $\phi_i(t)$ describes the motion of the wave packet with vanishing initial interaction V_i. From the normalisation $(\phi_i(t), \phi_i(t)) = 1$ follows

$$1 = (2\pi)^3 \int d\mathbf{K}|A(\mathbf{K})|^2 . \quad (1.5a)$$

Assuming a momentum distribution $|A(\mathbf{K})|^2$ as a Gaussian concentrated near $\mathbf{K}_i$ means

$$A(K) = \left\{ \frac{1}{(2\pi)^3} \frac{e^{-(\mathbf{K}-\mathbf{K}_i)^2/\varkappa^2}}{(\pi)^{3/2}\varkappa^3} \right\}^{1/2}$$

$$\cong \frac{1}{(2\pi)^{3/2}} \{\delta(\mathbf{K}-\mathbf{K}_i)\}^{1/2}{}^{\star} \tag{1.5b}$$

where $\varkappa$ is the width being small with respect to $\mathbf{K}_i$. Then (1.5) with (1.5b) is a wave packet moving with momentum $\mathbf{K}_i$ and describing maximum overlap of the colliding systems at $t = 0$.

The *Schroedinger* equation (1.1) with $\Psi(t \rightarrow -\infty) = \phi_i(t)$ is equivalent to the integral equation

$$\Psi(t) = \phi_i(t) + \int\limits_{-\infty}^{\infty} G_i(t-t')\, V_i \Psi(t')\mathrm{d}t' \tag{1.7}$$

where $G_i(t)$ is the time dependent *Green*'s operator to be calculated from

$$(-H_i + i\hbar\partial_t)\, G_i(t) = \delta(t). \tag{1.8}$$

With (1.6), (1.8) and $H = H_i + V_i$ (1.7) can be verified by differentiating with respect to time.

By Fouriertransformation of (1.8) with

$$G_i(t) = \frac{1}{2\pi} \int G_i(\omega)e^{-i\omega t}\mathrm{d}\omega$$

$$\delta(t) = \frac{1}{2\pi} \int e^{-i\omega t}\mathrm{d}\omega \tag{1.9}$$

one obtains

$$G_i(\omega) = \frac{1}{\hbar\omega - H_i + i\eta}, \qquad \eta \rightarrow 0 \tag{1.10}$$

and

$$G_i(t) = \Theta(t)\frac{1}{i\hbar}\, e^{-\frac{i}{\hbar}H_i t}, \qquad \Theta(t) = \begin{cases} 1 & t>0 \\ 0 & t<0. \end{cases} \tag{1.10a}$$

By adding $i\eta \rightarrow 0$ in (1.10) the integration in (1.9) has been defined in such a way, that $G_i(t)$ is the retarded *Green*'s operator, i.e. $G_i(t<0) = 0$. This implies in (1.7) that the scattered wave vanishes for $t \rightarrow -\infty$ and the initial condition holds.

With the convolution theorem (1.7) can be Fourier transformed

$$\Psi(\omega) = \phi_i(\omega) + G_i(\omega)\, V_i \Psi(\omega) \tag{1.11}$$

$\star$ This is a short hand notation to indicate that $A^2(k)$ behaves like a δ-function.

where

$$\Psi(\omega) = \int \Psi(t)\, e^{i\omega t}\, dt$$

$$\Psi(t) = \frac{1}{2\pi} \int \Psi(\omega)\, e^{-i\omega t}\, d\omega \tag{1.12}$$

and with (1.5)

$$\phi_i(\omega) = \int \phi_i(t)\, e^{i\omega t}\, dt$$

$$= 2\pi \int dK\, A(K)\, \delta\!\left(\omega - \frac{E_i(K)}{\hbar}\right)\phi_i(K). \tag{1.13}$$

Solving (1.11) for $\Psi(\omega)$ and using (1.13) results in

$$\Psi(\omega) = 2\pi \int dK\, A(K)\, \delta\!\left(\omega - \frac{E_i(K)}{\hbar}\right)\Psi(K) \tag{1.14}$$

with

$$\Psi(K) = \frac{1}{1 - G_i\!\left(\dfrac{E_i(K)}{\hbar}\right)V_i}\, \phi_i(K) = \phi_i(K) + G_i V_i \Psi(K). \tag{1.15}$$

(1.15) is an integral equation for $\Psi(K)$, called the *Lippmann-Schwinger* equation.

With the backtransformation according to (1.12) one obtains from (1.14), (1.15) and (1.5)

$$\Psi(t) = \phi_i(t) + \int dK\, A(K)\, e^{-\frac{i}{\hbar} E_i(K) t}\, G_i(K)\, V_i \Psi(K) \tag{1.16}$$

with

$$G_i(K) = \frac{1}{E_i(K) - H_i + i\eta}, \qquad \eta \to 0. \tag{1.16a}$$

The integral equation (1.16) for $\Psi(t)$ is in a suitable form for the calculation of the differential cross section. Now we have to distinguish between direct and rearrangement collisions.

1. Direct Collisions

The probability for finding the system at time t in any excited or ionized state $\varphi_f(r)^\star$ with relative momentum K' of v is

$$w_{if}(t)\, dK' = |(\phi_f(K'), \Psi(t))|^2\, \frac{dK'}{(2\pi)^3} \tag{1.17}$$

$\star$ For excitation into a bound state the normalisation is $(\varphi_{f'} \varphi_f) = 1$. For ionisation into a free state $\varphi_{k'}$ with energy $\varepsilon_{k'} = \hbar^2 k'^2/2\mu$ the differential dK' in the following formulas has to be replaced by $dK'\, dk'$, and the normalisation is $(\varphi_{k_1}\, \varphi_{k_2}) = \delta(k_1' - k_2')$.

where

$$\phi_f(K') = e^{iK'R}\varphi_f(r) \tag{1.17a}$$

$$H_i\phi_f(K') = E_f(K')\,\phi_f, E_f(K') = \frac{\hbar^2 K'^2}{2v} + \varepsilon_f. \tag{1.17b}$$

For the amplitude

$$F(t) = (\phi_f(K'), \Psi(t)) \tag{1.18}$$

one obtains from (1.16) with (1.5) and

$$G_i(K)\,\phi_f(K') = \frac{1}{E_i(K) - E_f(K') + i\eta}\,\phi_f(K')$$

$$F(t) = e^{-\frac{i}{\hbar}E_f(K')t}\left\{(2\pi)^3\,\delta_{if}A(K')\right.$$

$$\left. + \int dK\,A(K)\,T_{if}(K',K)\frac{e^{-\frac{i}{\hbar}(E_i(K) - E_f(K'))t}}{E_i(K) - E_f(K') + i\eta}\right\} \tag{1.19}$$

where

$$T_{if}(K', K) = (\phi_f(K'), V_i\Psi(K)) \tag{1.20}$$

is the *T*-matrix for direct collisions. Since we are interested only in times long after the collision, we obtain (Appendix 1) with

$$\frac{e^{-\frac{i}{\hbar}(E_i(K) - E_f(K'))t}}{E_i(K) - E_f(K') + i\eta} \xrightarrow{t\to\infty} -2\pi i\,\delta(E_i(K) - E_f(K')) \tag{1.21}$$

and (1.17), (1.19)

$$w_{if}(t\to\infty)dK' = \frac{dK'}{(2\pi)^3}\,|(2\pi)^3\,\delta_{if}A(K')$$

$$- 2\pi i \int dK\,A(K)\,T_{if}(K',K)\,\delta(E_i(K) - E_f(K'))|^2$$

$$= dK'\left\{(2\pi)^3\,\delta_{if}|A(K')|^2 + 2\pi i\,\delta_{if}\int dK_1\,dK_2\,A^*(K_1)\,A(K_2)\right.$$

$$\cdot [T_{if}^*(K',K_1)\,\delta(E_i(K_1) - E_f(K'))\,\delta(K_2 - K') \tag{1.22}$$

$$- T_{if}(K',K_2)\,\delta(E_i(K_2) - E_f(K'))\,\delta(K_1 - K')]$$

$$+ \frac{1}{2\pi}\int dK_1\,dK_2\,A(K_1)\,A(K_2)\,T_{if}^*(K',K_1)\,T_{if}(K',K_2)$$

$$\cdot \delta(E_i(K_1) - E_f(K'))\,\delta(E_i(K_2) - E_f)\right\}.$$

In the second term in (1.22)

$$\delta_{if}\delta(E_i(K_{1,2}) - E_f(K'))\,\delta(K_{2,1} - K') = \delta_{if}\delta(E_i(K_1) - E_i(K_2))\,\delta(K_{2,1} - K')$$

and in the third

$$\delta(E_i(\mathbf{K}_1) - E_f(\mathbf{K}')) \, \delta(E_i(\mathbf{K}_2) - E_f(\mathbf{K}'))$$
$$= \delta(E_i(\mathbf{K}_1) - E_i(\mathbf{K}_2)) \, \delta(E_i(\mathbf{K}_2) - E_f(\mathbf{K}')) \, .$$

For a wave packet that is concentrated in momentum space near $\mathbf{K}_i$ we obtain with (1.5b) in (1.22) provided $T_{if}(\mathbf{K}', \mathbf{K})$ is slowly varying for $|\mathbf{K} - \mathbf{K}_i| \lesssim \varkappa$:

$$w_{if} = w_{if}(t \to \infty) = \delta_{if} \delta(\mathbf{K}' - \mathbf{K}_i)$$
$$- 2\pi i \delta_{if} \delta(\mathbf{K}' - \mathbf{K}_i) \, (T_{ii}(\mathbf{K}_i, \mathbf{K}_i) - T_{ii}^*(\mathbf{K}_i, \mathbf{K}_i)) \qquad (1.23)$$
$$+ \frac{1}{2\pi} C |T_{if}(\mathbf{K}', \mathbf{K}_i)|^2 \, \delta(E_i(\mathbf{K}_i) - E_f(\mathbf{K}'))$$

with

$$C = \int d\mathbf{K}_1 \, d\mathbf{K}_2 \, A^*(\mathbf{K}_1) \, A(\mathbf{K}_2) \, \delta(E_i(\mathbf{K}_1) - E_i(\mathbf{K}_2)) \, .$$

The first and second term in (1.23) only contribute for elastic ($i = f$) scattering in the forward direction, while the genuine scattering is contained in the third term.

The differential cross section $d\sigma_{if}(\mathbf{K}_i, \mathbf{K}')$ for an internal transition from $i \to f$ and a momentum change of v from $\mathbf{K}_i \to \mathbf{K}'$ is defined by

$$J_i(\mathbf{R} = 0) \, d\sigma_{if}(\mathbf{K}_i, \mathbf{K}') = w_{if} \, d\mathbf{K}' \qquad (1.24)$$

where $J_i(\mathbf{R} = 0)$ is the time integrated probability flux of v per unit area attached to the bound state system, i.e.

$$J_i(\mathbf{R} = 0) = \int dt \, j_i(\mathbf{R} = 0, t) \qquad (1.25)$$

with

$$j_i(\mathbf{R}, t) = \frac{\hbar}{2vi} \int d\mathbf{r}(\phi_i^* \, \partial_{\mathbf{R}} \phi_i - \text{c.c.}) \, .$$

From (1.25) with (1.5) one gets

$$J_i(\mathbf{R} = 0) = \frac{\hbar}{2v} \int dt \int d\mathbf{K}_1 \, d\mathbf{K}_2 \, A^*(\mathbf{K}_1) \, A(\mathbf{K}_2)$$

$$\cdot e^{\frac{i}{\hbar}(E_i(\mathbf{K}_1) - E_i(\mathbf{K}_2))t} \qquad (\mathbf{K}_1 + \mathbf{K}_2)$$

$$= \frac{\pi \hbar^2}{v} \int d\mathbf{K}_1 \, d\mathbf{K}_2 \, A^*(\mathbf{K}_1) \, A(\mathbf{K}_2)$$

$$\cdot \delta(E_i(\mathbf{K}_1) - E_i(\mathbf{K}_2)) (\mathbf{K}_1 + \mathbf{K}_2) \, .$$

With a concentrated wave packet according (1.5b) we obtain

$$J_i(\mathbf{R} = 0) = \frac{2\pi \hbar^2}{v} C \mathbf{K}_i \qquad (1.26)$$

and from (1.23) and (1.24) for the genuine scattering $((K', f) \neq (K_i, i))$

$$d\sigma_{if} = \frac{v}{(2\pi)^2 \hbar^2 K_i} |T_{if}(K', K_i)|^2 \, \delta(E_i(K_i) - E_f(K')) dK'. \quad (1.27)$$

This is identical with the result of the common approach without wave packets. The δ-function in (1.27) takes care of energy conservation. Formula (1.27) shows that the T-matrix (1.20) is the quantity to be calculated for the differential cross section. That however means to find a solution of the *Lippmann-Schwinger* equation (1.15) for $\Psi(K)$.

From the normalisation of the probability w_{if} in (1.23) one obtains

$$\sum_f \int dK' w_{if} = 1 = 1 - 2\pi i C(T_{ii}(K_i, K_i) - T_{ii}^*(K_i, K_i))$$

$$+ \frac{1}{2\pi} C \sum_f \int dK' |T_{if}(K', K_i)|^2 \, \delta(E_i(K_i) - E_f(K')).$$

With (1.24) and (1.26) one arrives at the optical theorem

$$\sigma_{\text{tot}} = \sum_f \int d\sigma_{if} = \frac{v}{\hbar^2 K_i} i(T_{ii}(0) - T_{ii}^*(0)), \quad (1.28)$$

i.e. the total cross section is determined by the imaginary part of the T-matrix for elastic $(i = f)$ scattering in the forward direction.

2. Rearrangement Collisions (Charge Exchange)

For a rearrangement collision we have to calculate the probability of finding at time t m attached to M' in a bound state $\varphi'_f(r')$ moving with momentum K' with respect to M, i.e.

$$w'_{if}(t) dK' = |(\phi'_f(K'), \Psi(t))|^2 \frac{dK'}{(2\pi)^3} \quad (1.29)$$

$$\phi'_f = e^{iK'R'} \varphi'_f(r'), (\varphi'_f, \varphi'_f) = 1;$$

$$H_f \phi'_f = E'_f(K') \phi'_f; E'_f = \frac{\hbar^2 K'^2}{2v'} + \varepsilon'_f.$$

For atomic collisions this process is called charge exchange. The amplitude

$$F'(t) = (\phi'_f, \Psi(t))$$

is with (1.5), (1.15) and (1.16)

$$F'(t) = \int dK \, A(K) e^{-\frac{i}{\hbar} E_i(K)t} (\phi'_f(K'), \Psi(K)). \quad (1.30)$$

The representation (1.15) for $\Psi(K)$ is inconvenient in (1.30) as ϕ'_f is not eigenstate of G_i. We therefore write $\Psi(K)$ proportional to the total Green's operator

$$G(K) = \frac{1}{E_i(K) - H + i\eta} \qquad (1.31)$$

which is a solution of the integral equations

$$G = G_i + G_i V_i G = \frac{1}{1 - G_i V_i} G_i , \qquad (1.32a)$$

$$G = G_f + G_f V_f G = \frac{1}{1 - G_f V_f} G_f . \qquad (1.32b)$$

The final Green's operator G_f is defined by

$$G_f(K) = \frac{1}{E_i(K) - H_f + i\eta} . \qquad (1.33)$$

The relations (1.32a, b) can easily be verified by inserting G_i, G and G_f from (1.16a), (1.31) and (1.33).

As $G_i(E_i(K) - H_i + i\eta) = 1$ we obtain from (1.15)

$$\Psi(K) = \frac{1}{1 - G_i(K) V_i} G_i(E_i(K) - H_i + i\eta) \phi_i(K)$$

and with (1.32a), (1.32b) and $H_i \phi_i(K) = E_i(K) \phi_i(K)$

$$\Psi(K) = i\eta G \phi_i(K) = i\eta G_f \phi_i + G_f V_f \Psi(K) . \qquad (1.34)$$

The form (1.34) is the one desired for the scalar product in (1.30)

As

$$G_f(K) \phi'_f(K') = \frac{1}{E_i(K) - E'_f(K') + i\eta} \phi'_f$$

we obtain from (1.30)

$$F'(t) = \int dK \, A(K) \frac{e^{-\frac{i}{\hbar} E_i(K) t}}{E_i(K) - E'_f(K') + i\eta}$$
$$\cdot \{i\eta(\phi'_f(K'), \phi_i(K)) + (\phi'_f(K'), V_f \Psi(K))\} .$$

The first term vanishes for $\eta \to 0$. For $t \to \infty$ we have from (1.29) with (1.21)

$$w'_{if}(t \to \infty) = w'_{if} = (2\pi)^{-1} \int dK_1 \, dK_2 \, A^*(K_1) \, A(K_2)$$
$$\cdot T'^*_{if}(K', K_1) \, T'_{if}(K', K_2) \, \delta(E_i(K_1) - E'_f) \, \delta(E_i(K_2) - E'_f) . \qquad (1.35)$$

In (1.35) we have introduced the T-matrix for rearrangement collisions

$$T'_{if}(K', K) = (\phi'_f(K'), V_f \Psi(K)) . \qquad (1.36)$$

In contrast to T_{if} of (1.20) for direct collisions the final interaction V_f occurs in T'_{if}.

For an incoming wave packet concentrated near K_i we can simplify (1.35) in the same way as (1.22) for direct collisions:

$$w'_{if} = \frac{1}{2\pi} C |T'_{if}(K', K_i)|^2 \, \delta(E_i(K_i) - E'_f(K')). \tag{1.37}$$

With (1.24) and (1.26) the differential cross section for charge exchange is from (1.37)

$$d\sigma'_{if} = \frac{v}{(2\pi)^2 \hbar^2 K_i} |T'_{if}(K', K_i)|^2 \, \delta(E_i(K_i) - E'_f(K')) \, dK. \tag{1.38}$$

As for the direct collisions the solution of the *Lippmann-Schwinger* equation (1.15) is needed to evaluate the T-matrix in (1.36) and the differential cross section in (1.38).

By φ'_f in (1.29) we mean a bound state between M' and m, while in the case of direct collisions φ_f in (1.17a) also can be an ionized state. In principle one also could have charge exchange into the continuum. Then the rearranged states form a complete set as the excited plus ionized do, and either description would be complete. Physically however charge exchange into the continuum and ionisation mean the same, namely a transition of m from a bound into a free state. So one usually describes electron ejection by ions or electrons as an ionisation process. If however one is particularly interested in ejected electrons with velocity nearly equal to the velocity of the projectile, a charge exchange description would be the natural one.

3. Born Series for Direct and Rearrangement Collisions

In contrast to potential scattering one can produce several Born series of the T-matrices (1.20) and (1.36). With (1.15) one immediately has

$$\Psi(K) = \phi_i + \frac{1}{1 - G_i V_i} G_i V_i \phi_i \tag{1.39}$$

and with (1.32a)

$$\Psi(K) = \phi_i + G V_i \phi_i. \tag{1.40}$$

By iteration of the total Green's operator G in (1.32a, b) the following series expansions of G result

$$G = \sum_{n=0}^{\infty} G_i V_i)^n G_i = \sum_{n=0}^{\infty} (G_f V_f)^n G_f. \tag{1.41}$$

One also can expand G in powers of the total potential:

$$G = \sum_{n=0}^{\infty} (G_0 V)^n G_0 \tag{1.42}$$

with

$$G_0 = \frac{1}{E_i(K) - H_0 + i\eta}, \quad H_0 = -\frac{\hbar^2}{2\nu}\partial_{\mathbf{R}}^2 - \frac{\hbar^2}{2\mu}\partial_r^2 \tag{1.42a}$$

and

$$V = w + w' + W. \tag{1.42b}$$

(1.42) is obvious from (1.41) by replacing $V_{i,f} \to V$ and $G_{i,f} \to G_0$. Introducing the series expansions (1.41), (1.42) in (1.40), and $\Psi(K)$ of (1.40) in T_{if} and T'_{if} of (1.20) and (1.36), one arrives at the corresponding Born series of the T-matrices. The first Born approximation is the first term in these expansions. It is the same no matter which of the expansions (1.41), (1.42) are used and means to replace $\Psi(K)$ in (1.20) or (1.36) by $\phi_i(K)$. For the expansion (1.42) e.g. the Born series for T_{if} and T'_{if} are with (1.20), (1.36), (1.40)

$$T_{if}(\mathbf{K}', \mathbf{K}_i) = (\phi_f(\mathbf{K}'), V_i \phi_i(\mathbf{K}_i))$$
$$+ \sum_{n=0}^{\infty} (\phi_f(\mathbf{K}'), V_i(G_0 V)^n G_0 V_i \phi_i(\mathbf{K}_i)) \tag{1.43}$$

$$T'_{if}(\mathbf{K}', \mathbf{K}_i) = (\phi'_f(\mathbf{K}'), V_f \phi_i(\mathbf{K}_i))$$
$$+ \sum_{n=0}^{\infty} (\phi'_f(\mathbf{K}'), V_f(G_0 V)^n G_0 V_i \phi_i(\mathbf{K}_i)). \tag{1.44}$$

We can introduce $V_f = H - H_f$ in the first Born approximation of T'_{if} in (1.44). In the differential cross section (1.38) T'_{if} occurs only on the energy shell $(E_i(\mathbf{K}_i) = E'_f(\mathbf{K}'))$, where

$$(\phi'_f(\mathbf{K}'), H_f \phi_i(\mathbf{K}_i)) = (\phi'_f(\mathbf{K}'), H_i \phi_i(\mathbf{K}_i)) = E_i(\mathbf{K}_i)(\phi'_f(\mathbf{K}'), \phi_i(\mathbf{K}_i)).$$

Then is

$$(\phi'_f, V_f \phi_i) = (\phi'_f, (H - H_f)\phi_i) = (\phi'_f, (H - H_i)\phi_i) = (\phi'_f, V_i \phi_i) \tag{1.45}$$

on the energy shell. Therefore it is possible to evaluate the first Born approximation T'_{if} in the differential cross section (1.38) with V_i or V_f causing the transition from $\phi_i(\mathbf{K}_i)$ to $\phi'_f(\mathbf{K}')$.

II. Wave Packet Theory of Coulomb Scattering

From the derivation of the differential cross section for direct collisions it is obvious that (1.27) also holds for potential scattering of a particle with mass ν from a potential $V(R)$. The T-matrix for the transition from a state $\phi(K_i) = e^{iK_iR}$ to a state $\phi(K') = e^{iK'R}$ is corresponding to (1.20)

$$T(K', K_i) = (\phi(K'), V\Psi(K_i)).\tag{2.1}$$

The scattering state $\Psi(K_i)$ is a solution of the *Lippmann-Schwinger* equation for potential scattering, which is in analogy to (1.15)

$$\Psi(K_i) = \phi(K_i) + G_0 V\Psi(K_i)\tag{2.2}$$

with

$$G_0 = \frac{1}{E - H_0 + i\eta}; \quad E = \frac{\hbar^2 K_i^2}{2\nu}; \quad H_0 = -\frac{\hbar^2}{2\nu}\partial_R^2.$$

For atomic collisions the most important potential and strictly spoken the only one occurring is the Coulomb potential

$$V(R) = \frac{a}{R}.\tag{2.3}$$

Usually scattering from the Coulomb potential (2.3) is treated by solving the stationary *Schroedinger* equation

$$-\frac{\hbar^2}{2\nu}\Delta\Psi(R) + \frac{a}{R}\Psi(R) = E\Psi(R), \quad E > 0.\tag{2.4}$$

Fortunately this is possible exactly and the solution is [5]

$$\Psi(R) = e^{-\frac{\pi\alpha}{2}}\Gamma(1 + i\alpha)\, e^{iK_iR}\,{}_1F_1(-i\alpha; 1; i(K_iR - K_iR))\tag{2.5}$$

with

$$\alpha = \frac{a\nu}{\hbar^2 K_i}.$$

${}_1F_1$ is the confluent hypergeometric function. Asymptotically in leading order of $1/R$ (2.5) has the form

$$\Psi(R) \sim e^{iK_iR + i\alpha\ln(K_iR - K_iR)}$$

$$+\frac{\Gamma(1 + i\alpha)}{i\Gamma(-i\alpha)}\frac{e^{-i\alpha\ln\frac{1}{2}(1 - \cos\vartheta)}}{K_i(1 - \cos\vartheta)}\frac{e^{iK_iR + i\alpha\ln 2K_iR}}{R}.\tag{2.5a}$$

ϑ is the scattering angle, $K_iR = K_iR\cos\vartheta$.

One notices in (2.5a) that in contrast to the scattering state $\Psi(K_i)$ in the *Lippmann-Schwinger* equation (2.2) the incoming (and scattered) wave is modified by a logarithmic phase factor. This is a consequence of the long range Coulomb potential forbidding a particle to move according $H_0\Psi = E\Psi$ even for large distances from the scattering centre. Calculating the differential cross section from the scattered (j_s) and incoming (j_i) asymptotic fluxes by

$$\mathrm{d}\sigma = \frac{j_s}{j_i} R^2 \, \mathrm{d}\Omega$$

the *Rutherford* formula

$$\frac{\mathrm{d}\sigma}{\mathrm{d}\Omega} = \frac{\alpha^2}{4K_i^2 \sin^4 \vartheta/2} \tag{2.6}$$

results in agreement with classical theory.

The logarithmic phase factors in (2.5a) suggest that for the Coulomb potential the integral equation (2.2) and the T-matrix (2.1) have to be reconsidered. This becomes evident when one evaluates $T(K, K_i)$ in (2.1) with the exact scattering state $\Psi(K_i)$ of (2.5). The scalar product in (2.1) is an integral over R which is convergent only after multiplying the integrand with $\mathrm{e}^{-\varepsilon R}$, $\varepsilon \to 0$. The result is (Appendix 2)*

$$T_{KK_i} = \frac{-4\pi a e^{-\frac{\pi\alpha}{2}} \Gamma(1 + i\alpha)}{(K_i - K)^2 + \varepsilon^2} \left\{ \frac{K^2 - K_i^2 - 2i\varepsilon K_i}{(K_i - K)^2 + \varepsilon^2} \right\}^{i\alpha}; \quad \varepsilon \to 0. \tag{2.7}$$

On the energy shell $(K^2 = K_i^2) T_{KK_i}$ of (2.7) is divergent and therefore meaningless for the differential cross section. So the scattering theory of Chapter I breaks down for the Coulomb potential. We shall see in the following that already (1.7), the starting point of the wave packet theory, is wrong for the Coulomb potential, since (1.7) implies, that for $t \to -\infty \Psi(t) = \phi_i(t)$ moves as a free wave packet. On the other hand we saw in (2.5a) that the incoming wave is disturbed even at infinity.

One usually tries to avoid the long range difficulties in the T-matrix theory by replacing the Coulomb potential by $V_{R_0} = \frac{a}{R} \mathrm{e}^{-R/R_0}$ and letting $R_0 \to \infty$ at the end. However the *Schroedinger*- or *Lippmann-Schwinger* equation cannot be solved exactly for a screened Coulomb potential V_{R_0}. Surprisingly the Born approximation with V_{R_0}, $\Psi(K_i) = \phi(K_i)$ in (2.1), provides with (1.27) and $R_0 \to \infty$ the exact result (2.6). This however seems dubious since the second Born approximation of T [36] goes to infinity for $R_0 \to \infty$.

In the following we shall modify the wave packet theory of Chapter I in order to derive the *Rutherford* formula (2.6) from the T-matrix formulation. We now start with a wave packet $\phi_i(t)$ at a finite time $t = t_i$ instead of $t = -\infty$ as in (1.7). In analogy to (1.7) we then have for $V = a/R$

$$\Psi(t) = \phi_i(t) + \int_{t_i}^{\infty} \mathrm{d}t' \, G_0(t - t') V \Psi(t') \tag{2.8}$$

---

* We shall see in the following that (2.1) does not hold for Coulomb scattering. Therefore T_{KK_i} in (2.7) is not the correct T matrix (cp. 2.27).

with

$$G_0(t) = \frac{1}{2\pi} \int \frac{e^{-i\omega t}}{\hbar\omega - H_0 + i\eta} \, d\omega = \Theta(t)\frac{1}{i\hbar} e^{-\frac{i}{\hbar}H_0 t} \tag{2.8a}$$

$$\phi_i(t) = \int dK \, A(K) e^{-\frac{i}{\hbar}E(K)t} \phi(K); \tag{2.8b}$$

$$E(K) = \frac{\hbar^2 K^2}{2\nu}; \quad \phi(K) = e^{iKR}$$

$$H_0 \phi_i(t) = i\hbar \dot{\phi}_i. \tag{2.8c}$$

Differentiating (2.8) with respect to time one sees with (2.8a) and (2.8c) that the *Schroedinger* equation $(H_0 + V)\Psi = i\hbar\dot{\Psi}$ holds for $\Psi(t)$ of (2.8). Because of the step function in G_0 also the initial condition $\Psi(t_i) = \phi_i(t_i)$ is fullfilled. The lower integration limit t_i in (2.8) can be replaced by $-\infty$ when we multiply the integrand by $\Theta(t' - t_i)$. Multiplying Eq. (2.8) by $\Theta(t - t_i)$ results in

$$\Psi(t, t_i) = \phi_i(t, t_i) + \int_{-\infty}^{\infty} dt' \, G_0(t - t') V\Psi(t', t_i) \tag{2.9}$$

with

$$\Psi(t, t_i) = \Theta(t - t_i)\Psi(t), \quad \phi_i(t, t_i) = \Theta(t - t_i)\phi_i(t). \tag{2.9a}$$

In the second term of (2.9) $\Theta(t - t_i)$ drops out because the integrand contains a factor $\Theta(t - t') \Theta(t' - t_i)$ and $\Theta(t - t_i) \Theta(t - t') \Theta(t' - t_i) = \Theta(t - t') \Theta(t' - t_i)$. Fouriertransforming (2.9) we obtain with the Fourierrepresentation (2.8a) of $G_0(t)$ and

$$\Psi(t, t_i) = \frac{1}{2\pi} \int \Psi(\omega, t_i) e^{-i\omega t} d\omega$$

$$\phi_i(t, t_i) = \frac{1}{2\pi} \int \phi_i(\omega, t_i) e^{-i\omega t} d\omega \tag{2.10}$$

$$\Psi(\omega, t_i) = \phi_i(\omega, t_i) + G_0(\omega) V\Psi(\omega, t_i) \tag{2.11}$$

where

$$G_0(\omega) = \frac{1}{\hbar\omega - H_0 + i\eta} \tag{2.11a}$$

is the Fouriertransformed of $G_0(t)$. Multiplying (2.11) with $(2\pi)^{-1}e^{-i\omega t}$ and integrating over ω leads with (2.10) to

$$\Psi(t, t_i) = \phi_i(t, t_i) + \frac{1}{2\pi} \int d\omega \, e^{-i\omega t} G_0(\omega) V\Psi(\omega, t_i). \tag{2.12}$$

The probability for particle v moving at time t with momentum (K', dK') is

$$w_{K'}(t, t_i)dK' = |F(t, t_i)|^2 \frac{dK'}{(2\pi)^3} \tag{2.13}$$

with the amplitude

$$F(t, t_i) = (\phi(K'), \Psi(t, t_i)) \tag{2.13a}$$

and

$$\phi(K') = e^{iK'R}, \quad H_0\phi(K') = E(K')\phi(K')$$
$$E(K') = \frac{\hbar^2 K'^2}{2v}. \tag{2.13b}$$

Assuming the incoming wave packet $A(K)$ concentrated near K_i as in (1.5b), $\phi_i(t, t_i)$ in (2.12) does not contribute in (2.13a), when we confine ourselves to the genuine scattering, i.e. $K' \neq K_i$. This can be realized easily with (2.8b) and the orthogonality relation

$$(\phi(K'), \phi(K)) = (2\pi)^3 \delta(K - K').$$

With (2.12) and

$$G_0(\omega)\phi(K') = \frac{1}{\hbar\omega - E(K') + i\eta}\phi(K')$$

we obtain from (2.13a) for $K' \neq K_i$:

$$F(t, t_i) = \frac{e^{-\frac{i}{\hbar}E(K')t}}{2\pi} \int d\omega\, e^{-i\left(\omega - \frac{E(K')}{\hbar}\right)t}$$
$$\cdot \frac{1}{\hbar\omega - E(K') + i\eta}(\phi(K'), V\Psi(\omega, t_i)). \tag{2.14}$$

Solving the integral equation (2.11) for $\Psi(\omega, t_i)$ results in

$$\Psi(\omega, t_i) = \frac{1}{1 - G_0(\omega)V}\phi_i(\omega, t_i). \tag{2.15}$$

With the inverse Fouriertransformation of (2.10)

$$\phi_i(\omega, t_i) = \int \phi_i(t, t_i)e^{i\omega t}dt$$

and (2.8b), (2.9a) the incoming wave packet is

$$\phi_i(\omega, t_i) = \int dt\, dK\, A(K)\, \Theta(t - t_i)e^{i\omega t}e^{iKR - \frac{i}{\hbar}E(K)t}$$

The t-integration can be performed using (Appendix 3)

$$\Theta(t) = \frac{i}{2\pi}\int d\xi \frac{e^{-i\xi t}}{\xi + i\varepsilon}; \quad \varepsilon \to 0 \tag{2.16}$$

and then integrating over ξ:

$$\phi_i(\omega, t_i) = i \int dK \, d\xi \, A(K) \, \delta\left(\omega - \xi - \frac{E(K)}{\hbar}\right) e^{iKR} \frac{e^{i\xi t_i}}{\xi + i\varepsilon}$$

$$= i \int dK \, A(K) e^{iKR} \frac{e^{i\left(\omega - \frac{E(K)}{\hbar}\right)t_i}}{\omega - \frac{E(K)}{\hbar} + i\varepsilon} . \tag{2.17}$$

Introducing (2.17) in (2.15) and (2.15) in (2.14) results in

$$F(t, t_i) = -\frac{e^{-\frac{i}{\hbar}E(K')t}}{2\pi\hbar i} \int d\omega \, dK \, A(K)$$

$$\cdot \frac{e^{-i\left(\omega - \frac{E(K')}{\hbar}\right)t}}{\omega - \frac{E(K')}{\hbar} + i\eta} \frac{e^{i\left(\omega - \frac{E(K)}{\hbar}\right)t_i}}{\omega - \frac{E(K)}{\hbar} + i\varepsilon} T_{K'K}(\omega) \tag{2.18}$$

with the off energy-shell T-matrix

$$T_{K'K}(\omega) = \left(\phi(K'), V \frac{1}{1 - G_0(\omega)V} \phi(K)\right). \tag{2.19}$$

For $t_i \to -\infty$ one recognizes with (1.21) that (2.18) is reduced to the scattering term in (1.19). The T-matrix occurring for $t_i \to -\infty$ is (2.19) with $\omega = E(K)/\hbar$, which is half on the energy-shell. However in the following we will show that even half on the energy shell the T-matrix for the Coulomb potential is divergent. So the limit $t_i \to -\infty$ is not possible at this stage of the calculation. If in addition $t \to \infty$, $T_{K'K}$ is completely on the energy-shell because of energy conservation $E(K) = E(K') = \omega$ and twofold divergent for the Coulomb potential (see (2.27)).

For the evaluation of $F(t, t_i)$ in (2.18) and the transition probability $w_{K'}(t, t_i)$ in (2.13) we have to calculate the T-operator for the Coulomb potential:

$$T(\omega) = V \frac{1}{1 - G_0(\omega)V} . \tag{2.20}$$

T can be related to the free (2.11a) and the total Green's operator

$$G(\omega) = \frac{1}{\hbar\omega - H + i\eta} = \frac{1}{1 - G_0(\omega)V} G_0(\omega); \quad H = H_0 + V. \tag{2.21}$$

(2.21) can be verified with the definition (2.11a) of $G_0(\omega)$. Multiplying (2.21) with V from the left and (2.20) with $G_0(\omega)$ from the right results in

$$VG(\omega) = T(\omega) \, G_0(\omega) . \tag{2.22}$$

Applying $1 - G_0(\omega)V$ from the left on Eq. (2.21) leads to

$$G = G_0 + G_0 V G. \qquad (2.21a)$$

(2.21a) can be written with (2.22) as

$$G = G_0 + G_0 T G_0. \qquad (2.21b)$$

Solving (2.21b) for T gives the desired relation for T

$$T(\omega) = G_0^{-1} G G_0^{-1}(\omega) - G_0^{-1}(\omega). \qquad (2.23)$$

The total Green's operator G can be calculated from its definition in (2.21) by solving the equation $(\hbar\omega - H)G = 1$ or in space representation

$$\frac{\hbar^2}{2v} \partial_R^2 G(R, R') + \left(\hbar\omega - \frac{a}{R}\right) G(R, R') = \delta(R - R'). \qquad (2.24)$$

This has been done by several authors [37, 38] and the result in momentum space as *Schwinger* [37] presents it is

$$
\begin{aligned}
G_{K'K}(E) &= (\phi(K') G(\omega) \phi(K)) \\
&= (2\pi)^3 \frac{\delta(K - K')}{E - E(K)} \\
&\quad + \frac{4\pi a}{(E - E(K))(E - E(K'))} \int_0^1 dx \, x^{i\alpha(E)} \\
&\quad \cdot \frac{d}{dx} \frac{x}{(K - K')^2 x - \frac{v}{2E\hbar^2}(E - E(K))(E - E(K'))(1 - x)^2}
\end{aligned}
\qquad (2.25)
$$

with

$$E = \hbar\omega = \frac{\hbar^2 k^2}{2v}, \qquad \alpha(E) = \frac{av}{\hbar^2 k}, \qquad E(K) = \frac{\hbar^2 K^2}{2v}.$$

With

$$
\begin{aligned}
G_{0K'K}^{-1}(E) &= (\phi(K'), (\hbar\omega - H_0 + i\eta) \phi(K)) \\
&= (2\pi)^3 (E - E(K)) \delta(K - K')
\end{aligned}
$$

we obtain from (2.23) and (2.25)

$$
\begin{aligned}
T_{K'K}(E) &= 4\pi a \int_0^1 dx \, x^{i\alpha(E)} \\
&\quad \cdot \frac{d}{dx} \frac{x}{(K - K')^2 x + \frac{v}{2E\hbar^2}(E - E(K))(E - E(K'))(1 - x)^2}.
\end{aligned}
\qquad (2.26)
$$

Since we are interested in $F(t, t_i)$ of (2.18) only for t_i being great, the ω-integration in (2.18) contributes because of (1.21) only for $\hbar\omega \cong E(\mathbf{K})$. Therefore we need $T_{\mathbf{K}'\mathbf{K}}(E)$ only in the neighbourhood of the energy-shell. Then in (2.26) $E = E(\mathbf{K})$ and apart from $x \ll 1$ the second term in the denominator can be neglected and the integrand vanishes. So only small x contribute in (2.26) and we replace $(1 - x)^2$ by 1. We now perform the integration in (2.26) by introducing a branch cut from 0 to $+\infty$ along the real axis of the complex x-plane (Fig. 2). Choosing the phase of x as 0 in the upper and as 2π in the lower half plane we have

$$\int_0^1 = \frac{1}{1 - e^{-2\pi\alpha}} \oint_{1-i0}^{1+i0}$$

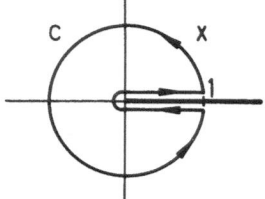

Fig. 2. Integration path C for evaluating integral (2.26)

for the integral in (2.26). As only the neighbourhood of $x = 0$ contributes, we may close the integration path $\circlearrowright$ along the unit circle and have for $E \cong E(\mathbf{K})$:

$$T_{\mathbf{K}'\mathbf{K}}(E) \cong \frac{-4\pi a}{1 - e^{-2\pi\alpha}}$$

$$\cdot \int_C dx\, x^{i\alpha} \frac{d}{dx} \frac{x}{(\mathbf{K} - \mathbf{K}')^2\, x - \dfrac{v}{2E\hbar^2}\, (E - E(\mathbf{K}))\, (E - E(\mathbf{K}'))}$$

with the integration path C according Fig. 2. Integrating by parts results in

$$T_{\mathbf{K}'\mathbf{K}}(E) \cong -\frac{4\pi a i\alpha}{1 - e^{-2\pi\alpha}}$$

$$\cdot \int_C dx\, \frac{x^{i\alpha}}{(\mathbf{K} - \mathbf{K}')^2\, x - \dfrac{v}{2E\hbar^2}\, (E - E(\mathbf{K}))\, (E - E(\mathbf{K}'))}.$$

With E having a small imaginary part the integral has a simple pole within C near $x = 0(E \cong E(K))$ and we obtain with *Cauchy*'s theorem

$$T_{K'K}(E) \cong \frac{8\pi^2 a\alpha(E)}{1 - e^{-2\pi\alpha(E)}} \frac{1}{(K - K')^2} \left(\frac{v}{2E\hbar^2(K - K')^2} \right)^{i\alpha(E)}$$
$$\cdot (E - E(K))^{i\alpha} (E - E(K'))^{i\alpha} \tag{2.27}$$

for the Coulomb T-matrix near the energy-shell on the "right hand side", i.e. $E \cong E(K)$. In (2.27) the small imaginary part of E to define the contour integral along C can be taken as zero. From (2.27) we see that the T-matrix for the Coulomb potential has a logarithmic singularity on the energy-shell from the right-, $E = E(K)$, and the left hand side, $E = E(K')$, as well. We also see that the proper Coulomb T-matrix (2.27) is different from the naive one (2.7), which is on the energy-shell from the right hand side already by the definition (2.1).

Before continuing with the calculation of $F(t, t_i)$ in (2.18) we now can investigate the behaviour of the generalized *Lippmann-Schwinger* equation in (2.11), (2.15) for $t_i \to -\infty$.

With the identity

$$\frac{1}{1 - G_0 V} = 1 + G_0 V \frac{1}{1 - G_0 V} = 1 + G_0 T$$

we obtain from (2.15)

$$\Psi(\omega, t_i) = \phi_i(\omega, t_i) + G_0 T(\omega) \phi_i(\omega, t_i). \tag{2.28}$$

Applying in (2.28) $T(\omega)$ on $\phi_i(\omega, t_i)$ of (2.17), we see with (1.21) that for $t_i \to -\infty$ in (2.17) $\omega \cong E(K)$. Representing (2.28) in momentum space by multiplying with $\phi(K')$ from the left, matrix elements $T_{K'K}(\omega)$ are formed which are on the energy-shell on the right hand side. This means that the second term in (2.28), the scattered wave, diverges for $t_i \to -\infty$. Since (2.28) is identical with the generalized *Lippmann-Schwinger* equation (2.11), this divergence means that a solution of the genuine *Lippmann-Schwinger* equation for the Coulomb potential does not exist. In Appendix 2 we show that the solution (2.5) of the *Schroedinger* equation fullfills the homogeneous *Lippmann-Schwinger* equation $\Psi = G_0 V \Psi$, provided we define the space – integrations involved by convergence factors $e^{-\varepsilon R}$, $\varepsilon \to 0$.

With the T-matrix (2.27) we now proceed with the calculation of the transition probability. It is convenient to write $T_{K'K}(E)$ for $E \cong E(K)$ in (2.27)

$$T_{K'K}(E) \cong \frac{8\pi^2 a\alpha(K)}{1 - e^{-2\pi\alpha(K)}} \frac{1}{(K - K')^2} \left(\frac{v}{2E(K)\hbar^2(K - K')^2} \right)^{i\alpha(K)}$$
$$\cdot (E - E(K))^{i\alpha(K)} (E(K) - E(K'))^{i\alpha(K)} \tag{2.27a}$$

with

$$\alpha(K) = \frac{av}{\hbar^2 K}.$$

For large $|t_i|$, $|t_i| \gg \dfrac{\hbar}{|E(K_i) - E(K_i + \varkappa)|}$, we can write (2.18):

$$F(t, t_i) \cong - \frac{e^{-\frac{i}{\hbar} E(K')t}}{2\pi\hbar i} \int d\omega \, dK \, A(K) \frac{e^{-i\Delta\omega(K, K')t}}{\Delta\omega(K, K') + i\eta}$$

$$\cdot \frac{e^{i\left(\omega - \frac{E(K)}{\hbar}\right)t_i}}{\omega - \frac{E(K)}{\hbar} + i\varepsilon} T_{K'K}(\omega) \qquad (2.28a)$$

with

$$\Delta\omega(K, K') = \frac{E(K) - E(K')}{\hbar}.$$

From (2.13) and (2.18a) we have for large $|t_i|$ with (2.27a)

$$w_{K'}(t, t_i) \cong \frac{1}{(2\pi)^5 \hbar^2} \int dK_1 \, dK_2 \, A(K_1) \, A^*(K_2) \, I(K_1) \, I^*(K_2) \qquad (2.29)$$

with the integral $I(K)$ after substituting $\xi = \omega - E(K)/\hbar$ in (2.18)

$$I(K) = \frac{8\pi^2 a\alpha(K)}{1 - e^{-2\pi\alpha(K)}} \frac{1}{(K - K')^2} \left(\frac{v}{2E(K)(K - K')^2}\right)^{i\alpha(K)}$$

$$\cdot (\Delta\omega(K, K'))^{i\alpha(K)} \frac{e^{-i\Delta\omega(K, K')t}}{\Delta\omega(K, K') + i\eta} \int d\xi \, \xi^{i\alpha(K)} \frac{e^{i\xi t_i}}{\xi + i\varepsilon}. \qquad (2.29a)$$

The remaining integral in (2.29a) is calculated in Appendix 4. The result is $(t_i < 0)$

$$\int d\xi \, \xi^{i\alpha} \frac{e^{i\xi t_i}}{\xi + i\varepsilon} = - \frac{i}{(-t_i)^{i\alpha}} \frac{1 - e^{-2\pi\alpha}}{\alpha} e^{\pi\alpha/2} \Gamma(1 + i\alpha). \qquad (2.30)$$

Substituting (2.29a) and (2.30) would yield the transition probability at arbitrary t. Since we are interested only in t being large,

$$t \gg \frac{\hbar}{|E(K_i) - E(K')|},$$

we can simplify (2.29a) with the relation

$$\omega^{i\alpha} \frac{e^{-i\omega t}}{\omega + i\eta} \xrightarrow{t \to \infty} \frac{1}{t^{i\alpha}} \int x^{i\alpha} \frac{e^{-ix}}{x + i\eta} \, dx \, \delta(\omega). \qquad (2.31)$$

The δ-function in (2.31) is obvious, as for $t \to \infty$ only $\omega = 0$ contributes. The factor in front of $\delta(\omega)$ provides the integrals over ω of both sides of (2.31) to be equal. The integral in (2.31) is given by (2.30).

A further simplification follows by choosing a wave packet $A(K)$ concentrated with small width at K_i. Then $K_{1,2} = K_i$ in (2.29) and we obtain with (2.29a), (2.30) and (2.31) from (2.29) for large t and $|t_i|$

$$w_{K'}(t, t_i) \cong \frac{8a^2}{\pi} \frac{\sinh^2 \pi \alpha}{\alpha^2} |\Gamma(1 + i\alpha)|^4 \frac{1}{(K_i - K')^4}$$
$$\cdot \int dK_1 \, dK_2 \, A(K_1) \, A^*(K_2) \, \delta(E(K_1) - E(K')) \, \delta(E(K_2) - E(K')) ; \qquad (2.32)$$
$$\alpha = \alpha(K_i) .$$

The time dependent phase factors of (2.30), (2.31) have dropped now and (2.32) is the asymptotic expression for $w_{K'}(t, |t_i| \to \infty) = w_{K'}$. For a wave packet sharply concentrated according (1.5b) we obtain with

$$|\Gamma(1 + i\alpha)|^2 = \frac{\alpha \pi}{\sinh \alpha \pi}$$

[39] as in (1.23)

$$w_{K'} = \frac{a^2 \pi C}{2 K_i^4 \sin^4 \vartheta/2} \delta(E(K') - E(K_i)) \qquad (2.33)$$

with

$$C = \int dK_1 \, dK_2 \, A(K_1) \, A(K_2) \, \delta(E(K_1) - E(K_2)) .$$

The definition for the differential cross section (1.24) yields with the current (1.26)

$$d\sigma(K_i, K') = \frac{a^2 v}{4\hbar^2 K_i^5 \sin^4 \vartheta/2} \delta(E(K') - E(K_i)) dK'$$

and with $E(K) = \dfrac{\hbar^2 K^2}{2v}$ and $\alpha = \dfrac{av}{\hbar^2 K_i}$

$$\frac{d\sigma}{d\Omega} = \int d\sigma(K_i, K') K'^2 dK' = \frac{\alpha^2}{4 K_i^2 \sin^4 \vartheta/2} . \qquad (2.34)$$

(2.34) is in agreement with the *Rutherford* formula (2.6). It has been derived here by starting at an early but finite time t_i with a long wave packet containing many wave lengths and analyzing it at a late time. Early time means in this context $|t_i| \gg \dfrac{\hbar}{|E(K_i) - E(K_i + \varkappa)|}$ and late time $t \gg \dfrac{\hbar}{|E(K_i) - E(K')|}$. The usual divergencies in a T-matrix treatment of Coulomb scattering have been avoided by letting $t_i \to -\infty$ not in a scattering-state or amplitude but in the transition probability for a wave packet scattering. Finally the wave packet has been smeared out to a plane wave.

III. Impact Parameter Treatment of Atomic Collisions

In the following we will formulate a semiclassical theory, which treats the motion of M and M' (Fig. 1) classically but the transition of m quantum mechanically. This restricts the applications to collisions of ions with atomic systems where one would expect a classical treatment of the heavy nuclei to be valid. Furthermore the impact parameter theory assumes for the nuclei a straight path motion with constant velocity which is completely defined by the impact parameter b and velocity v (Fig. 3).

Intuitively one expects a classical motion of the nuclei, if the wavelength of $v \cong \dfrac{MM'}{M+M'}$ is small compared with a_i, the radius of the electron orbit corresponding to the initial bound state φ_i; i.e. $1/K_i \ll a_i$ or

$$\frac{\hbar^2 K_i^2}{2v} \gg \frac{\hbar^2}{2ma_i^2} \frac{m}{v}. \tag{3.1}$$

Eq. (3.1) shows that the small ratio m/v provides the classical behaviour. For a proton-hydrogen collision for instance a_i is the Bohr radius and $\hbar^2/2ma_i^2 = -\varepsilon_i = 13.6\,\text{eV}$ the binding energy of the hydrogen atom. The right hand side of (3.1) is then of the order of magnitude of 0.01 eV, so that several eV of the incoming ion would be sufficient to guarantee a classical motion of the nuclei.

This energy is raised by the condition of straight path motion with constant velocity. This condition holds, if the average deflection $\langle \vartheta \rangle$ of M' is small. With $\langle \vartheta \rangle \cong W(a_i) \Big/ \dfrac{\hbar^2 K_i^2}{2v}$ and $E_i(K_i) \cong \dfrac{\hbar^2 K_i^2}{2v}$ in order of magnitude, we obtain

$$\frac{W(a_i)}{E_i(K_i)} \ll 1. \tag{3.2}$$

For a proton-hydrogen collision (3.2) yields $|\varepsilon_i|/E_i \ll 1$, which holds for several hundred eV of the incoming proton. Since the maxima of proton excitation or ionisation cross sections are usually at 100 KeV in order of magnitude, the restrictions (3.1), (3.2) hold even for rather low energies. So the impact parameter theory is a high energy treatment in a very weak sense.

In the following we will set up the formalism of the impact parameter treatment. We again will stick to the three body system of Fig. 1. The generalisation to more complicated systems is straightforward. For a one dimensional model of a three body collision we then will justify this semi-classical approach and check the conditions (3.1), (3.2).

1. Formulation of the Impact Parameter Treatment

The three particle system of Fig. 1 is now described by only one coordinate r of the electron with mass m, because the internuclear coordinate is given by (Fig. 3)

$$R(t) = vt + b. \tag{3.3}$$

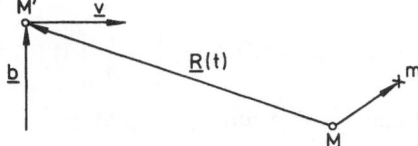

Fig. 3. Three body collision in impact parameter treatment

The *Schroedinger* equation is

$$h\psi(r, t) = i\hbar\dot{\psi}(r, t) \tag{3.4}$$

with the initial condition

$$\psi(t_i) = \varphi_i(t_i) = \varphi_i(r)e^{-\frac{i}{\hbar}\varepsilon_i t_i} \tag{3.4a}$$

for m being in the bound state φ_i with binding energy ε_i at an initial time t_i.

The Hamiltonian h in (3.4) is

$$
h = \underbrace{-\frac{\hbar^2}{2m}\partial_r^2 + w(r)}_{h_i} + \underbrace{w'(r - R(t)) + W(R(t))}_{V_i(r, t)}
$$
$$
= \underbrace{-\frac{\hbar^2}{2m}\partial_r^2 + w'}_{h_f} + \underbrace{w + W}_{V_f} \tag{3.5}
$$

and the bound state $\varphi_i(t)$ is a stationary solution of the *Schroedinger* equation with vanishing initial interaction V_i:

$$h_i\varphi_i(rt) = i\hbar\dot{\varphi}_i(r, t). \tag{3.6}$$

In contrast to the quantum-mechanical three body problem with time independent interactions in Chapter I we have to deal here with a one particle problem with time dependent interactions. The internuclear potential W is even a function of time alone.

It is convenient to split from ψ a phase according to

$$\psi = \tilde{\psi} \exp\left\{-\frac{i}{\hbar} \int\limits_{t_i}^{t} W(t')\mathrm{d}t'\right\}. \tag{3.7}$$

By differentiating (3.7) with respect to time we obtain from (3.4), (3.5)

$$i\hbar\dot{\psi} = W(t)\,\psi + i\hbar\dot{\tilde{\psi}} \exp\left\{-\frac{i}{\hbar} \int\limits_{t_i}^{t} W(t')\mathrm{d}t'\right\}$$

$$= (h_i + w' + W)\,\tilde{\psi} \exp\left\{-\frac{i}{\hbar} \int\limits_{t_i}^{t} W(t')\,\mathrm{d}t'\right\}.$$

The W-terms drop and for $\tilde{\psi}$ results the equation

$$i\hbar\dot{\tilde{\psi}} = (h_i + w')\tilde{\psi} \tag{3.8}$$

which differs from the original *Schroedinger* equation (3.4) by the missing internuclear potential W. So W contributes only by a phase factor in the wave function and does not affect any transition probabilities. This means that apart from the low energy range ($E_i \gtrsim 100$ eV), where (3.2) is violated, the internuclear potential can be dropped for all inelastic collisions. This is also true for the wave treatment of Chapter I since it is equivalent with the impact parameter theory for not too low energies. W has to be kept of course for the wave treatment of the elastic cross section, since one investigates here the, though small, deflection of M'. In the impact parameter theory the elastic collisions cannot be treated because of the straight path assumption.

For convenience we drop the tilde in (3.8) in the following calculation of the cross section and deal with (3.4)–(3.6) with $W = 0$. As in Chapter I we now distinguish between direct and rearrangement collisions.

a) Excitation and Ionisation

The probability for finding the electron at time t in an excited state φ_f after a collision with impact parameter b is

$$w_b(t) = |f_b(t)|^2 \tag{3.9}$$

with the amplitude

$$f_b(t) = (\varphi_f(t), \psi(t)) \tag{3.9a}$$

and the excited state

$$\varphi_f(t) = \varphi_f(r)\mathrm{e}^{-\frac{i}{\hbar}\varepsilon_f t} \tag{3.9b}$$

which is like the initial state a stationary solution of the *Schroedinger* equation with vanishing initial interaction V_i:

$$h_i\varphi_f = i\hbar\dot{\varphi}_f = \varepsilon_f\varphi_f. \tag{3.9c}$$

ε_f is the binding energy in the excited state. For ionisation into a state with energy $\varepsilon_{k'} = \hbar^2 k'^2/2m$ (3.9) has to be multiplied with dk'.

The cross section for excitation into φ_f is

$$\sigma_f = \int w_b(t \to \infty) d\boldsymbol{b}. \tag{3.10}$$

Therefore we have to calculate the amplitude $f_b(t \to \infty) = f_b$ for a time long after the collision.

In analogy to Eq. (1.7) the following integral equation is equivalent to the *Schroedinger* equation (3.4) with initial condition (3.4a)

$$\psi(t) = \varphi_i(t) + \frac{1}{i\hbar} \int_{t_i}^{t} U_i(t, \vartheta) \, w'(\vartheta) \, \psi(\vartheta) d\vartheta \tag{3.11}$$

with the time evolution operator $U_i(t, \vartheta)$ defined by the relation

$$\varphi_i(t) = U_i(t, \vartheta) \, \varphi_i(\vartheta). \tag{3.12}$$

From (3.6) and (3.12) we obtain the differential equation

$$h_i U_i(t, \vartheta) = i\hbar \partial_t U_i(t, \vartheta) \tag{3.13}$$

and with $U_i(\vartheta, \vartheta) = 1$ the integral equation

$$U_i(t, \vartheta) = 1 - \frac{i}{\hbar} \int_{\vartheta}^{t} h_i U_i(t', \vartheta) dt' \tag{3.13a}$$

which can be verified by differentiation with respect to time. The solution of (3.13) with $U_i(\vartheta, \vartheta) = 1$ or (3.13a) is

$$U_i(t, \vartheta) = e^{-\frac{i}{\hbar} h_i(t - \vartheta)} \tag{3.14}$$

With (3.14) and (3.6) $\psi(t)$ of (3.11) can be shown to fulfill the *Schroedinger* equation (3.4) by differentiation with respect to time.

Calculating $f_b(t)$ of (3.9a) with (3.11) the incoming wave does not contribute because of the orthogonality of $\varphi_{i,f}$, $(\varphi_i, \varphi_f) = 0$, and we obtain

$$f_b(t) = -\frac{i}{\hbar} \int_{t_i}^{t} (U_i^+(t, \vartheta) \, \varphi_f(t), w'(\vartheta) \, \psi(\vartheta)) d\vartheta. \tag{3.15}$$

It is evident from (3.14) that $U_i^+(t, \vartheta) = U_i(\vartheta, t)$. Then (3.15) yields with $t_i \to -\infty, \ t \to \infty$

$$f_b = -\frac{i}{\hbar} \int_{-\infty}^{\infty} (\varphi_f(\vartheta), w'(\vartheta) \, \psi(\vartheta)) d\vartheta. \tag{3.16}$$

In order to calculate f_b and the cross section $\sigma = \int |f_b|^2 d\boldsymbol{b}$ for excitation into φ_f one has to find a solution of the integral equation (3.11) or the equivalent time dependent *Schroedinger* equation.

b) Charge Exchange

For charge exchange we have to calculate the probability $w'(t)$ for finding the electron at time t in a bound state φ'_f centered around M'. For a fixed impact parameter b is

$$w'_b(t) = |f'_b(t)|^2 \qquad (3.17)$$

with amplitude

$$f'_b(t) = (\varphi'_f(t), \psi(t)) \qquad (3.17a)$$

and the *Galilei* transformed bound state φ'_f. Since the *Galilei* transformed charge density is $|\varphi'_f(r - R(t))|^2$, the state has the form $\varphi'_f(t) = \varphi'_f(r - R(t)) e^{i\lambda(r,t)}$ The phase $\lambda(r, t)$ can be found by considering a free particle wave $\psi = \exp\left\{ik_i r - \dfrac{i}{\hbar}\dfrac{\hbar^2 k_i^2}{2m} t.\right\}$ *Galilei* transforming ψ with velocity v turns it into a wave

$$\psi_v = \exp\left\{i(k_i + k) r - \frac{i}{\hbar}\frac{\hbar^2(k_i + k)^2}{2m} t\right\} = e^{i k_i (r - vt)} e^{i\lambda(r,t)}.$$

Comparing the phases yields $\lambda = \dfrac{mv}{\hbar} r - \dfrac{i}{\hbar}\dfrac{mv^2}{2} t - \dfrac{i}{\hbar}\dfrac{\hbar^2 k_i^2}{2m} t$, which means an additional momentum mv and kinetic energy $mv^2/2$ of m. Therefore in (3.17a)★

$$\varphi'_f(t) = \varphi'_f(r - R(t)) \exp\left\{i\frac{mv}{\hbar} r - \frac{i}{\hbar}\frac{mv^2}{2} t - \frac{i}{\hbar}\varepsilon'_f t\right\} \qquad (3.18)$$

$$h_f \varphi'_f = i\hbar \dot{\varphi}'_f.$$

ε'_f is the binding energy of m in the bound state φ'_f. The cross section for charge exchange into φ'_f is

$$\sigma'_f = \int w'_b(t \to \infty) d b = \int |f'_b(t \to \infty)|^2 d b \qquad (3.19)$$

For the calculation of $f'_b(t)$ in (3.17a) it is convenient to represent $\psi(t)$ as

$$\psi(t) = U_f(t, t_i)\, \varphi_i(t_i) - \frac{i}{\hbar} \int_{t_i}^{t} U_f(t, \vartheta)\, w\psi(\vartheta) d\vartheta \qquad (3.20)$$

$U_f(t, \vartheta)$ is the time evolution operator for $\varphi'_f(t)$, i.e.

$$\varphi'_f(t) = U_f(t, \vartheta)\, \varphi'_f(\vartheta). \qquad (3.21)$$

From (3.21) we obtain with $i\hbar\dot{\varphi}'_f = h_f \varphi_f$

$$i\hbar\, \partial_t U_f(t, \vartheta) = h_f\, U_f(t, \vartheta) \qquad (3.22)$$

★ $\varphi'_f(t)$ in (3.18) can also be determined as a solution of $h_f(t) \varphi'_f = i\hbar\dot{\varphi}'_f$.

and from (3.22a) with $U_f(\vartheta, \vartheta) = 1$

$$U_f(t, \vartheta) = 1 - \frac{i}{\hbar} \int_\vartheta^t h_f(t')\, U_f(t', \vartheta)\mathrm{d}t' . \qquad (3.22a)$$

Iteration of (3.22b) yields

$$U_f(t, \vartheta) = T\exp\left\{ -\frac{i}{\hbar} \int_\vartheta^t h_f(\vartheta')\mathrm{d}\vartheta' \right\} \qquad (3.23)$$

with Dyson's time ordering operator T, ordering products of $h_f(t_n)$ with respect to t_n with smaller times on the right of later ones.

Introducing (3.20) in (3.17a) results in

$$f_b'(t) = (U_f^+(t, t_i)\, \varphi_f'(t), \varphi_i(t_i)) - \frac{i}{\hbar} \int_{t_i}^t (U_f^+(t_i, \vartheta)\, \varphi_f'(t), w\psi(\vartheta))\mathrm{d}\vartheta .$$

With $U_f^+(t, \vartheta) = U_f^{-1}(t, \vartheta) = U_f(\vartheta, t)$ from unitarity and definition (3.21) of U_f we obtain

$$f_b'(t) = (\varphi_f'(t_i), \varphi_i(t_i)) - \frac{i}{\hbar} \int_{t_i}^t (\varphi_f'(\vartheta), w\psi(\vartheta))\mathrm{d}\vartheta . \qquad (3.24)$$

For $t_i \to -\infty$ the first term drops because the bound states φ_i and φ_f' are then well separated in space and the overlap vanishes. For the cross section (3.19) we need $f_b'(t \to \infty) = f_b'$. With (3.24) the result is

$$f_b' = -\frac{i}{\hbar} \int_{-\infty}^\infty (\varphi_f'(\vartheta), w\psi(\vartheta))\mathrm{d}\vartheta . \qquad (3.25)$$

As for excitation and ionisation f_b' and σ_f' can be calculated after having solved (3.11) or the equivalent *Schroedinger* equation for $\psi(t)$.

c) Born Series in the Impact Parameter Treatment

Similar to the wave treatment of Chapter I we can derive several Born series for f_b (3.16) and f_b' (3.25) by expanding in powers of $V_i = w'$, $V_f = w$ or $V = w + w'$. To this end a convenient representation of ψ is for $t_i \to -\infty$

$$\psi(t) = \varphi_i(t) - \frac{i}{\hbar} \int_{-\infty}^t U(t, \vartheta)\, w'(\vartheta)\, \varphi_i(\vartheta)\mathrm{d}\vartheta \qquad (3.26)$$

with the time evolution operator U defined by

$$\psi(t) = U(t, \vartheta)\, \psi(\vartheta) . \qquad (3.27)$$

From the *Schroedinger* equation $h\psi = i\hbar\dot{\psi}$ follows

$$i\hbar\, \partial_t U(t, \vartheta) = h(t)\, U(t, \vartheta) . \qquad (3.28)$$

10*

By differentiating (3.26), using (3.28) and $h_i \varphi_i = i\hbar \dot{\varphi}_i$, the *Schroedinger* equation $h\psi = (h_i + w')\psi = i\hbar \dot{\psi}$ can be shown to hold. For $t \to -\infty$ the scattered wave in (3.26) vanishes and the initial condition $\psi(t \to -\infty) = \varphi_i(t)$ is fulfilled.

As for $U_{i,f}$ in (3.13) and (3.22) the following integral equation follows for U from (3.28) with $U(\vartheta, \vartheta) = 1$

$$U(t, \vartheta) = 1 - \frac{i}{\hbar} \int_\vartheta^t h(t') \, U(t', \vartheta) \mathrm{d}t' \qquad (3.28a)$$

and after iteration of (3.28a)

$$U(t, \vartheta) = T \exp\left\{ - \frac{i}{\hbar} \int_\vartheta^t h(\vartheta') \mathrm{d}\vartheta' \right\}. \qquad (3.29)$$

Introducing (3.26) in (3.16) and (3.25) yields

$$f_b = - \frac{i}{\hbar} \int_{-\infty}^\infty \mathrm{d}\vartheta (\varphi_f(\vartheta), w'(\vartheta) \, \varphi_i(\vartheta))$$

$$- \frac{1}{\hbar^2} \int_{-\infty}^\infty \mathrm{d}\vartheta \int_{-\infty}^\vartheta \mathrm{d}\vartheta' (\varphi_f(\vartheta), w'(\vartheta) \, U(\vartheta, \vartheta') \, w'(\vartheta') \, \varphi_i(\vartheta')) \qquad (3.30)$$

$$f_b' = - \frac{i}{\hbar} \int_{-\infty}^\infty \mathrm{d}\vartheta (\varphi_f'(\vartheta), w\varphi_i(\vartheta))$$

$$- \frac{1}{\hbar^2} \int_{-\infty}^\infty \mathrm{d}\vartheta \int_{-\infty}^\vartheta \mathrm{d}\vartheta' (\varphi_f'(\vartheta), w U(\vartheta, \vartheta') \, w'(\vartheta') \, \varphi_i(\vartheta')). \qquad (3.31)$$

The desired expansions of f_b, f_b' follow by expanding U in (3.30), (3.31). Therefore we will set up integral equations for U with kernels proportional to that potential, in powers of which we wish to expand.

We introduce an interaction picture by

$$U(t, \vartheta) = U_0(t, \vartheta) \, \tilde{U}(t, \vartheta) \qquad (3.32)$$

with

$$U_0(t, \vartheta) = \mathrm{e}^{-\frac{i}{\hbar} h_0 (t - \vartheta)}, \qquad h_0 = - \frac{\hbar^2}{2m} \partial_r^2.$$

From (3.32) and (3.28) we obtain with $h = h_0 + V$

$$i\hbar \, \partial_t U = h_0 U + i\hbar U_0 \, \partial_t U = (h_0 + V) U.$$

The term $h_0 U$ cancels and multiplying with U_0^{-1} from the left yields

$$i\hbar \, \partial_t \tilde{U} = U_0^{-1} V U \qquad (3.33)$$

and with $\tilde{U}(\vartheta, \vartheta) = 1$ (3.32) the equivalent integral equation

$$\tilde{U}(t, \vartheta) = 1 - \frac{i}{\hbar} \int_\vartheta^t U_0^{-1}(t', \vartheta) \, V(t') \, U(t', \vartheta) \mathrm{d}t'. \qquad (3.33a)$$

By multiplying (3.33a) with $U_0(t, \vartheta)$ from the left we obtain with (3.32) and $U_0(t, \vartheta)\, U_0^{-1}(t', \vartheta) = U_0(t, \vartheta)\, U_0(\vartheta, t') = U_0(t, t')$

$$U(t, \vartheta) = U_0(t, \vartheta) - \frac{i}{\hbar} \int_\vartheta^t U_0(t, t')\, V(t')\, U(t', \vartheta)\mathrm{d}t' \,. \qquad (3.34)$$

Replacing in (3.34) $U_0 \to U_i$, $V \to V_i = w'$ or $U_0 \to U_f$, $V \to V_f = w$ similar integral equations for U result which could be derived by the same procedure as (3.34).

We obtain the Born series of f_b and f_b' by expanding U in (3.30) and (3.31) in powers of V, V_i or V_f. This can be done easily by iterating (3.34) or the corresponding integral equations with V_i and V_f instead of V. In all cases the first term in (3.30), (3.31) is the first Born approximation. The second Born approximation corresponding to an expansion of U in powers of V means to replace U by U_0 in (3.30), (3.31).

The vital question is under which circumstances the Born series can be truncated after a few terms, possibly after the first term. From potential scattering it is known that the first Born approximation is valid for high energies, because then the scattering is small and the scattering state in the T-matrix can be approximated by the incoming wave. So one hopes that also for inelastic scattering a few terms in the Born series are sufficient for high energy of the incoming particle. In the wave treatment it seems plausible that the high energy criterion is $E_i \gg |\varepsilon_i|$. In the impact parameter treatment however instead of E_i the projectile velocity v occurs. The relevant velocity of the three particle system under consideration is the average velocity of m in the initial bound state φ_i. So a Born approximation approach can be expected to be valid for $v \gg v_i$ or, in energy, $E_i \gg \dfrac{v}{m} |\varepsilon_i|$ instead of $E_i \gg |\varepsilon_i|$.

Because of the great factor v/m this means $E_i = 100\,\mathrm{KeV}$ and more for a proton-hydrogen collision.

2. Justification of the Impact Parameter Treatment

In the following we shall derive the impact parameter treatment from the wave theory of Chapter I for a one dimensional model of the three particle collision in Fig. 1. As an example we consider the probability for excitation of m into $\varphi_f(x)$ with the internuclear potential $W = 0$. The general case with arbitrary W can be treated in the same way, also for ionisation and charge exchange.

Instead of a differential cross section we have to consider in a one dimensional collision the probability for excitation with M' moving in the forward or backward direction. In contrast to the three dimensional case the normalisation in (1.17) is $dK'/(2\pi)^2$, so that one arrives at (1.23) multiplied with $(2\pi)^2$:

$$(i, K_i) \neq (f, K'):$$
$$w_{if} = 2\pi C |T_{if}(K', K_i)|^2\, \delta(E_i(K_i) - E_f(K')). \tag{3.35}$$

The constant C in (1.23) can easily be calculated for a one dimensional model with

$$A(K) \cong \left\{ \frac{1}{2\pi}\, \delta(K - K_i) \right\}^{1/2}$$

corresponding to (1.5b). We obtain

$$C = \frac{v}{2\pi\hbar^2 K_i}. \tag{3.35a}$$

The probability for forward and backward scattering of M' with excitation of m is

$$w_{if}(+1) = \int\limits_0^\infty w_{if}\, dK' \qquad w_{if}(-1) = \int\limits_{-\infty}^0 w_{if}\, dK' \tag{3.36}$$

and we have with (3.35), (3.35a)

$$w_{if}(\pm 1) = \frac{v^2}{\hbar^4 K_i K_f} |T_{if}(\pm|K_f|, K_i)|^2,$$
$$K_f = \pm \sqrt{\frac{2v}{\hbar^2} (E_i(K_i) - \varepsilon_f)}. \tag{3.37}$$

The T-matrix in (3.37) for $W = 0$ is with (1.20)

$$T_{if} = (\phi_f, w'\Psi) = \int e^{-iK_f X} \varphi_f^*(x)\, w'(-X + \zeta x)\, \Psi(x, X)\, dx\, dX. \tag{3.38}$$

We have replaced r, R by the one dimensional coordinates x, X and written in $w'(x'): x' = -X + \zeta x$ according to (1.3).

In Fourierspace (3.38) has the form

$$T_{if} = (2\pi)^2 \int dk_1\, dk_2\, \tilde{\varphi}_f(k_1)\, \tilde{w}'(k_2)\, \tilde{\Psi}(k_1 + \zeta k_2; K_f - k_2) \tag{3.38a}$$

with

$$\varphi_f(x) = \int \tilde{\varphi}_f(k) e^{ikx}\, dk \quad \text{etc.}$$
$$\Psi(x, X) = \int \tilde{\Psi}(k, K) e^{i(kx + KX)}\, dk\, dK.$$

Eqs. (3.9) and (3.16) of the impact parameter treatment hold in one dimension when we put $b = 0$:

$$w_{b=0} = |f_{b=0}|^2;$$
$$f_{b=0} = f = -\frac{i}{\hbar} \int\limits_{-\infty}^\infty (\varphi_f(t), w'(x - vt)\, \psi(t))\, dt. \tag{3.39}$$

The amplitude in Fourierspace is

$$f = \frac{(2\pi)^2}{i\hbar} \int dk_1\, dk_2\, \tilde{\varphi}_f(k_1)\, \tilde{w}'(k_2)\, \tilde{\psi}\left(k_1 + k_2, -\frac{\varepsilon_f}{\hbar} - k_2 v\right) \quad (3.39a)$$

with

$$\psi(x, t) = \int \tilde{\psi}(k, \omega)\, e^{i(kx + \omega t)}\, dk\, d\omega .$$

We will show in the following, that for $v \to \infty$ in forward direction $(K_f > 0)$

$$\tilde{\Psi}(k_1 + \zeta k_2, |K_f| - k_2) = \frac{1}{v}\, \tilde{\psi}\left(k_1 + k_2, -\frac{\varepsilon_f}{\hbar} - k_2 v\right) \quad (3.40)$$

and therefore with (3.38a), (3.39a)

$$f = -\frac{i}{\hbar v}\, T(|K_f|, K_i) . \quad (3.40a)$$

With $\hbar|K_{i,f}| \cong vv$ for $v \to \infty$ we obtain from (3.37) and (3.40a)

$$w_{if}(+1) \cong |f|^2 \cong w_{b=0} . \quad (3.41)$$

For backward scattering and $v \to \infty$ T_{if} vanishes. Then we see from (3.41) that for $v \to \infty$ the total probability of excitation from $i \to f$ in the wave treatment, $w_{if} = w_{if}(+1) + w_{if}(-1)$, agrees with the result of the impact parameter theory. It will turn out that $v \to \infty$ means $m/v \ll 1$ and $|\varepsilon_{i,f}|/E_i \ll 1$. Therefore the conditions (3.1) and (3.2) are necessary but not sufficient, as they are fulfilled even for $v \cong m$ or $E_i \cong \varepsilon_f$ provided $E_i \gg \varepsilon_i$.

Equation (3.40) can be proved to hold by showing that $\tilde{\Psi}$ and $\frac{1}{v}\tilde{\psi}$ are solutions of integral equations which become identical for m/v, $|\varepsilon_{i,f}|/E_i \ll 1$. The *Lippmann-Schwinger* equation (1.15) is in space representation

$$\Psi(x, X) = \phi_i(x, X)$$
$$+ \int G_i(x, X; x_1 X_1)\, w'(\zeta x_1 - X_1)\, \Psi(x_1, X_1)\, dx_1\, dX_1 \quad (3.42)$$

where

$$G_i(x, X; x_1 X_1) = \frac{1}{2\pi} \int dk\, dK\, \varphi_k(x)\, \varphi_k^*(x_1) \frac{e^{iK(X - X_1)}}{E_i + i\eta - \dfrac{\hbar^2 K^2}{2v} - \varepsilon_k}$$

is the initial Green's function $G_i = \dfrac{1}{E_i + i\eta - H_i}$ represented in space by the eigenfunctions

$$\phi_k(x, X) = e^{iKX} \varphi_k(x) \quad \text{of} \quad H_i = -\frac{\hbar^2}{2v}\, \partial_X^2 - \frac{\hbar^2}{2\mu}\, \partial_x^2 + w(x) .$$

The integral over k means a sum over the bound states and an integration over the ionized states, the energies ε_k being the binding energies $\varepsilon_n < 0$ and ionisation energies $\varepsilon_k = \hbar^2 k^2 / 2\mu$ respectively. Representing Ψ, ϕ_i and w' in Fourierspace as in (3.38a) a straightforward calculation yields

$$\tilde{\Psi}(k, K) = \tilde{\phi}_i(k, K) + \int dk_1 \, dK_1 \, I(k, K; k_1 K_1) \, \tilde{\Psi}(k_1 K_1) \qquad (3.43)$$

with the kernel

$$I(k, K; k_1 K_1) = \frac{1}{2\pi} \int dx_1 \, dx_2 \, dk_3 \, dk_4 \, \delta(K_1 - K - k_4) \, \varphi_{k_3}(x_1)$$

$$\cdot \varphi_{k_3}^*(x_2) \, \tilde{w}'(k_4) \frac{e^{i(k_4 \zeta x_2 + k_1 x_2 - k x_1)}}{E_i + i\eta - \dfrac{\hbar^2 K^2}{2v} - \varepsilon_{k_3}} \qquad (3.43a)$$

and the incoming wave

$$\tilde{\phi}_i(k, K) = \frac{1}{(2\pi)^2} \int dx \, dX \, e^{iK_i X} \, \varphi_i(x) \, e^{-i(kx + KX)} \qquad (3.43b)$$

$$= \tilde{\varphi}_i(k) \, \delta(K - K_i).$$

With $k = k_1 + \zeta k_2$ and $K = K_f - k_2$ and substituting $k_1 = k_1' + \zeta k_2'$ and $K_1 = K_f - k_2'$ we obtain from (3.43)

$$\tilde{\Psi}(k_1 + \zeta k_2, K_f - k_2) = \tilde{\phi}_i(k_1 + \zeta k_2, K_f - k_2)$$

$$+ \int dk_1' \, dk_2' \, I(k_1 + \zeta k_2, K_f - k_2; k_1' + \zeta k_2', K_f - k_2') \qquad (3.44)$$

$$\cdot \tilde{\Psi}(k_1' + \zeta k_2', K_f - k_2').$$

This is an integral equation for $\tilde{\Psi}$ in (3.40). By an analogous procedure we can derive an integral equation for $\tilde{\psi}$ in (3.40). Eq. (3.11) for $t_i \to -\infty$ is in space representation

$$\psi(x, t) = \varphi_i(x, t) - \frac{i}{\hbar} \int_{-\infty}^{t} dt_1 \int dx_1 \, U_i(t - t_1; x, x_1)$$

$$\cdot w'(x_1 - vt_1) \, \psi(x_1, t_1) \qquad (3.45)$$

where

$$U_i(t - t_1; x, x_1) = \int dk \, \varphi_k(x) \, \varphi_k^*(x_1) \, e^{-\frac{i}{\hbar} \varepsilon_k(t - t_1)}$$

is the time evolution operator $U_i(t; t_1) = e^{-\frac{i}{\hbar} h_i(t - t_1)}$ in space representation with the eigenfunctions $\varphi_k(x)$ and eigenvalues ε_k of $h_i = -\dfrac{\hbar^2}{2m} \partial_x^2 + w(x)$. For the bound states ε_n the integral over k means a sum over n. As $\mu = m + O(m/v)$, φ_k and ε_k are identical with those in G_i of (3.42) in leading order of m/v. After a straightforward calculation we obtain

(3.45) in Fourierspace with $\tilde{\psi}(k, \omega)$ etc. defined in (3.39a)

$$\tilde{\psi}(k, \omega) = \tilde{\varphi}_i(k, \omega) + \int dK_1 \, d\omega_1 \, i(k, \omega; k_1 \omega_1) \, \tilde{\psi}(k_1, \omega_1) \qquad (3.46)$$

with the kernel

$$i(k, \omega; k_1 \omega_1) = \frac{1}{2\pi} \int dx_1 \, dx_2 \, dk_3 \, dk_4 \, \delta(k_2 v - \omega_1 + \omega) \, \varphi_{k_3}(x_1)$$

$$\cdot \varphi_{k_3}^*(x_2) \, \tilde{w}'(k_4) \frac{e^{i(k_4 x_2 + k_1 x_2 - k x_1)}}{-\varepsilon_{k_3} - \hbar\omega} \qquad (3.46a)$$

and the incoming wave

$$\tilde{\varphi}_i(k, \omega) = \frac{1}{(2\pi)^2} \int \varphi_i(x) \, e^{-\frac{i}{\hbar}\varepsilon_i t} \, e^{-i(kx + \omega t)} \, dx \, dt$$

$$= \tilde{\varphi}_i(k) \, \delta\left(\omega + \frac{\varepsilon_i}{\hbar}\right). \qquad (3.46b)$$

With $k = k_1 + k_2$ and $\omega = -\varepsilon_f/\hbar - k_2 v$ and substituting $k_1 = k_1' + k_2'$ and $\omega_1 = -\varepsilon_f/\hbar - k_2'v$ we obtain from (3.46)

$$\tilde{\psi}\left(k_1 + k_2, -\frac{\varepsilon_f}{\hbar} - k_2 v\right) = \tilde{\varphi}_i\left(k_1 + k_2, -\frac{\varepsilon_f}{\hbar} - k_2 v\right)$$

$$+ v \int dk_1' \, dk_2' \, i\left(k_1 + k_2, -\frac{\varepsilon_f}{\hbar} - k_2 v; k_1' + k_2'; -\frac{\varepsilon_f}{\hbar} - k_2'v\right) \qquad (3.47)$$

$$\cdot \tilde{\psi}\left(k_1' + k_2', -\frac{\varepsilon_f}{\hbar} - k_2'v\right).$$

We now have to compare the integral equations (3.44) and (3.47). The inhomogeneous term in (3.44) is with (3.43b)

$$\tilde{\varphi}_i(k_1 + \zeta k_2; K_f - k_2) = \tilde{\varphi}_i(k_1 + \zeta k_2) \, \delta(K_f - K_i - k_2). \qquad (3.48)$$

With K_f according to (3.37) we obtain for forward scattering

$$|K_f| - K_i = K_i\left(\left\{\frac{1 - \varepsilon_f/E_i}{1 - \varepsilon_i/E_i}\right\}^{1/2} - 1\right)$$

and expanding the square root in powers of $1/E_i$ with $\hbar K_i = vv$, $v \cong MM'/(M + M')$

$$|K_f| - K_i = \frac{\varepsilon_i}{\hbar v} - \frac{\varepsilon_f}{\hbar v} \qquad (3.49)$$

for $m/v \ll 1$ and $|\varepsilon_{i, f}|/E_i \ll 1$. Since $\zeta = M/(M + m) \cong 1$ for $m/M \ll 1$, (3.48) becomes with (3.49)

$$\tilde{\varphi}_i(k_1 + \zeta k_2; |K_f| - k_2) \cong \tilde{\varphi}_i(k_1 + k_2) \, \delta\left(\frac{\varepsilon_i}{\hbar v} - \frac{\varepsilon_f}{\hbar v} - k_2\right). \qquad (3.48a)$$

Comparing (3.48a) with the inhomogeneity $\tilde{\varphi}_i \left(k_1 + k_2, -\dfrac{\varepsilon_f}{\hbar} - k_2 v \right)$ in (3.47) we see with (3.46b), that

$$\tilde{\varphi}_i \left(k_1 + k_2, -\frac{\varepsilon_f}{\hbar} - k_2 v \right) \cong \frac{1}{v} \, \tilde{\phi}_i(k_1 + \zeta k_2; |K_f| - k_2) \qquad (3.50)$$

in leading order of m/v and $|\varepsilon_{i,f}|/E_i$.

The integral kernel in (3.44) is with (3.43a) for forward scattering

$$I(k_1 + \zeta k_2, |K_f| - k_2; k_1' + \zeta k_2', |K_f| - k_2')$$

$$= \frac{1}{2\pi} \int dx_1 \, dx_2 \, dk_3 \, dk_4 \, \delta(k_2 - k_2' - k_4) \, \varphi_{k_3}(x_1) \, \varphi_{k_3}^*(x_2) \, \tilde{w}'(k_4) \qquad (3.51)$$

$$\frac{e^{i\{k_4 \zeta x_2 + (k_1' + \zeta k_2') x_1 - (k_1 + \zeta k_2) x_1\}}}{E_i + i\eta - \dfrac{\hbar^2 (|K_f| - k_2)^2}{2v} - \varepsilon_{k_3}}$$

As for $m/v \ll 1$, $|\varepsilon_{i,f}|/E_i \ll 1$ in (3.51):

$$E_i - \frac{\hbar^2 K_f^2}{2v} + \frac{\hbar^2 |K_f|}{v} k_2 - \frac{\hbar^2 k_2^2}{2v} \cong \varepsilon_f + \hbar v k_2$$

with $\hbar|K_f| \cong vv$ we obtain from (3.47) and (3.46a) with $\zeta \cong 1$

$$i \left(k_1 + k_2, -\frac{\varepsilon_f}{\hbar} - k_2 v; k_1' + k_2'; -\frac{\varepsilon_f}{\hbar} - k_2' v \right)$$

$$\cong \frac{1}{v} I(k_1 + \zeta k_2, |K_f| - k_2; k_1' + \zeta k_2', |K_f| - k_2'). \qquad (3.52)$$

With (3.50) and (3.52) we see from (3.44), (3.47) that Eq. (3.40) holds for m/v, $|\varepsilon_{i,f}|/E_i \ll 1$.

For backward scattering we see by introducing (3.48) and iteration terms of (3.44) in (3.38a) that $T_{if}(-|K_f|, K_i) \sim \tilde{w}'(|K_f| + K_i)$ and therefore vanishes for $v \to \infty$.

IV. High Energy Limit for Inelastic Collisions

In this chapter we calculate the leading term of the high energy expansion for the differential and total cross sections in the three particle collision of Fig. 1. For sake of generality we will set up the asymptotic expansion without specifying the potentials and masses. As an application we will investigate a proton-hydrogen collision by employing the asymptotic expansion for the Coulomb potential and m/M, $m/M' \ll 1$.

1. Wave Treatment of Direct Collisions

a) High Energy Expansion

The differential cross section for exciting or ionizing m from φ_i to φ_f with M' being deflected in the solid angle $(\Omega, d\Omega)$ is with (1.27) and (1.20)

$$\frac{d\sigma_{if}}{d\Omega} = \frac{v}{(2\pi)^2 \hbar^2 K_i} \int |T_{if}(K', K_i)|^2 \, \delta(E_i(K_i) - E_f(K')) K'^2 \, dK'$$

$$E_i(K_i) = \frac{\hbar^2 K_i^2}{2v} + \varepsilon_i; \quad E_f(K') = \frac{\hbar^2 K'^2}{2v} + \varepsilon_f.$$

The K'-integration can be performed because of the δ-function and the result is

$$\frac{d\sigma_{if}}{d\Omega} = \frac{v^2}{(2\pi)^2 \hbar^4} \frac{K_f}{K_i} |T_{if}(K_f, K_i)|^2, \quad K_{f,i} = \sqrt{\frac{2v}{\hbar^2}(E_i - \varepsilon_{f,i})} \quad (4.1)$$

when $\varepsilon_f < E_i$. For $\varepsilon_f > E_i$ $\dfrac{d\sigma_{if}}{d\Omega}$ vanishes. This is obvious as for $\varepsilon_f > E_i$ the incoming energy is too small to supply the energy ε_f of m in the ionized state φ_f. A suitable starting point for a high energy expansion of $\dfrac{d\sigma_{if}}{d\Omega}$ in (4.1) is the Born series (1.43) for the T-matrix, in which the total Green's operator has been expanded in powers of the total interaction $V = w + w' + W$. The Born approximation is the first term in (1.43).

$$T_B = (\phi_f, (w' + W)\phi_i) = T_B^{w'} + T_B^{W}, \tag{4.2}$$

$$T_B^{w'} = \int e^{-iK_f R} \varphi_f^*(r) \, w'(-R + \zeta r) \, e^{iK_i R} \varphi_i(r) \, dr \, dR, \tag{4.2a}$$

$$T_B^{W} = \int e^{-iK_f R} \varphi_f^*(r) \, W(R + (1 - \zeta)r) \, e^{iK_i R} \varphi_i(r) \, dr \, dR. \tag{4.2b}$$

For the high energy expansion it is convenient to write T_B in Fourier space. The w'-term for instance is then

$$T_B^{w'} = (2\pi)^6 \, \tilde{w}'(K_i - K_f) \int dk \, \tilde{w}_f^*(k) \, \tilde{\varphi}_i(-k + \zeta(K_i - K_f)) \tag{4.3}$$

with

$$\varphi_i(r) = \int \tilde{\varphi}_i(k) e^{ikr} \, dk \quad \text{etc.}$$

To expand $T_B^{w'}$ in (4.3) for high energies we need for large E_i "semi-asymptotic" expansions of the Fourier coefficients $\tilde{w}'(k) = \tilde{w}'(\sqrt{E_i}\,k')$ etc.[*], which are not only correct for $k' = 0$ but which also cover k'-values near $k' = 0$ because an integration over k' has to be carried out. For simplicity we assume that w and w' are potentials of the same kind, for

---

[*] By substitution $k = \sqrt{E_i}\,k'$ all arguments of the Fourier-coefficients become proportional to $\sqrt{E_i}$.

instance both Coulomb potentials or Yukawa potentials, and that φ_i is the groundstate and φ_f any excited s-state. Then φ_i and φ_f are isotropic and depend only on $|r| = r$. In order to establish the expansion for

$$\tilde{w}(k) = \tilde{w}(\sqrt{E_i}\,k') = \frac{1}{(2\pi)^3} \int d\mathbf{r}\, w(r)e^{-i\mathbf{k}\mathbf{r}} \qquad (4.4)$$

we expand $w(r)$ in powers of r,

$$w(r) = w^{(-1)}\frac{1}{r} + w^{(0)} + w^{(1)}r + w^{(2)}r^2 + \cdots . \qquad (4.5)$$

If one inserts (4.5) into (4.4) and adds a factor $e^{-\varepsilon r}$, $\varepsilon \to 0$, for convergence one has

$$\tilde{w}(k) = \frac{1}{(2\pi)^3} \int d\mathbf{r}\, e^{-i\mathbf{k}\mathbf{r}-\varepsilon r}\left\{ w^{(-1)}\frac{1}{r} + w^{(0)} + w^{(1)}r + w^{(2)}r^2 + \cdots \right\}$$

$$= \frac{w^{(-1)}}{2\pi^2 k^2} + w^{(0)}\,\delta(k) + \frac{w^{(1)}}{2\pi^2 ik}\left\{ \frac{1}{(\varepsilon - ik)^3} - \frac{1}{(\varepsilon + ik)^3} \right\} \qquad (4.6)$$

$$+ w^{(2)}\partial_k^2\,\delta(k) + \cdots .$$

The even terms contain only δ-functions which drop out for large k. Therefore the odd terms with $\varepsilon = 0$ give the genuine asymptotic expansion of w:

$$w(k) \sim \frac{w^{(-1)}}{2\pi^2 k^2} - \frac{w^{(1)}}{\pi^2 k^4} + \cdots . \qquad (4.6a)$$

The quantity $\varepsilon \to 0$ in (4.6) is needed in order to define integrations in k-space.

As an example let us consider

$$w(r) = \frac{e^{-\varkappa r}}{r} = \frac{1}{r} - \varkappa + \frac{\varkappa^2}{2}r - \frac{\varkappa^3}{3!}r^2 + \cdots$$

$$w^{(-1)} = 1, \quad . \quad w^{(0)} = -\varkappa, \qquad w^{(1)} = \frac{\varkappa^2}{2}, \qquad w^{(2)} = -\frac{\varkappa^3}{3!}$$

$$\tilde{w}(k) = \frac{1}{2\pi^2(k^2 + \varkappa^2)} \sim \frac{1}{2\pi^2 k^2} - \frac{\varkappa^2}{2\pi^2 k^4} + \cdots$$

which agrees with (4.6a). Eq. (4.6) becomes

$$\tilde{w}(k) \sim \frac{1}{2\pi^2 k^2} - \varkappa\delta(k) + \frac{\varkappa^2}{4\pi^2 ik}\left\{ \frac{1}{(\varepsilon - ik)^3} - \frac{1}{(\varepsilon + ik)^3} \right\} + \cdots .$$

By differentiation with respect to $-\varkappa$ we obtain the expansion for $w(r) = -\partial_\varkappa \frac{e^{-\varkappa r}}{r} = e^{-\varkappa r}$. We see that the first term in the semiasymptotic expansion of $\tilde{w}(k)$ vanishes and that the leading term is a δ-function. This can also be realized by considering the Fourier transform of $e^{-\varkappa r}$, i.e. $\tilde{w}(k) = \dfrac{2\varkappa}{[k^2 + \varkappa^2]^2}$, which for great E_i is concentrated near $k = \sqrt{E_i} k' = 0$ with a width of $\varkappa/\sqrt{E_i}$ in k'-space and which is a representation of $\delta(k)$ for $E_i \to \infty$. This δ-function property is a consequence of $w(r) = e^{-\varkappa r}$ being finite at $r = 0$.

For the following we have to assume, that not all odd terms in (4.6) vanish, i.e. we confine ourselves to potentials with Fouriertransforms $\tilde{w}(k)$ having a genuine asymptotic expansion in powers of $1/k^\beta$. Thereby we exclude potentials like $w = e^{-\varkappa^2 r^2}$ which depend only on r^2 and for which the expansion (4.6) contains only δ-functions and derivatives of δ-functions. But the common potentials like $w = 1/r$ or $w = e^{-\varkappa r}/r$, $e^{-\varkappa r}$ are included. Also included are potentials depending on $\sqrt{r}$ rather than on r itself, such as $w = e^{-\sqrt{\varkappa r}}$ or $e^{-\sqrt{\varkappa r}}/\sqrt{r}$, where one has to expand w in powers of $\sqrt{r}$ and (4.6) is an expansion in powers of $k^{-1/2}$ rather than k^{-1}.

For these potentials we can prove the following properties of the semiasymptotic expansions of $\tilde{\varphi}_i(k)$ or $\tilde{\varphi}_f(k)$:

1. the expansion (4.6) always starts with a δ-function ($\varphi_{i,f}(0) \neq 0$ for s-states),

2. the asymptotic expansion (4.6a) of $\tilde{\varphi}_{i,f}$ is by a factor k^{-2} smaller than that of $\tilde{w}(k)$, i.e. $\tilde{\varphi}_{i,f}$ vanishes faster asymptotically than $\tilde{w}$.

For simplicity we show 1. and 2. for potentials with an expansion (4.5) and $w^{(-1)} \neq 0$, e.g. Coulomb- or Yukawa potential. If in the *Schroedinger* equation (1.5) with r separated from R

$$\left(-\frac{\hbar^2}{2\mu} \Delta + w\right) \varphi_i(r) = -\frac{\hbar^2}{2\mu}\left(\varphi_i''(r) + \frac{2}{r}\varphi_i'(r)\right) + w\varphi_i = \varepsilon_i \varphi_i \quad (4.7)$$

one introduces (4.5) and a corresponding expansion for $\varphi_i(r)$, one obtains from comparing equal powers of r

$$\varphi_i^{(1)} = -\frac{\mu}{\hbar^2} w^{(-1)} \varphi_i(0), \quad \text{etc.} \tag{4.8}$$

This shows that $\varphi_i(0)$ and $\varphi_i^{(1)}$ are both nonvanishing as otherwise $\varphi_i(r)$ in (4.7) would be identically zero. The expansions (4.6) and (4.6a) become for $\tilde{\varphi}_i$

$$\tilde{\varphi}_i(k) = \varphi_i(0)\, \delta(k) + \frac{\varphi_i^{(1)}}{2\pi^2 ik}\left\{\frac{1}{(\varepsilon - ik)^3} - \frac{1}{(\varepsilon + ik)^3}\right\} + \cdots \tag{4.9}$$

$$\tilde{\varphi}_i(k) \sim -\frac{\varphi_i^{(1)}}{\pi^2 k^4} + \cdots = \frac{\mu}{\hbar^2}\frac{w^{(-1)}}{\pi^2 k^4}\varphi_i(0) + \cdots \tag{4.9a}$$

or considering only the leading terms in the asymptotic expansions (4.6a), (4.9a)

$$\tilde{\varphi}_i(\boldsymbol{k}) \sim -\frac{2\mu}{\hbar^2 k^2} \varphi_i(0)\, \tilde{w}(\boldsymbol{k}).$$ (4.9b)

Using $\varphi_i(0) \neq 0$ for s-states one also can derive (4.9b) from the Fourier-transform of (4.7)

$$\frac{\hbar^2 k^2}{2\mu}\, \tilde{\varphi}_i(\boldsymbol{k}) + \int w(\boldsymbol{k} - \boldsymbol{k}')\, \tilde{\varphi}_i(\boldsymbol{k}')\, \mathrm{d}\boldsymbol{k}' = \varepsilon_i \tilde{\varphi}_i(\boldsymbol{k})$$ (4.7a)

by neglecting ε_i asymptotically and by using only the first term in (4.6) and (4.9):

$$\frac{\hbar^2 k^2}{2\mu}\, \tilde{\varphi}_i(\boldsymbol{k}) + \tilde{w}(\boldsymbol{k})\, \varphi_i(0) \sim 0$$

which is equivalent to (4.9b).

The statements 1. and 2. can be proved for the excited s-state φ_f in the same way replacing w by w' in (4.7) to (4.9).

With (4.9) we can evaluate the asymptotic expression for $T_{\mathrm{B}}^{w'}$. For $\boldsymbol{K}_f$ being outside the forward direction $\boldsymbol{K}_i/K_i$ the term resulting from the δ-function for $\tilde{\varphi}_i$ and $\tilde{\varphi}_f$ vanishes. The next order term is

$$T_{\mathrm{B}}^{w'} \sim (2\pi)^6\, \tilde{w}'(\boldsymbol{K}_i - \boldsymbol{K}_f)\, \{\varphi_f(0)\, \tilde{\varphi}_i(\zeta(\boldsymbol{K}_f - \boldsymbol{K}_i)) + \varphi_i(0)\, \tilde{\varphi}_f(\zeta(\boldsymbol{K}_f - \boldsymbol{K}_i))\}^\star.$$ (4.10)

In (4.10) $\tilde{w}'$, $\tilde{\varphi}_i$ and $\tilde{\varphi}_f$ mean the leading terms in the genuine asymptotic expansion, which we will not distinguish from $\tilde{w}'$ etc. of (4.4) if there is no danger of misunderstanding. For forward scattering is with (4.1)

$$K_i - K_f = K_i\left(1 - \sqrt{\frac{1 - \varepsilon_f/E_i}{1 - \varepsilon_i/E_i}}\right) \cong K_i\left(\frac{\varepsilon_f}{2E_i} - \frac{\varepsilon_i}{2E_i}\right).$$ (4.11)

Then we obtain from (4.3)

$$T_{\mathrm{B}}^{w'} = (2\pi)^3\, \tilde{w}'(\boldsymbol{K}_i - \boldsymbol{K}_f) \int \varphi_i(r)\, \varphi_f(r)\, e^{i\zeta(\boldsymbol{K}_i - \boldsymbol{K}_f)r}\, \mathrm{d}r$$

by expanding the exponential

$$T_{\mathrm{B}}^{w'} \sim -\frac{\zeta^2}{6}\, (2\pi)^3\, \tilde{w}'(\boldsymbol{K}_i - \boldsymbol{K}_f)\, (K_i - K_f)^2 \int \mathrm{d}r\, r^2\, \varphi_i(r)\, w_f(r).$$ (4.12)

In (4.12) we have used that $\int \varphi_i(r)\, \varphi_f(r)\, \mathrm{d}r = 0$ because of the orthogonality of φ_i and φ_f, and that $\int \varphi_i(r)\, \varphi_f(r)\, r\, \mathrm{d}r = 0$ because of inversion symmetry of $\varphi_{i,f}$.

★ Here and in (4.14) one could insert (4.9b) for $\tilde{\varphi}_i$ and $\tilde{\varphi}_f$ so that the asymptotic behaviour of $T_{\mathrm{B}}^{w'}$ and $T_2^{w'w'}$ outside the forward direction is given by the potentials w, w'.

The term T_B^W of (4.2b) can be expanded in the same way as $T_B^{w'}$. The w'-contribution of the second Born approximation is from (1.43)

$$T_2^{w'w'} = (\phi_f, w' G_0 w' \phi_i) \tag{4.13}$$

or in Fourierspace

$$
\begin{aligned}
T_2^{w'w'} &= (2\pi)^{12} \int d\mathbf{k} \, d\mathbf{K} \, G_0(k, K) \\
&\quad \cdot \tilde{\phi}_f(\mathbf{k} + \zeta(\mathbf{K} - \mathbf{K}_f)) \, \tilde{w}'(\mathbf{K} - \mathbf{K}_f) \, \tilde{w}'(\mathbf{K} - \mathbf{K}_i) \, \tilde{\phi}_i(\mathbf{k} + \zeta(\mathbf{K} - \mathbf{K}_i))
\end{aligned}
$$

$$G_0(k, K) = \frac{1}{(2\pi)^6} \frac{1}{E_i + i\eta - \dfrac{\hbar^2 K^2}{2v} - \dfrac{\hbar^2 k^2}{2\mu}} . \tag{4.13a}$$

Asymptotically we obtain from (4.13a) outside the forward direction with (4.9)

$$T_2^{w'w'} \sim (2\pi)^6 \{\varphi_f(0) \, \tilde{\phi}_i(\zeta(\mathbf{K}_f - \mathbf{K}_i)) \, I_f + \varphi_i(0) \, \tilde{\phi}_f(\zeta(\mathbf{K}_f - \mathbf{K}_i)) \, I_i\}$$

$$I_{i,f} = \int d\mathbf{K} \frac{\tilde{w}'(\mathbf{K} - \mathbf{K}_i) \, \tilde{w}'(\mathbf{K} - \mathbf{K}_f)}{E_i + i\eta - \dfrac{\hbar^2 K^2}{2v} - \dfrac{\hbar^2 \zeta^2 (\mathbf{K}_{i,f} - \mathbf{K})^2}{2\mu}} . \tag{4.14}$$

In the integrals of (4.14) the numerators become large at $\mathbf{K} = \mathbf{K}_{i,f}$ and the denominator small at $\mathbf{K} = \mathbf{K}_f$ in I_f and at $\mathbf{K} = \mathbf{K}_i$ in I_i. Therefore the leading terms in (4.14) are provided by $\mathbf{K} = \mathbf{K}_f$ in I_f and by $\mathbf{K} = \mathbf{K}_i$ in I_i. Then we can replace in $I_f \tilde{w}'(\mathbf{K} - \mathbf{K}_i)$ by $\tilde{w}'(\mathbf{K}_f - \mathbf{K}_i)$ and in $I_i \tilde{w}'(\mathbf{K} - \mathbf{K}_f)$ by $\tilde{w}'(\mathbf{K}_i - \mathbf{K}_f)$. After rearranging the denominator the integrals can then be evaluated by the convolution theorem and we obtain for I_f for instance

$$I_f \cong \tilde{w}'(\mathbf{K}_f - \mathbf{K}_i) \int d\mathbf{K} \frac{\tilde{w}'(\mathbf{K})}{E_i + i\eta - \dfrac{\hbar^2}{2v}(\mathbf{K} + \mathbf{K}_f)^2 - \dfrac{\hbar^2 \zeta^2 K^2}{2\mu}}$$

$$\cong \frac{-\mu' \tilde{w}'(\mathbf{K}_f - \mathbf{K}_i)}{2\pi\hbar^2}$$

$$\cdot \int d\mathbf{r} \, w'(r) \frac{\exp\left\{ i \sqrt{\dfrac{2\mu'}{\hbar^2}\left(E_i - \dfrac{\hbar^2 K_f^2}{2v'}\zeta^2\right)} \, r \right\}}{r} \exp\left\{ i \frac{\mu}{v'} \mathbf{K}_f r \right\} .$$

Introducing $E_i = \hbar^2 K_f^2 / 2v + \varepsilon_f$, expanding the square root for small ε_f, and performing the angular integration results in

$$I_f \cong \frac{-v \tilde{w}'(\mathbf{K}_f - \mathbf{K}_i)}{\hbar^2 K_f}$$

$$\cdot \int_0^\infty dr \exp\left\{ i\left(\frac{\mu}{v'} K_f + \frac{v\varepsilon_f}{\hbar^2 K_f} \right) r \right\} \sin\left(\frac{\mu}{v'} K_f r \right) w'(r) . \tag{4.15}$$

For $w'(r)$ being finite or at least integrable at $r = 0$ (4.15) yields

$$I_f \cong \frac{-v\tilde{w}'(K_f - K_i)}{2i\hbar^2 K_f} \int_0^\infty dr\, w'(r) \tag{4.15a}$$

and for $w'(r) = \frac{A}{r} e^{-\varkappa r}$

$$I_f = \frac{vA\tilde{w}'(K_f - K_i)}{2i\hbar^2 K_f} \ln \frac{2\frac{\mu}{v'}K_f + i\varkappa}{\frac{v\varepsilon_f}{\hbar^2 K_f} + i\varkappa}. \tag{4.15b}$$

I_f in (4.15b) with $\varkappa = 0$ corresponds to the Coulomb potential. The integral I_i can be evaluated in the same way.

Introducing (4.15a, b) and the analogous result for I_i in (4.14) shows by comparison with the first Born approximation (4.10) that the second Born approximation can be neglected for high energies outside the forward direction.

In the forward direction $K_i - K_f \to 0$ for $E_i \to \infty$ according to (4.11) and we have to reconsider $T_2^{w'w'}$ in (4.13a). The k-integration can be separated from the K-integration, as only $K = K_{i,f}$ contributes and then $k = 0$ in $G_0(k, K)$. Therefore we obtain from (4.13a)

$$\begin{aligned}
T_2^{w'w'} &\sim (2\pi)^6\, I \int dk\, \tilde{\varphi}_f(k)\, \tilde{\varphi}_i(k - \zeta(K_i - K_f)) \\
&\sim (2\pi)^3\, I \int dr\, \varphi_i(r)\, \varphi_f(r)\, e^{i\zeta(K_i - K_f)r} \\
I &= \int dK\, \frac{\tilde{w}'(K - K_i)\, \tilde{w}'(K - K_f)}{E_i + i\eta - \frac{\hbar^2 K^2}{2v}}.
\end{aligned} \tag{4.16}$$

In the integral I all factors become rapidly varying at $K = K_{i,f}$ for $E_i \to \infty$. We can estimate I by replacing the denominator by the minimum ε_f near $K = K_{i,f}$:

$$\begin{aligned}
|I| &\leqq \frac{1}{|\varepsilon_f|} \int dK\, \tilde{w}'(K)\, \tilde{w}'(K - (K_i - K_f)) \\
&= \frac{1}{(2\pi)^3 |\varepsilon_f|} \int w'^2(r)\, e^{i(K_i - K_f)r}\, dr.
\end{aligned}$$

The convolution integrals in (4.16) have already been calculated in the first Born approximation $T_B^{w'}$ in (4.12). Then we obtain from (4.16)

$$\begin{aligned}
|T_2^{ww'}| &\leqq \frac{\zeta^2}{6} \left| \int dr\, r^2 \varphi_i(r)\, \varphi_f(r) \right| \\
&\quad \cdot \frac{(K_i - K_f)^2}{|\varepsilon_f|} \int w'^2(r)\, e^{i(K_i - K_f)r}\, dr.
\end{aligned} \tag{4.17}$$

For any screened potential $\int w'^2(r) e^{i(K_i - K_f)r} \, d\mathbf{r} \cong \int w'^2(r) \, d\mathbf{r}$ in (4.17) and $\tilde{w}'(\mathbf{K}_i - \mathbf{K}_f) \cong \tilde{w}'(0)$ in (4.12) as in the forward direction $K_i - K_f \to 0$ for $E_i \to \infty$. According to (4.17) and (4.12) $T_2^{w'w'}$ is then of higher or at most the same order of magnitude as $T_B^{w'}$. As only an angular integration over the forward direction is meaningfull and the second Born approximation has been proved to be neglegible for any scattering angle greater than zero, we can neglect $T_2^{w'w'}$ for high energies.

For the Coulomb potential $\int w'^2(r) e^{i(K_i - K_f)r} \, d\mathbf{r}$ is proportional to $(K_i - K_f)^{-1} \sim \sqrt{E_i}$ and according to (4.17) $T_2^{w'w'}$ is smaller or of the order $1/\sqrt{E_i}$ and therefore neglegible with respect to $T_B^{w'}$, which is of the order E_i^0 according to (4.12).

In order to check (4.17) we have calculated I in (4.16) asymptotically for $E_i \to \infty$ for the screened Coulomb potential $V = A e^{-\varkappa r}/r$ (Appendix 6). The result is for $\varkappa > 0$

$$I \sim \frac{\nu A^2}{4 i \pi^2 \hbar^2 K_f (\varepsilon_f - \varepsilon_i)} \ln\left(1 + \frac{2\nu(\varepsilon_f - \varepsilon_i)}{\hbar^2 \varkappa^2}\right) \tag{4.18a}$$

and for the Coulomb case $\varkappa = 0$

$$I \sim \frac{\nu i A^2}{8 \hbar^2 \pi^2 K_f (\varepsilon_f - \varepsilon_i)} \ln \frac{2\nu^3 \varepsilon_f^6}{\hbar^6 K_i^6 (\varepsilon_f - \varepsilon_i)^3} . \tag{4.18b}$$

Inserting (4.18a), (4.18b) in (4.16) we see that for the screened Coulomb potential $T_2^{w'w'} \sim E_i^{-3/2}$, while $T_B^{w'} \sim 1/E_i$. Therefore we can exclude the possibility in (4.17) that $T_2^{w'w'}$ is of the same order small as $T_B^{w'}$ in the forward direction. For the Coulomb potential $T_2^{w'w'} \sim E_i^{-3/2} \ln E_i$, while (4.17) only guaranteed $|T_2^{w'w'}| < c/\sqrt{E_i}$, which was already sufficient to neglect the second Born approximation.

Eqs. (4.18b), (4.15b) show that in contrast to potential scattering the Coulomb potential does not lead to divergencies in the second Born approximation for inelastic collisions. This is intuitively clear as a finite energy transfer cannot be supplied by collisions in great distances, which are responsible for the divergencies in potential scattering from a Coulomb potential. This can formally be realized in (4.18b), where I would diverge for a vanishing energy transfer $\varepsilon_f - \varepsilon_i$.

We have shown that for high energies E_i the asymptotic behaviour of the inelastic cross sections is determined by T_B in (4.2), (4.3). Furthermore we see in (4.10), (4.12) that only the forward direction contributes for great E_i, as then $K_i - K_f \sim E_i^{-1/2}$. For excitation into $l \neq 0$ states $\varphi_f(r)$ in (4.3a) is not isotropic and in the expansion of the exponential the first nonvanishing term is proportional to $K_i - K_f$, so that $T_B^{w'}$ in forward direction is by $\sqrt{E_i}$ greater than (4.12) with isotropic $\varphi_f(r)$. Therefore $l \neq 0$ transitions occur predominantly for high energies.

b. Cross Sections and Angular Dispersion for Proton Hydrogen Collisions

We have shown in the preceding chapter that we can confine ourselves to the first Born approximation to obtain the correct high energy limit for direct collisions. As an example we will investigate now a collision of a proton with a hydrogen-like atom. Then in Fig. 1 w, w' and W are Coulomb potentials and $m/M, m/M' \ll 1$.

The cross section for direct collisions is given by (4.1), and the high energy limit results from the Born approximation (4.2) for the T-matrix. In (4.2b) is $1 - \zeta \cong m/M \ll 1$ neglegible and the r, R integrations are separated. For excitation and ionisation φ_i and φ_f are orthogonal and the r-integration yields zero. So the internuclear potential W does not contribute in inelastic collisions. This is consistent with the impact parameter treatment where W is only a function of time and can be discarded by a unitary transformation.

For elastic collisions W has to be kept, as $\varphi_i = \varphi_f$ and $\int \varphi_i^2(r)\, dr = 1^\star$. We obtain from (4.2) with $\zeta \cong 1$

$$T_B = \int dR\, e^{-i(K_f - K_i)R}\, V_{\text{eff}}(R) \tag{4.19}$$

$$V_{\text{eff}}(R) = W(R) + \int dr\, \varphi_i^2(r)\, w'(r - R). \tag{4.19a}$$

Eq. (4.19) shows that the Born approximation for an elastic atomic collision can be considered as a Born approximation for potential scattering from an effective potential $V_{\text{eff}}(R)$. This is the electrostatic potential of a point charge together with a charge density φ_i^2. In a proton-hydrogen collision is

$$W = \frac{e^2}{R}; \quad w' = -\frac{e^2}{|R - r|}; \quad \varphi_i = \frac{1}{\sqrt{\pi a_0^3}}\, e^{-\frac{r}{a_0}};$$

$$a_0 = \frac{\hbar^2}{2me^2}; \quad \mu \cong m \tag{4.20}$$

and the integral in (4.19a) can be calculated conveniently in Fourier space.

$$\int dr\, \varphi_i^2(r)\, w'(r - R) = (2\pi)^3 \int dk\, \tilde{\varphi}_i^2(k)\, \tilde{w}'(k)\, e^{ikR}$$

$$\tilde{\varphi}_i^2(k) = \frac{1}{(2\pi)^3} \int \varphi_i^2(r)\, e^{-ikr}\, dr = \frac{2}{\pi^3 a_0^4}\, \frac{1}{\left(\frac{4}{a_0^2} + k^2\right)^2}$$

$$\tilde{w}'(k) = \frac{1}{(2\pi)^3} \int w'(r)\, e^{-ikr}\, dr = \frac{-e^2}{2\pi^2 k^2}.$$

The result is

$$V_{\text{eff}}(R) = \frac{e^2}{R}\, e^{-\frac{2R}{a_0}} + \frac{e^2}{a_0}\, e^{-\frac{2R}{a_0}}. \tag{4.19b}$$

This is a screened potential as the hydrogen atom is neutral.

$\star$ The groundstate φ_i is real.

Introducing (4.19b) in (4.19) yields

$$T_{\text{B}} = 4\pi e^2 \left\{ \frac{4}{a_0^2} \frac{1}{\left(\frac{4}{a_0^2} + K^2\right)^2} + \frac{1}{\frac{4}{a_0^2} + K^2} \right\}, \quad K = K_f - K_i. \quad (4.21)$$

The angular dependence of the elastic differential cross section is given by

$$\frac{1}{\frac{4}{a_0^2} + K^2} = \frac{1}{2K_i^2} \frac{1}{2/a_0^2 K_i^2 + 1 - \cos\vartheta} \cong \frac{1}{K_i^2} \frac{1}{4\frac{m}{v} \frac{|\varepsilon_i|}{E_i} + \vartheta^2} \quad (4.21\text{a})$$

where we have used $|\varepsilon_i| = \dfrac{\hbar^2}{2ma_0^2}$, $E_i \cong \dfrac{\hbar^2}{2v} K_i^2$ and $1 - \cos\vartheta \cong \vartheta^2/2$ near the forward direction $\vartheta = 0$. Eq. (4.21a) shows that the elastic scattered protons are confined to a small angle near the forward direction, $\vartheta^2 \lesssim \dfrac{4m}{v} \dfrac{|\varepsilon_i|}{E_i}$.

Inserting (4.21) in (4.1) and performing the Ω-integration yields the total elastic cross section for a proton-hydrogen collision. The leading term for large E_i is

$$\sigma_{\text{el}} \sim \frac{7}{3} \frac{v}{m} \frac{|\varepsilon_i|}{E_i} \pi a_0^2 \quad (4.22)$$

i.e. the elastic cross section vanishes proportional to $1/E_i$ for large E_i. Eq. (4.22) shows that large E_i means $\dfrac{v}{m} \dfrac{|\varepsilon_i|}{E_i} \ll 1$, so that the velocities of the incoming proton and the electron in the Bohr orbit are the relevant quantities (cp. Impact Parameter Treatment, Chapter III). As $|\varepsilon_i| = 13.6\,\text{eV}$ and $v/m \cong 10^3$ the proton energy has to be substantially larger than 25 KeV for (4.22) to be valid.

The total cross section for inelastic collisions is

$$\sigma_{\text{inel.}} = \sum_{f \neq i} \int d\sigma_{if} \quad (4.23)$$

with $d\sigma_{if}$ according to (4.1). The summation has to be performed over all excited and ionized states φ_f with $\varepsilon_f < E_i$. In the high energy limit (4.23) has to be calculated with $T_{\text{B}}^{w'}$ and $\zeta \cong 1$ in (4.2a), as $T_{\text{B}}^{W} = 0$ for $\varphi_f \neq \varphi_i$. Again high energy will mean

$$\frac{m}{v} \frac{E_i}{|\varepsilon_i|} \gg 1. \quad (4.23\text{a})$$

For a proton-hydrogen collision we obtain with (4.20) and (4.2a)

$$T_B^{w'} = - \frac{4\pi e^2}{K^2} F_f(K), \quad K = K_f - K_i$$

$$F_f(K) = (\varphi_f, e^{iKx}\varphi_i)$$

(4.24)

where we have chosen the x-coordinate in direction of K. With $d\Omega = 2\pi \sin\vartheta \, d\vartheta$ we have from (4.24) and (4.1)

$$d\sigma_{if} \sim \frac{4\pi v^2 e^4}{\hbar^4 K_i} \frac{K_f}{K^4} |F_f(K)|^2 \sin\vartheta \, d\vartheta.$$

(4.25)

It is convenient to introduce the momentum transfer K instead of ϑ by $K^2 = K_f^2 + K_i^2 - 2K_i K_f \cos\vartheta$, $2K \, dK = 2K_i K_f \sin\vartheta \, d\vartheta$. Then (4.25) yields

$$d\sigma_{if} \sim \frac{8\pi v^2 e^4}{\hbar^4 K_i^2} |F_f(K)|^2 \frac{dK}{K^3}$$

(4.25a)

and the total inelastic cross section is with (4.23)

$$\sigma_{\text{inel.}} \sim \frac{8\pi v^2 e^4}{\hbar^4 K_i^2} \sum_{f \neq i} \int_{K_i - K_f}^{K_i + K_f} \frac{dK}{K^3} |F_f(K)|^2.$$

(4.26)

The integration limits can be realized in Fig. 4a: For fixed ε_f, K_f varies on a circle with radius $K_f = \sqrt{\frac{2v}{\hbar^2}(E_i - \varepsilon_f)}$. Forward scattering corresponds to a momentum transfer $K = K_i - K_f$ and backward scattering $K = K_i + K_f$. Fig. 4b shows the area over which the sum and integral in (4.26) has to be carried out. It is bounded by $K = K_i \pm K_f$, which is given, after taking the square, by the parabola $\varepsilon_f - \varepsilon_i = \frac{\hbar^2}{2v}(2K_i K - K^2)$.

In order to approximate (4.26) we have to investigate the function $|F_f(K)|^2$. With (4.24) is

$$\sum_f |F_f(K)|^2 = \sum_f (A\varphi_i, \varphi_f)(\varphi_f, A\varphi_i), \quad A = e^{-iKx}$$

and by the closure relation and the normalisation of φ_i

$$\sum_f |F_f(K)|^2 = (A\varphi_i, A\varphi_i) = (\varphi_i, A^+ A \varphi_i) = (\varphi_i, \varphi_i) = 1$$

(4.27)

i.e. $|F_f(K)|^2$ is for fixed K a normalized state density function.

We can calculate the average transferred energy by

$$\overline{\varepsilon(K)} = \sum_f (\varepsilon_f - \varepsilon_i) |F_f(K)|^2 = \sum_f (\varepsilon_f - \varepsilon_i)(A\varphi_i, \varphi_f)(\varphi_f, A\varphi_i).$$

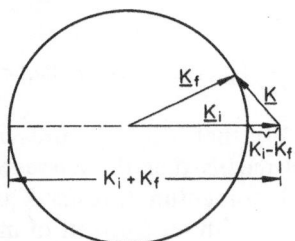

Fig. 4a. Initial (K_i)-final (K_f)- and transferred momentum (K) for inelastic collisions

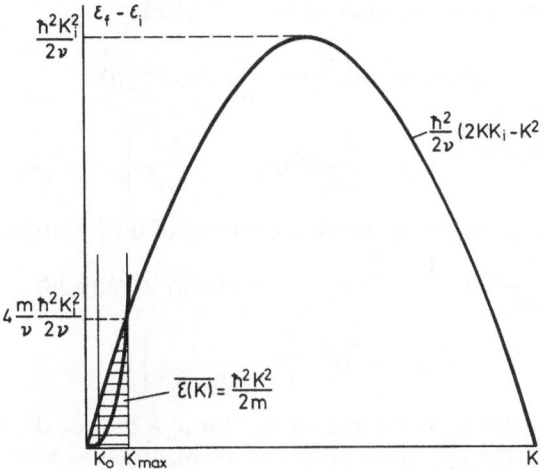

Fig. 4b. Integration area for the calculation of the total inelastic cross section and angular dispersion

With

$$h\varphi_{i,f} = \varepsilon_{i,f}\varphi_{i,f}, \quad h = -\frac{\hbar^2}{2m}\partial_r^2 - \frac{e^2}{r}, \quad \mu \cong m$$

we have

$$\overline{\varepsilon(K)} = \sum_f (A\varphi_i, h\varphi_f)(\varphi_f, A\varphi_i) - (Ah\varphi_i, \varphi_f)(\varphi_f, A\varphi_i)$$

and by closure

$$\overline{\varepsilon(K)} = (\varphi_i, (A^+ hA - h)\varphi_i) = (\varphi_i, (A^+ hA - A^+ Ah)\varphi_i)$$
$$= (\varphi_i, A^+ [hA]\varphi_i). \tag{4.28}$$

As e^2/r commutes with $A = \mathrm{e}^{-iKx}$ we obtain for the commutator

$$[h, A] = -\frac{\hbar^2}{2m}\{\partial_x^2 \mathrm{e}^{-iKx} - \mathrm{e}^{-iKx}\partial_x^2\}$$
$$= -\frac{\hbar^2}{2m}\{-K^2 \mathrm{e}^{-iKx} - 2iK\mathrm{e}^{-iKx}\partial_x\}$$

and therefore with (4.28)

$$\overline{\varepsilon(K)} = \frac{\hbar^2}{2m}(\varphi_i, (K^2 + 2iK\partial_x)\varphi_i) = \frac{\hbar^2 K^2}{2m}. \tag{4.29}$$

We have used in (4.29) that $\partial_x\varphi_i$ is antisymmetric and therefore $(\varphi_i, \partial_x\varphi_i) = 0$. Eq. (4.29) means that the average energy transfer to the hydrogen atom for fixed momentum transfer K is the same as the energy transfer in a free collision with an electron of mass m.

In order to obtain the width of $|F_f(K)|^2$ we have to evaluate $\overline{\varepsilon^2(K)}$ $= \sum_f (\varepsilon_f - \varepsilon_i)^2 |F_f(K)|^2$.

An analogous manipulation as for $\overline{\varepsilon(K)}$ yields

$$\overline{\varepsilon^2(K)} = \overline{\varepsilon(K)}^2 \left\{ 1 - \frac{4}{K^2}(\varphi_i, \partial_x^2 \varphi_i) \right\} \tag{4.30}$$

with

$$(\varphi_i, \partial_x^2 \varphi_i) = -\frac{2m}{3\hbar^2}\left(\varphi_i, -\frac{\hbar^2}{2m}\partial_r^2 \varphi_i \right) = -\frac{2m}{3\hbar^2}\left\{ \varepsilon_i + e^2\left(\varphi_i, \frac{1}{r}\varphi_i \right) \right\}.$$

With $\varepsilon_i = -e^2/2a_0$ for the groundstate energy of the hydrogen atom and $\left(\varphi_i, \frac{1}{r}\varphi_i \right) = \frac{1}{\pi a_0^3}\int d r \frac{1}{r} e^{-\frac{2r}{a_0}} = \frac{1}{a_0}$ we obtain from (4.30)

$$\overline{\varepsilon^2(K)} = \overline{\varepsilon(K)}^2 \left\{ 1 + \frac{4}{3a_0^2 K^2} \right\}. \tag{4.31}$$

This means $|F_f(K)|^2$ is well concentrated for $a_0 K \gg 1$, or the collision of the proton with the hydrogen atom can be regarded as a collision with a free electron for sufficiently large momentum transfer.

We have plotted in Fig. 4b $\overline{\varepsilon(K)} = \hbar^2 K^2/2m$ intersecting the integration boundary at $K_{max} \cong 2mK_i/v$. As

$$K_{max} a_0 = 2\frac{m}{v}K_i a_0 = 2\sqrt{\frac{m}{v}\frac{E_i}{|\varepsilon_i|}} \gg 1$$

with (4.23a), $|F_f(K)|^2$ is sharply concentrated for $K \gtrsim K_{max}$, so that $K > K_{max}$ does not contribute in (4.26). Therefore only the small dashed part of the integration area in Fig. 4b has to be taken into account. K_{max} is in good approximation the maximum momentum transfer occurring in (4.26), and it corresponds to the momentum transfer in a free head on collision between M and $m \ll M$. From Fig. 4a it is obvious that $K_{max}/K_i = 2m/v \ll 1$ is an upper limit for the deflection angle of inelastically scattered protons. As $K \le K_{max}$ we can write (4.26)

$$\sigma_{inel.} \sim \frac{8\pi v^2 e^4}{\hbar^4 K_i^2} \sum_{f \ne i} \int_{K_i - K_f}^{K_{max}} \frac{d K}{K^3} |F_f(K)|^2. \tag{4.32}$$

The sum and integral has to be carried out over the dashed area in Fig. 4b. We now introduce an intermediate momentum $K_0 = |\varepsilon_i| K_i/2E_i$, which corresponds to $\varepsilon_f - \varepsilon_i = |\varepsilon_i|$ on the integration boundary, and divide $\sigma_{\text{inel.}}$ in two parts:

$$\sigma_{\text{inel.}} \sim \frac{8\pi v^2 e^4}{\hbar^4 K_i^2} \left\{ \sum_{f \neq i} \int_{K_0}^{K_{\max}} \frac{dK}{K^3} |F_f(K)|^2 + \sum_{f \neq i} \int_{K_i - K_f}^{K_0} \frac{dK}{K^3} |F_f(K)|^2 \right\}. \quad (4.33)$$

In the first term of (4.33) we sum over $f \neq i$ up to infinity instead of the dashed area in Fig. 4b. This means we add contributions from ionisation processes which are not accessible energetically. One immediately realizes that this is correct for $Ka_0 \gg 1$, since then only the neighbourhood of the parabola $\varepsilon(K) = \hbar^2 K^2/2m$ contributes. But even near $K = K_0$, where $K_0 a_0 = \frac{1}{2} \left(\frac{v|\varepsilon_i|}{mE_i} \right)^{1/2} \ll 1$ with (4.23a), $|F_f(K)|^2$ is well concentrated within the integration area, as with (4.31), (4.23a), and $K_0 = |\varepsilon_i| K_i/2E_i$

$$\cdot \overline{\varepsilon(K_0)} + \{\overline{\varepsilon(K_0)^2} - \overline{\varepsilon(K_0)}^2\}^{1/2} = \frac{\overline{\varepsilon(K_0)}}{a_0 K_0} = 2|\varepsilon_i| \left(\frac{v|\varepsilon_i|}{mE_i} \right)^{1/2} \ll |\varepsilon_i|.$$

With $F_f(K)$ of (4.24) and the closure relation we then obtain for the first term in (4.33)

$$\sum_{f \neq i} \int_{K_0}^{K_{\max}} \frac{dK}{K^3} |F_f(K)|^2 = \int_{K_0}^{K_{\max}} \frac{dK}{K^3} (1 - |F_i(K)|^2) \quad (4.34)$$

with

$$F_i(K) = (\varphi_i, e^{-iKx} \varphi_i) = \frac{16}{a_0^4} \frac{1}{\left(\frac{4}{a_0^2} + K^2 \right)^2} .$$

The integration in (4.34) can be performed easily and the asymptotic result for $mE_i/v|\varepsilon_i| \gg 1$ is

$$\sum_{f \neq i} \int_{K_0}^{K_{\max}} \frac{dK}{K^3} |F_f(K)|^2 \sim \frac{a_0^2}{2} \ln \frac{16 m E_i}{v|\varepsilon_i|}. \quad (4.35)$$

In the second term of (4.33) we extend the sum from $\varepsilon_f - \varepsilon_i \lesssim |\varepsilon_i|$ to $\varepsilon_f - \varepsilon_i \lesssim 2|\varepsilon_i|$. Thereby we substract contributions which we have added in the first term and which are neglegible anyway. The greatest K which occurs in the integral is then about $2K_0$. As $2K_0 a_0 \ll 1$, we can expand e^{-iKx} in $F_f(K) = (\varphi_f, e^{-iKx} \varphi_i)$, where φ_i cuts off the integration at $x \cong a_0$:

$$F_f(K) \cong (\varphi_f, (1 - iKx) \varphi_i) = -iK(\varphi_f, x\varphi_i) = -iK x_{fi}. \quad (4.36)$$

With (4.34) the second term in (4.33) is

$$\sum_{\substack{f \neq i \\ \varepsilon_f - \varepsilon_i \leq 2|\varepsilon_i|}} \int_{K_i - K_f}^{K_0} \frac{dK}{K^3} |F_f(K)|^2 \cong \sum_{\varepsilon_f - \varepsilon_i \leq 2|\varepsilon_i|} |x_{fi}|^2 \ln \frac{K_0}{K_i - K_f}$$

$$\cong -a_0^2 \sum_{\varepsilon_f - \varepsilon_i \leq 2|\varepsilon_i|} |x_{fi}|^2/a_0^2 \ln \frac{\varepsilon_f - \varepsilon_i}{|\varepsilon_i|}. \tag{4.37}$$

We have dropped the restriction $f \neq i$, as $x_{ii} = 0$.

Along the lines of (4.27), (4.29), (4.31) one can realize the following properties of the distribution $|x_{fi}|^2/a_0^2$:

$$\sum_f |x_{fi}|^2/a_0^2 = 1 ; \quad \bar{\varepsilon} = \sum_f (\varepsilon_f - \varepsilon_i) |x_{fi}|^2/a_0^2 = |\varepsilon_i| ,$$

$$(\overline{\varepsilon^2} - \bar{\varepsilon}^2)^{1/2} = \frac{|\varepsilon_i|}{\sqrt{3}} \tag{4.38}$$

$\bar{\varepsilon} = |\varepsilon_i|$ means that $|x_{fi}|^2/a_0^2$ is centred where $\ln \dfrac{\varepsilon_f - \varepsilon_i}{|\varepsilon_i|}$ in (4.37) vanishes.

This makes the second term in (4.33) small and justifies the choice of K_0. We can estimate the second term in (4.33) by expanding $\ln \dfrac{\varepsilon_f - \varepsilon_i}{|\varepsilon_i|}$ in (4.37) in powers of $\varepsilon_f - \varepsilon_i - |\varepsilon_i|$. The first order term vanishes and the second order term is

$$\sum_{\substack{f \neq i \\ \varepsilon_f - \varepsilon_i \leq 2|\varepsilon_i|}} \int_{K_i - K_f}^{K_0} \frac{dK}{K^3} |F_f(K)|^2 = \frac{a_0^2}{2\varepsilon_i^2} \sum_f (\varepsilon_f - \varepsilon_i - |\varepsilon_i|)^2 \frac{|x_{fi}|^2}{a_0^2} = \frac{a_0^2}{6}. \tag{4.39}$$

We have extended in (4.39) the sum over all f, as the region $\varepsilon_f - \varepsilon_i > 2|\varepsilon_i|$ does not contribute with the width of $|x_{fi}|^2/a_0^2$ in (4.38).

Introducing (4.35) and (4.39) in (4.33) yields

$$\sigma_{\text{inel.}} \sim \frac{4v}{m} \frac{|\varepsilon_i|}{E_i} \ln \frac{22mE_i}{v|\varepsilon_i|} \pi a_0^2 ; \quad \frac{mE_i}{v|\varepsilon_i|} \gg 1 \tag{4.40}$$

or substituting $E_i \sim vv^2/2$ with the velocity v of the incoming proton

$$\sigma_{\text{inel.}} \sim \frac{8|\varepsilon_i|}{mv^2} \ln \frac{11mv^2}{|\varepsilon_i|} \pi a_0^2 ; \quad \frac{mv^2}{|\varepsilon_i|} \gg 1 . \tag{4.40a}$$

The error introduced by the expansion of $\ln \dfrac{\varepsilon_f - \varepsilon_i}{|\varepsilon_i|}$ in (4.39) can be assessed by comparing with the result of *Bethe*, who calculated the second term in (4.33) numerically. The difference is the argument of the logarithm in (4.40a) with a factor 3.14 instead of 11. This is neglegible for $mv^2/|\varepsilon_i| \gg 1$.

Comparing (4.40a) with (4.22) we see that $\sigma_{inel.}$ dominates for high energies.

Bethe [41] and later *Bates* and *Griffing* [16] calculated the cross section σ_{ion} for ionisation of hydrogen by protons in the Born approximation. Fig. 5 shows that the agreement with experimental data is good for energies greater than 100 KeV. Asymptotically for high energies $\sigma_{ion} \sim 0.285\, \sigma_{inel.}$.

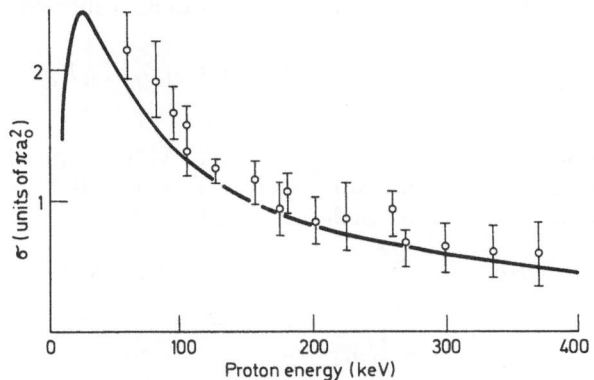

Fig. 5. Cross sections for $H^+ + H(1s) \rightarrow H^+ + H^+ + e$. Solid curve: Born approximation (*Bates* and *Griffing* 1953) ⌀ : Absolute measurements (*Gilbody* and *Ireland*, 1963)

The stopping power dE/dz for fast protons passing through hydrogen is defined as the energy loss per unit path z. We have seen that in the high energy region only inelastic collisions occur. Therefore the stopping power is due to excitation and ionisation. The calculation of dE/dz is contained in many text books [6, 7]. We rather calculate here a related quantity, the angular dispersion $\overline{d\theta^2}/dz$ of fast protons passing through hydrogen defined as the increase of average angular divergence per unit path z.

The average deflection $\overline{\theta^2}(1)$ of a proton after one collision is

$$\overline{\theta^2}(1) = \zeta^2 \int d\Omega\, \vartheta^2\, \frac{d\sigma}{d\Omega}\Big/ \sigma \qquad (4.41)$$

with the scattering angle $\theta = \zeta\vartheta$ in the laboratory system and the total cross section σ. Travelling a distance z in a gas with density ϱ the proton suffers $\varrho\sigma z$ collisions and the average angular divergence is $\overline{\theta^2}(z) = \varrho\sigma z\overline{\theta^2}(1)$, as all scattering events are supposed to be independent.

We obtain from (4.41)

$$\frac{d\overline{\theta^2}}{dz} = \varrho\zeta^2 \int d\Omega\, \vartheta^2\, \frac{d\sigma}{d\Omega}. \tag{4.42}$$

The differential cross section for inelastic collisions is with (4.1)

$$\frac{d\sigma}{d\Omega} = \frac{v^2}{(2\pi)^2\hbar^4 K_i} \sum_{\substack{f \neq i \\ \varepsilon_f \leq E_i}} K_f |T_{if}(\boldsymbol{K}_f, \boldsymbol{K}_i)|^2$$

and (4.42) yields for the angular dispersion due to inelastic collisions

$$\frac{d\overline{\theta^2}}{dz} = \frac{\varrho\zeta^2 v^2}{(2\pi)^2\hbar^4 K_i} \sum_{\substack{f \neq i \\ \varepsilon_f \leq E_i}} \int d\Omega\, \vartheta^2 K_f |T_{if}(\boldsymbol{K}_f, \boldsymbol{K}_i)|^2. \tag{4.43}$$

In the high energy limit is in (4.43) $T_{if} = T_{\mathrm{B}}^{\mathrm{w}'}$ with $T_{\mathrm{B}}^{\mathrm{w}'}$ given by (4.24). Introducing in (4.43) the momentum transfer $K = |\boldsymbol{K}_i - \boldsymbol{K}_f|$ as integration variable, we have with only small ϑ occuring in (4.43)

$$\cos\vartheta \cong 1 - \frac{\vartheta^2}{2} = \frac{1}{2K_i K_f}(K_f^2 + K_i^2 - K^2)$$

$$K_f = \sqrt{K_i^2 - \frac{2v}{\hbar^2}(\varepsilon_f - \varepsilon_i)}.$$

Expanding K_f for small $\varepsilon_f - \varepsilon_i$ results in

$$\vartheta^2 \cong \frac{K^2}{K_i^2} - \frac{v^2}{\hbar^4 K_i^2}(\varepsilon_f - \varepsilon_i)^2; \quad \vartheta \ll 1; \quad \frac{\varepsilon_f - \varepsilon_i}{E_i} \ll 1. \tag{4.44}$$

One recognizes in Fig. 6a, b that in good approximation $\varepsilon_f - \varepsilon_i \lesssim 4mE_i/v$ and $\vartheta < K_{\max}/K_i = 2m/v$. Therefore (4.44) is justified. With (4.24) and $2K\,dK = 2K_i K_f \sin\vartheta\,d\vartheta$ we obtain from (4.43)

$$\frac{d\overline{\theta^2}}{dz} = \frac{8\pi\varrho\zeta^2 v^2 e^4}{\hbar^4 K_i^4} \sum_{\varepsilon_i < \varepsilon_f \leq E_i} \int_{K_i - K_f}^{K_i + K_f} \frac{dK}{K^3} |F_f(K)|^2$$
$$\cdot \left\{ K^2 - \frac{v^2}{\hbar^4 K_i^2}(\varepsilon_f - \varepsilon_i)^2 \right\}. \tag{4.45}$$

The evaluation of (4.45) is completely analogous to that of (4.26). The upper limit of integration can be replaced by $K_{\max} = 4mK_i/v$. By the intermediate momentum $K_0 = |\varepsilon_i| K_i/2E_i$ (4.45) is divided in two parts

$$\frac{d\overline{\theta^2}}{dz} = \frac{8\pi\varrho\zeta^2 v^2 e^4}{\hbar^4 K_i^4} \left\{ \sum_{\varepsilon_i < \varepsilon_f} \int_{K_0}^{K_{\max}} \frac{dK}{K^3} |F_f(K)|^2 \left[K^2 - \frac{v^2}{\hbar^4 K_i^2}(\varepsilon_f - \varepsilon_i)^2 \right] \right.$$
$$\left. + \sum_{\varepsilon_i < \varepsilon_f < |\varepsilon_i|} \int_{K_i - K_f}^{K_0} \frac{dK}{K^3} K^2 |x_{fi}|^2 \left[K^2 - \frac{v^2}{\hbar^4 K_i}(\varepsilon_f - \varepsilon_i)^2 \right] \right\}. \tag{4.46}$$

As in (4.33) we have omitted in the first term the restriction $\varepsilon_f \lesssim E_i$ and in the second term expanded $F_f(K)$ as well as extended the sum from $\varepsilon_f < 0$ to $\varepsilon_f < |\varepsilon_i|$. With the closure relation and (4.31) we evaluate the sum in the first term and can then easily perform the K-integration. In the second term we first integrate over K and then perform the sum after expanding in powers of $\varepsilon_f - \varepsilon_i - |\varepsilon_i|$ up to quadratic terms. For $mv^2/|\varepsilon_i| \gg 1$ the result for the second term in the curly brackets of (4.46) is $|\varepsilon_i|/6mv^2$ compared with $\frac{1}{2} \ln 3mv^2/3|\varepsilon_i|$ for the first term. Asymptotically for $mv^2/|\varepsilon_i| \gg 1$ we drop the second term and obtain finally

$$\frac{d\overline{\theta^2}}{dz} = \frac{4\pi\varrho\zeta^2 e^4}{v^2 v^4} \ln \frac{3mv^2}{2|\varepsilon_i|}. \tag{4.47}$$

2. Wave Treatment of Rearrangement Collisions

a) High Energy Expansion

The high energy treatment of rearrangement collisions is one of the fundamental problems in collision theory. Since 1930 it has been investigated for the charge exchange process in a proton-hydrogen collision, and even nowadays new proposals for the correct high energy cross section come up.

The historical development is roughly the following. *Brinkman* and *Kramers* (1930) [18] calculated the first Born approximation (1.44) neglecting the internuclear potential W in $V_f = w + W$. Their result is a E_i^{-6}-dependence for the leading term in the charge exchange cross section. In 1953 *Jackson* and *Schiff* [20] included W in their Born approximation calculation. Surprisingly they obtained a cross section asymptotically by a factor of 0.66 smaller than the *Brinkman Kramers* result. This was difficult to understand as for high energies and $m/M, M' \ll 1$ an impact parameter treatment should be valid, where W drops out after a unitary transformation. Furthermore it is intuitively clear that for high energies, where the proton follows a straight path, the electronic transitions should not be influenced by the internuclear repulsion. *Drisko* (1955) [19] calculated the second Born approximation (1.44) in the high energy limit. He showed that the W-contributions in the first and second Born approximation cancel, so that this problem was settled. On the other hand the second Born approximation provided also a $E_i^{-11/2}$-term which is for high energies dominant over the E_i^{-6}-term of the first Born approximation. This suggested that the Born series would diverge for all energies, and *Aaron, Amado* and *Lee* (1961) [28] produced a "proof" of the divergence of the Born series for rearrangement collisions. However in 1966 [30] we found a simple counter example, which showed the divergence proof not to be conclusive. *Corbett* (1967) [31] extended our argument by pointing out the ordering, matrix element series diverges implies vector series diverges implies operator series diverges, but not the inverse. In 1968 [24] and 1969 [25] we investigated the high energy behaviour of rearrangement collisions with arbitrary masses for potentials with Fourier transforms having an asymptotic expansion in powers of $k^{-\beta}$, e.g. the Coulomb potential. It turned out that depending on the mass ratio either the second, or the first and second, Born approximation provide the leading terms in a high energy expansion of the cross section. For $m/M, M' \ll 1$ the W-terms in the first and second Born approximation cancel as already *Drisko* had shown for the Coulomb case. This high energy expansion will be reproduced in the following.

The differential cross section for M' being scattered in the solid angle $(\Omega, d\Omega)$ having picked up m in the bound state φ'_f is with (1.38) and (1.36)

$$\frac{d\sigma'_{if}}{d\Omega} = \frac{v}{(2\pi)^2 \hbar^2 K_i} \int |T'_{if}(K', K_i)|^2 \, \delta(E_i(K_i) - E'_f(K')) K'^2 \, dK'$$

$$E_i(K_i) = \frac{\hbar^2 K_i^2}{2v} + \varepsilon_i ; \quad E_f(K') = \frac{\hbar^2 K'^2}{2v'} + \varepsilon'_f . \tag{4.48a}$$

After integrating over K' we obtain

$$\frac{d\sigma'_{if}}{d\Omega} = \frac{vv'}{(2\pi)^2 \hbar^4} \frac{K_f}{K_i} |T'_{if}(K_f, K_i)|^2, \quad K_f = \sqrt{\frac{2v'}{\hbar^2}(E_i - \varepsilon'_f)} \tag{4.48}$$

when $\varepsilon'_f < E_i$ and zero otherwise. For the high energy expansion of T'_{if} in (4.48) the Born series (1.44) is convenient. The first term is the Born approximation

$$T'_B = (\phi'_f, (w + W) \phi_i) = T'^w_B + T'^W_B \tag{4.49}$$

with the Fourier representation

$$T'^w_B = (2\pi)^6 \, \tilde{\varphi}'_f(K^{w'}) \int d\mathbf{k} \, w(K^w - k) \, \tilde{\varphi}_i(k), \tag{4.49a}$$

$$T'^W_B = (2\pi)^6 \int d\mathbf{k} \, \tilde{\varphi}'_f(K^w - k) \, W(k) \, \tilde{\varphi}_i(K^w + k) \tag{4.49b}$$

where

$$K^{w'} = K_i - \xi K_f, \quad K^w = K_f - \zeta K_i$$

and

$$\varphi_i(r) = \int \tilde{\varphi}_i(k) \, e^{ikr} \, dk \quad \text{etc.}$$

ξ and ζ are defined in (1.3). We can expand (4.49a, b) asymptotically with (4.6), (4.9), and (4.9b) assuming that w, w', and W are potentials of the same kind with Fourier transforms having an asymptotic expansion in powers of $k^{-\beta}$, like a Yukawa- or Coulomb potential. For simplicity we assume that φ_i, φ'_f are the groundstates and therefore isotropic. With (4.9) and (4.9b) we obtain for (4.49a)

$$T_B'^w \sim -(2\pi)^6 \frac{2\mu' \varphi_i(0) \varphi'_f(0)}{\hbar^2 (K^{w'})^2} \tilde{w}(K^w) \tilde{w}'(K^{w'}) \tag{4.50}$$

where $\tilde{w}$ and $\tilde{w}'$ mean the asymptotically leading terms. In the forward direction $K^{w'} \sim K_i(1 - \sqrt{\zeta\xi})$ and $K^w \sim K_f(1 - \sqrt{\zeta\xi})$ are nonvanishing, so that all terms in (4.50) exist even for forward scattering.

We can estimate the energy for the asymptotic formula (4.50) to be valid by calculating (4.49a) exactly. This can be done easily with (4.7a) and the result is

$$T_B'^w = -(2\pi)^6 \tilde{\varphi}_i(K^w) \tilde{\varphi}'_f(K^w) \left\{ \frac{\hbar^2}{2\mu} (K^w)^2 - \varepsilon_i \right\}. \tag{4.50a}$$

Obviously the validity of the asymptotic expansion requires that $\hbar^2 (K^w)^2/2\mu \sim \gtrsim \hbar^2 K_f^2 (1 - \sqrt{\zeta\xi})^2/2\mu \sim E_i v'(1 - \sqrt{\zeta\xi})^2/\mu \gg |\varepsilon_i|$. If the three masses are of comparable magnitude, this means $E_i \gg |\varepsilon_i|$. If, on the other hand, the exchanged mass is small ($m \ll M, M'$) as for instance in proton hydrogen scattering, one obtains $E_i \gg |\varepsilon_i| v/m$. This shows, as in the case of direct collisions, that the velocities and not the energies are the relevant quantities ($E_i/v \sim v^2/2$). This is also suggested by the impact parameter treatment. For proton hydrogen scattering v has to be great compared with the velocity of the electron in the Bohr orbit and means, as for direct collisions, proton energies substantially greater than 25 KeV.

If in (4.49b) one introduces the semiasymptotic expansions (4.9), (4.6), the leading term contains at least two, for finite $W(0)$ even three δ-functions peaked at different k-values, if $K^{w'} \neq -K^w$. The case $K^{w'} = -K^w$ will be discussed later. The next term containing two δ-functions vanishes for the same reason. The first non-vanishing terms contain only one δ-function. Of the three possible cases the δ-functions due to $\tilde{\varphi}'_f$ and $\tilde{\varphi}_i$ determine the two leading terms, because W is assumed of the same kind as w, w' and is with (4.9b) asymptotically larger than $\tilde{\varphi}_{i,f}$. Consequently is with $K^{ww'} = K^w + K^{w'} \neq 0$

$$T_B'^W \sim -(2\pi)^6 \left\{ \varphi'_f(0) \tilde{W}(K^{w'}) \tilde{\varphi}_i(K^{ww'}) + \varphi_i(0) \tilde{W}(K^w) \tilde{\varphi}'_f(K^{ww'}) \right\}$$

and with (4.9b)

$$T_B'^W \sim -\frac{2(2\pi)^6 \varphi_i(0) \varphi'_f(0)}{\hbar^2 (K^{ww'})^2} \left\{ \mu \tilde{W}(K^{w'}) \tilde{w}(K^{ww'}) + \mu' \tilde{w}'(K^{ww'}) \tilde{W}(K^w) \right\}. \tag{4.51}$$

Again the Fourier transforms are meant to be the asymptotically leading terms. From (4.50) and (4.51) one sees that both Born amplitudes have the same order of magnitude.

For

$$K^{ww'} = K^w + K^{w'} = (1 - \xi)\, K_f + (1 - \zeta)\, K_i = 0 \qquad (4.52)$$

the expansion (4.51) breaks down because then $\tilde\varphi_{i,f}$ are peaked at the same value $k = K^{w'} = -K^w \neq 0$. Eq. (4.52) can only be fulfilled if $(1 - \xi)\, K_f = (1 - \zeta)K_i$ or asymptotically $(1 - \xi)(v'/v)^{1/2} = (1 - \xi)(\zeta/\xi)^{1/2} = 1 - \zeta$, which implies $\zeta = \xi$ or $M' = M$. Condition (4.52) means back scattering $(K_f = -K_i)$ for equal masses and one has from (4.49b) with $\tilde\varphi_i(k) = \tilde\varphi_i(-k)$

$$T_B^W(-K_i, K_i) = (2\pi)^6 \int dk\, \tilde\varphi_f'(K^{w'} - k)\, \tilde\varphi_i(K^{w'} - k)\, \tilde W(k). \qquad (4.53)$$

Here one has to employ the semiasymptotic expansion for the product

$$\tilde\varphi_f'(k)\, \tilde\varphi_i(k) = \frac{1}{(2\pi)^3} \int dr\, e^{ikr}\, O(r) = \tilde O(k)$$

where

$$O(r) = \frac{1}{(2\pi)^3} \int dr'\, \varphi_f'(|r - r'|)\, \varphi_i(r')$$

is an overlap integral between the two ground states which can be expanded in powers of r

$$\tilde O(k) = \delta(k)\, O(0) + \cdots.$$

Then the leading term in (4.53) becomes with $K^{w'} = (1 + \xi)\, K_i$

$$T_B'^W(-K_i, K_i) \sim (2\pi)^6\, O(0)\, \tilde W((1 + \xi)\, K_i) \qquad (4.54)$$

which is asymptotically much larger than $T_B'^w$ in (4.50). Therefore for $M = M'$ charge exchange in backward direction is dominant for sufficiently high energies. This is in agreement with the classical argument that charge exchange works best if after a certain collision, here a W-collision between M' and M, the exchanged charge and the capturing nucleus have the same velocity. In the laboratory system m is at rest initially. After a head-on collision between M' and M (corresponding to back scattering in the centre of mass system) M' replaces M and will be at rest after the collision. As $(1 + \xi)\, K_i$ is the momentum of the incoming particle M', $T_B'^W$ in (4.54) is the Born amplitude for internuclear potential scattering multiplied with the overlap of the initial and final bound states. If M and $M' = M$ are identical particles, back scattering could not be observed experimentally. In a nuclear reaction where an incoming neutron $(M' = 1)$ collides with a light ^3_2He nucleus, replaces the proton $(M = 1)$ and forms a tritium nucleus (exchange of the deuteron $(m = 2)$), one then should expect at high energies predominant proton emission in forward direction.

To discuss the angular dependence of (4.49b) for $M = M'$ near the backward direction, one can take $\tilde W(k) = \tilde W((1 + \xi)\, K_i)$ out of the integral because it is slowly varying with respect to $\tilde{\varphi}_i \tilde{\varphi}'_f$:

$$T_B'^W \sim (2\pi)^6 \, \tilde{W}((1+\xi)\,K_i) \int dk \, \tilde{\varphi}'_f(k) \, \tilde{\varphi}_i(k + (1-\xi)\,(K_i + K_f))$$

$$= (2\pi)^3 \, \tilde{W}((1+\xi)\,K_i) \int dr \, \varphi'_f(r) \, \varphi_i(r) \, e^{i(1-\xi)(K_i + K_f)r}$$

$$= (2\pi)^3 \, \tilde{W}((1+\xi)\,K_i) \frac{4\pi}{u} \int_0^\infty r \, dr \, \varphi'_f \varphi_i \sin ur$$

with

$$u = (1-\xi)\,|K_i + K_f| \sim (1-\xi)\left\{\frac{4v}{\hbar^2} E_i (1 - \cos\vartheta)\right\}^{1/2}$$

$$\cos\vartheta = K_i K_f / K_i K_f = K_i K_f / K_i^2 \, .$$

The total cross section for charge exchange from φ_i to φ'_f due to T_B^W alone is obtained from (4.48) by integration over Ω, $\int d\Omega = \int_{-1}^{+1} 2\pi \, d\cos\vartheta$.
If one introduces u instead of $\cos\vartheta$, one can replace the upper limit of u, which is $(1-\xi)\,(8vE_i/\hbar^2)^{1/2}$, by infinity and obtains

$$\sigma_B^W \sim \frac{2v(2\pi)^7}{\hbar^2(1-\xi)^2\,E_i} |\tilde{W}((1+\xi)\,K_i)|^2 \int_0^\infty \frac{du}{u} \left\{\int_0^\infty r \, dr \, \varphi'_f \varphi_i \sin ur\right\}^2 . \quad (4.55)$$

Comparing (4.55) with (4.54) one sees that the width of the angular distribution $d\sigma_B^W$ is proportional to $1/E_i$ and is sharply peaked at high energies.

The second order Born amplitude in (1.44)

$$T_2' = (\phi'_f, (w + W)\, G_0 (w' + W)\, \phi_i) \quad (4.56)$$

contains four terms $(T_2'^{ww'}, T_2'^{wW}, \ldots)$ which represent two subsequent collisions between free particles, e.g. $T_2'^{ww'}$ corresponds to a process where first M' and m interact by w' and then m and M by w. T_2' is in Fourier representation

$$T_2' = (2\pi)^{12} \int dk \, dK \, G_0(k, K)$$

$$ww' \times \tilde{\varphi}'_f(\xi K_f - K) \, \tilde{w}(K_f - k - \zeta K)$$
$$\cdot \tilde{w}'(K - K_i) \, \tilde{\varphi}_i(k + \zeta(K - K_i)), \quad (4.56a)$$

$$wW \times \tilde{\varphi}'_f(\xi K_f - K) \, \tilde{w}(K_f - k - \zeta K)$$
$$\cdot \tilde{W}(K - K_i) \, \tilde{\varphi}_i(k - (1-\zeta)\,(K - K_i)), \quad (4.56b)$$

$$Ww' \times \tilde{\varphi}'_f(k - (1-\zeta)\,K - (1-\xi)\,K_f) \, \tilde{W}(K_f - k - \zeta K)$$
$$\cdot \tilde{w}'(K - K_i) \, \tilde{\varphi}_i(k + \zeta(K - K_i)), \quad (4.56c)$$

$$WW \times \tilde{\varphi}_f(k - (1-\zeta)\,K - (1-\xi)\,K_f) \, \tilde{W}(K_f - k - \zeta K)$$
$$\cdot \tilde{W}(K - K_i) \, \tilde{\varphi}_i(k - (1-\zeta)\,(K - K_i)) \quad (4.56d)$$

with

$$G_0(\mathbf{k}, \mathbf{K}) = \frac{1}{(2\pi)^6} \frac{1}{E_i + i\eta - \dfrac{\hbar^2 K^2}{2\nu} - \dfrac{\hbar^2 k^2}{2\mu}}.$$

Because $\tilde{\varphi}_{i,f}$ vanish faster asymptotically than the Fourier transforms of the potentials, the leading terms are obtained by inserting $\varphi_{i,f}(0)\,\delta(\mathbf{k}')$ for $\tilde{\varphi}_{i,f}(\mathbf{k}')$, with the result

$$\frac{T_2'}{(2\pi)^6\,\varphi_i(0)\,\varphi_f'(0)} = ww' : \tilde{w}(\mathbf{K}^w)\,\tilde{w}'(\mathbf{K}^{w'})/D^{ww'}$$

$$D^{ww'} = E_i \left\{ 1 - \eta' - \frac{\eta'}{1-\eta'}(\mathbf{u}_i - \sqrt{\eta'}\,\mathbf{u}_f)^2 \right\} \tag{4.57a}$$

with

$$\eta' = \zeta\xi, \qquad \mathbf{u}_{i,f} = \frac{\mathbf{K}_{i,f}}{k_{i,f}}, \qquad \mathbf{u}_{i,f}^2 = 1,$$

$$wW : \tilde{w}(\mathbf{K}^{ww'})\,\tilde{W}(\mathbf{K}^{w'})/D^{wW}$$

$$D^{wW} = E_i \left\{ 1 - \eta' - \frac{\xi(1-\zeta)^2}{\zeta(1-\eta')}(\mathbf{u}_i - \sqrt{\eta'}\,\mathbf{u}_f)^2 \right\} \tag{4.57b}$$

$$Ww' : \tilde{W}(\mathbf{K}^w)\,\tilde{w}'(\mathbf{K}^{ww'})/D^{Ww'}$$
$$D^{Ww'}(\xi,\zeta) = D^{wW}(\zeta,\xi), \tag{4.57c}$$

$$WW : 0, \text{ vanishes of higher order if } K^{ww'} \neq 0 \text{ or } M \neq M'. \tag{4.57d}$$

In this order T_2^{WW} vanishes because the two δ-functions in (4.56d) are peaked at different values $\mathbf{k} - (1-\zeta)\mathbf{K} = (1-\xi)\mathbf{K}_f$, $-(1-\zeta)\mathbf{K}_i$ if $K^{ww'} \neq 0$. In Eqs. (4.57) again only the asymptotically leading terms are meant. By comparison with (4.50), (4.51) one recognizes that the second order amplitudes (4.57a, b, c) have the same energy dependence as the Born approximation and are in contrast to the direct collisions not negligible in the high energy limit. The third and higher Born approximations will be shown of higher order small Γ. Consequently in general

$$T' \sim T_B' + T_2'. \tag{4.58}$$

The semiasymptotic expansion gives a divergent result for particular angles depending on the mass ratios. It appears that some transition amplitudes in these cases become infinite. Actually, the amplitudes stay finite; in these cases the semiasymptotic expansion has to be reconsidered, but the divergencies indicate that for these angles T_2' becomes particularly large and, as in the case of $T_B^{'W}$ for $M = M'$, becomes of lower order in $1/E_i$. Consequently, the asymptotic behavior of the total cross section is determined by the divergent terms if they are present.

One kind of divergence is established by vanishing denominators in (4.57) which occur at certain angles $(\cos\vartheta = u_i u_f, (u_i - \sqrt{\eta'} \, u_f)^2 = 1 + \eta' - 2\sqrt{\eta'}\cos\vartheta)$,

$$D^{ww'} = 0, \quad \cos\vartheta = \frac{3\zeta\xi - 1}{2(\zeta\xi)^{3/2}}, \quad 1 \leq 4\zeta\xi \leq 4, \quad (4.59\text{a})$$

$$D^{wW} = 0, \quad \cos\vartheta = \frac{1}{2(\zeta\xi)^{1/2}}\left\{1 + \zeta\xi - \frac{\zeta}{\xi}\frac{(1-\zeta\xi)^2}{(1-\zeta)^2}\right\}.$$
$$(4.59\text{b})$$
$$\zeta \leq \xi \leq \frac{\zeta}{(1-2\zeta)^2},$$

$$D^{Ww'} = 0, \ \zeta \text{ and } \xi \text{ interchanged in } (4.59\text{b}). \quad (4.59\text{c})$$

Eqs. (4.59) give the critical angles, ϑ_c, and the conditions for real angles $(-1 \leq \cos\vartheta \leq 1)$ which are shown in the $\zeta - \xi$ plane of Fig. 8 $(0 < \zeta, \xi < 1)$. In the hatched area of Fig. 6a the denominator $D^{ww'}$ can vanish. The lines for $\vartheta = \pi(\zeta\xi = 1/4)$ and for $\vartheta = \pi/2(\zeta\xi = 1/3)$ are indicated. Small angles belong to points in the right upper corner where $\zeta, \xi \cong 1$ or $m \ll M, M'$ and $\cos\vartheta \cong 1 - 3m^2(M + M')^2/8M^2M'^2$, e.g., for an electron exchange in a proton hydrogen collision $(M = M')$ one obtains $\vartheta \cong \sqrt{3}m/M = 10^{-3}$ or $3.5'$, for deuteron hydrogen collision $(M = M'/2)$ one obtains an angle smaller by a factor 3/4. Fig. 6b shows the corresponding areas for wW and Ww' collisions according to (4.59b, c). There the lines for $\vartheta = \pi, 0$ are also shown. However, a point near $\zeta = \xi$ does not necessarily indicate a critical angle close to π. In particular for small m one has for a wW collision $(M' > M)$ from (4.59b) $\cos\vartheta = 1 - (M + M')^2/2M'^2 + m(M + M')(M - M')^2/4MM'^3$, which shows that the angle is mainly determined by M'/M. Only for $M' > \sim M$ will one have critical back scattering $(\cos\vartheta \cong -1, \vartheta \cong \pi)$, for instance in a proton hydrogen collision. In a deuteron hydrogen collision one obtains $\cos\vartheta = -1/8$, $\vartheta = 97°$ corresponding to an angle of 28° in the Lab system. No singularities occur in the areas left free in Fig. 8a and b. For such values the asymptotic expressions are given by (4.57a–c). This corresponds roughly to masses m intermediate between M' and M, e.g., $M', m, M = 1, 4, 16; \zeta, \xi = 4/5, 1/5$; point 1 in Fig. 8b. In other cases one can have one critical angle, e.g., point 2 with $M', m, M = 1, 2, 2; \zeta, \xi = 1/2, 1/3; \cos\vartheta^{Ww'} = \sqrt{6}/16, \vartheta^{Ww'} \cong 80°$, or two critical angles, e.g., point 3 with $M', m, M = 1, 1, 2; \zeta, \xi = 2/3, 1/2, \zeta\xi = 1/3$ where the critical angles due to $D^{ww'}, D^{Ww'} = 0$ are both $\pi/2$. This corresponds to a nuclear exchange reaction where an incoming proton $(M' = 1)$ picks up a neutron $(m = 1)$ from a tritium $(m + M = 3, M = 2)$ nucleus and forms two

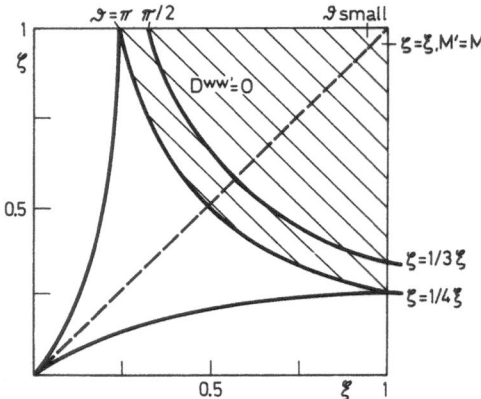

Fig. 6a. Area of the $\zeta - \xi$ plane where $D^{ww'}$ vanishes

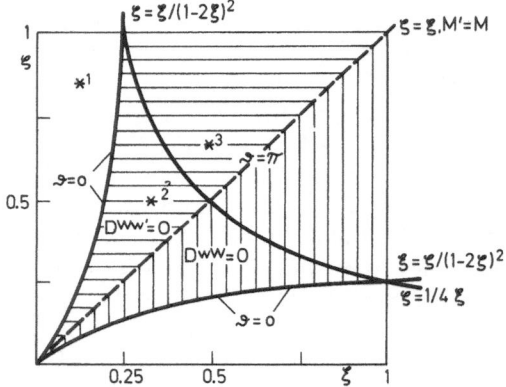

Fig. 6b. Areas of the $\zeta - \xi$ plane where D^{wW} and $D^{Ww'}$ vanish, indicated are the points
1: $M', m, M = 1, 4, 16$; 2: 1, 2, 2; 3: 1, 1, 1

deuterons. In this reaction 90° scattering in the center of mass system should prevail for large incoming energies. For small masses m one has always two critical angles, one small angle from (4.58a) and one angle depending on M'/M to be determined from either (4.59b) or (4.59c).

These critical angles in the 2nd Born approximation have already been discussed by *Thomas* [42] in his classical treatment of charge exchange. They can also be defined by energy-momentum conservation and by the additional requirement that M' and m have equal final velocities after the corresponding free collisions, for instance first a w'-collision between M' and m and a subsequent w-collision between m

and M as in (4.59a). This is in line with critical back scattering showing up in $T_B'^W$ for equal masses.

The behavior of T_2' near the critical angles can be investigated in the following way. As an example we discuss

$$T_2'^{ww'} = (\phi_f', w G_0 w' \phi_i). \tag{4.60}$$

Here one can take $\tilde{w}$ and $\tilde{w}'$ out of the integral (4.56a) at k values where $\tilde{\phi}_{f,i}$ are concentrated,

$$T_2'^{ww'} \sim (2\pi)^6 \, \tilde{w}(K^w) \, \tilde{w}'(K^{w'}) F^{ww'} \tag{4.61}$$
where

$$F^{ww'} = \int d\mathbf{K} \, d\mathbf{k} \, \tilde{\phi}_f'(\mathbf{K}_f - \mathbf{K}) \frac{1}{i\eta + E_i - \dfrac{\hbar^2 K^2}{2v} - \dfrac{\hbar^2 k^2}{2\mu}} \, \tilde{\phi}_i(\mathbf{k} + \zeta(\mathbf{K} - \mathbf{K}_i)) \ .$$

The semiasymptotic expansion for $\tilde{\phi}_{i,f}$ would result in

$$F^{ww'} = \varphi_f'(0) \, \varphi_i(0) \frac{1}{D^{ww'}} \tag{4.62}$$

and would diverge for the critical angle where $D^{ww'} = 0$. Actually, $F^{ww'}$ stays always finite and can be calculated when $\varphi_{f,i}$ are known. Asymptotically one can simplify the calculation by using the fact that $\tilde{\phi}_{i,f}(k)$ are concentrated near $k = 0$ and by expanding the denominator about the values $\mathbf{K}_0, \mathbf{k}_0$ where $\tilde{\phi}_{i,f}$ are peaked.

$$\mathbf{K}_0 = \xi \mathbf{K}_f, \quad \mathbf{k}_0 = \zeta(\mathbf{K}_i - \mathbf{K}_0) = \zeta(\mathbf{K}_i - \xi \mathbf{K}_f)$$
and

$$E_i - \frac{\hbar^2}{2v} K^2 - \frac{\hbar^2}{2\mu} k^2 = D^{ww'} + D_1^{ww'} + D_2^{ww'}$$
with

$$D_1^{ww'} = -\frac{\hbar^2}{v}(\mathbf{K} - \mathbf{K}_0)\mathbf{K}_0 - \frac{\hbar^2}{\mu}(\mathbf{k} - \mathbf{k}_0)\mathbf{k}_0$$

$$D_2^{ww'} = -\frac{\hbar^2}{2v}(\mathbf{K} - \mathbf{K}_0)^2 - \frac{\hbar^2}{2\mu}(\mathbf{k} - \mathbf{k}_0)^2 \ .$$

Since $\mathbf{K}_0, \mathbf{k}_0$ are of the order $\sqrt{E_i}$ and the quantities $(\mathbf{K} - \mathbf{K}_0)$, $(\mathbf{k} - \mathbf{k}_0)$ are confined to the width of $\tilde{\phi}_{i,f}$ the order of D, D_1, D_2 is $E_i, E_i^{1/2}, E_i^0$ and D_2 can be dropped asymptotically. Outside the critical angle also D_1 can be neglected and one comes back to (4.62). One can see here that at $\vartheta = \vartheta_c$ the quantity F is proportional to $E_i^{-1/2}$ rather than to E_i^{-1}. The amplitude therefore is by a factor $\sqrt{E_i}$ larger than normally. Because $D^{ww'} \propto (\vartheta - \vartheta_c)$ near the critical angle, one can also see that the angular widths of critical scattering will be proportional to $1/\sqrt{E_i}$. The contribution to the total cross section, therefore, will be larger by a factor $\sqrt{E_i}$

12*

as compared to the normal result. Consequently, the scattering into critical angles will be leading asymptotically. By introducing the arguments of $\tilde{\varphi}_{i,f}$ as new variables

$$K - K_0 = K', \quad k + \zeta K - \zeta K_i = k + \zeta K - k_0 - \zeta K_0 = k - k_0 + \zeta K' = k'$$

one obtains after dropping D_2

$$F^{ww'} = \int d K' d k' \, \tilde{\varphi}'_f(K') \frac{1}{i\eta + D^{ww'} - K' K'_0 - k' k'_0} \, \tilde{\varphi}_i(k')$$

where

$$K'_0 = \frac{\hbar^2}{v} K_0 - \frac{\hbar^2}{\mu} \zeta k_0, \quad k'_0 = \frac{\hbar^2}{\mu} k_0 \, .$$

The Hamiltonian $\mathscr{H}_0$ in G_0 has now been effectively linearized and one can easily see that $F^{ww'}$ can be written

$$F^{ww'} = -i \int_0^\infty e^{-\eta s + i D^{ww'} s} \varphi'_f(K'_0 s) \varphi_i(k'_0 s) d s \, . \tag{4.63}$$

As an example, let us consider the Coulomb case where

$$\varphi_{f,i}(r) = \varphi_{f,i}(0) e^{-\varkappa_{f,i} r} \tag{4.64}$$

and

$$F^{ww'} = \varphi'_f(0) \varphi_i(0) \frac{1}{D^{ww'} + i(\kappa_f K'_0 + \kappa_i k'_0)}, \quad \eta = 0$$

which exhibits clearly the properties of F pointed out above: $D^{ww'}$ is of the order of E_i and vanishes at ϑ_c, whereas $\kappa_f K'_0 + \kappa_i k'_0$ is of the order $\sqrt{E_i}$ and does not vanish at ϑ_c.

Other critical angles can be treated similarly, e.g.

$$T_2'^{wW} \sim (2\pi)^6 \, \tilde{w}(K^{ww'}) \, \tilde{W}(K^{w'}) F^{wW} \tag{4.65}$$

where

$$F^{wW} = \int d k \, d K \, \tilde{\varphi}'_f(\xi K_f - K) \frac{1}{E_i + i\eta - \dfrac{\hbar^2 K^2}{2v} - \dfrac{\hbar^2 k^2}{2\mu}}$$

$$\cdot \tilde{\varphi}_i(k - (1 - \zeta)(K - K_i)) \, .$$

By comparing (4.61) with (4.65) one sees that F^{wW} can be evaluated analogously to $F^{ww'}$ by replacing ζ by $-(1 - \zeta)$

$$K_0 \to \xi K_f, \quad k_0 \to -(1 - \zeta)(K_i - K_0),$$

$$K'_0 \to \frac{\hbar^2}{v} \xi K_f - \frac{\hbar^2}{\mu} (1 - \zeta)^2 (K_i - \xi K_f),$$

$$k'_0 \to -\frac{\hbar^2}{\mu} (1 - \zeta)(K_i - \xi K_f) \, .$$

The equal mass case ($M = M'$, $\zeta = \xi$) has to be treated separately. Here, as we have seen, the amplitude $T_B^{\prime W}$ of the 1st Born approximation is singular for back scattering, since then $K^{ww'} = 0$. For the same reason, T_2^{wW} and $T_2^{Ww'}$ diverge at $\vartheta = \pi$ because they contain $\tilde{w}(K^{ww'})$ as factors; furthermore, their denominators vanish for back scattering. $T_2^{ww'}$ shows no divergence at $\vartheta = \pi$ unless its critical angle is π corresponding to $\zeta\xi = \zeta^2 = 1/4$ or $M' = m = M$. Also T_2^{WW} becomes divergent employing the semi-asymptotic expansion because in back scattering the arguments of $\tilde{\varphi}_{i,f}$ are both the same, $k - (1 - \zeta)(K - K_i)$ for $K_f = -K_i$. However one can demonstrate that $T_B^{\prime W}$ is asymptotically larger than the 2nd order contributions and therefore becomes the leading term.

The situation is then the following. In general the first and second order Born terms give the asymptotically leading terms. The asymptotic expansion (4.57) can be used if no critical angles are present. Critical angles show up as divergencies in (4.57). Near the critical angles the expansion has to be reconsidered. The amplitudes containing critical angles are asymptotically leading; they are terms of the 2nd Born approximation unless $M = M'$. For $M = M'$ the first order Born amplitude $T_B^{\prime W}$ is the leading term.

If the exchanged mass m is small, $m \ll M, M'$, $1 - \xi, 1 - \zeta \ll 1$, the amplitudes are peaked in forward direction because there ($u_f = u_i$) the quantities $K^w = (1 - \xi)K_i$, $K^{w'} = (1 - \zeta)K_i$, $K^{ww'} = K^w + K^{w'}$ are very small. One can then show with $u_i = u_f$ in (4.57b, c) and (4.51) that

$$T_B^{\prime W} + T_2^{\prime wW} + T_2^{Ww'} \cong \frac{m(M + M')}{4MM'}(T_B^{\prime w} + T_2^{\prime ww'}) \qquad (4.66)$$

if the three potentials are of the same kind and have the same numerical values. In this case the potential W can be neglected

$$T' \sim T_B^{\prime w} + T_2^{\prime ww'} \qquad (4.67)$$

and the result agrees with the impact parameter version of the theory where the W contribution drops out naturally. This, of course, is only a numerical result and does not hold in general. If W were numerically much larger than w, w', its contribution could not be neglected. But if all potentials are of the same order, the transition amplitude near forward direction is given by (4.67) including the critical ww' scattering.

This cancellation of W terms fails for critical back scattering if $M = M'$ where $T' \cong T_B^{\prime W}$ for $\vartheta \cong \pi$ and, e.g., for $M' > M$ or $\zeta > \xi$, near the critical angle ϑ^{wW} where $T_2^{\prime wW}$ is predominant.

We will show now that the third and higher Born terms in (1.44) can be neglected for great E_i. As a example we investigate

$$T_3^{\prime www'} = (\phi_f', wG_0wG_0w'\phi_i). \qquad (4.68)$$

Other terms like $T_3'^{ww'w'}$ etc. can be treated in the same way. In Fourier representation (4.68) is

$$T_3'^{www'} = (2\pi)^{18} \int d\mathbf{k} \, d\mathbf{K} \, d\mathbf{k}' \, \tilde{\varphi}'_f(\mathbf{K} - \xi\mathbf{K}_f) \, \tilde{w}(\mathbf{k}' + \zeta\mathbf{K} - \mathbf{K}_f) \, G_0(\mathbf{K}, \mathbf{k}')$$

$$\cdot \tilde{w}(\mathbf{k}' - \mathbf{k}) \, G_0(\mathbf{K}, \mathbf{k}) \, \tilde{w}'(\mathbf{K} - \mathbf{K}_i) \, \tilde{\varphi}_i(\mathbf{k} + \zeta(\mathbf{K} - \mathbf{K}_i)) \, . \qquad (4.68a)$$

One can use the δ-function property of $\tilde{\varphi}'_f$ and $\tilde{\varphi}_i$, except in $G_0(\mathbf{K}, \mathbf{k})$ to include critical scattering. The $\tilde{w}'$-term can be taken out of the integral. The triple integral is factorized. The result can be compared with (4.56a) where $\tilde{w}$ and $\tilde{w}'$ can be taken out even for special mass ratios discussed above. The comparison yields

$$T_3'^{www'} \sim \frac{T_2'^{ww'}}{\tilde{w}(\mathbf{K}^w)} \int d\mathbf{k}' \, \tilde{w}(\mathbf{k}' - (1 - \zeta\xi)\mathbf{K}_f)$$

$$\cdot G_0(\xi\mathbf{K}_f, \mathbf{k}') \, \tilde{w}(\mathbf{k}' + \zeta\xi\mathbf{K}_f - \zeta\mathbf{K}_i) \, . \qquad (4.69)$$

In the integral (4.69) the leading contribution is provided by $\mathbf{k}' \cong (1 - \zeta\xi)\mathbf{K}_f$, as there $G_0(\xi\mathbf{K}_f, \mathbf{k}')$ is ε_f^{-1} and not of the order E_i^{-1}, while also $\tilde{w}(\mathbf{k}' - (1 - \zeta\xi)\mathbf{K}_f)$ is peaked there or has even a pole for the Coulomb case. The factor $\tilde{w}(\mathbf{k}' + \zeta\xi\mathbf{K}_f - \zeta\mathbf{K}_i) \cong \tilde{w}(\mathbf{K}^w)$ can be taken out of the integral and the denominator cancels. The remaining integral can be rewritten by the convolution theorem

$$T_3'^{www'} \sim - T_2'^{ww'} \frac{\mu}{2\pi\hbar^2} \int d\mathbf{r} \, w(r) \cdot \frac{\exp\left\{ i \sqrt{\dfrac{2\mu}{\hbar^2} \left(E_i - \dfrac{\hbar^2}{2\nu} \xi^2 K_f^2 \right)} r \right\}}{r} \qquad (4.70)$$

$$e^{i(1 - \zeta\xi)\mathbf{K}_f \mathbf{r}} \, .$$

With $E_i = \hbar^2 K_f^2 / 2\nu' + \varepsilon'_f$ the square root can be expanded for high energies and (4.70) yields after performing the angular integration

$$T_3'^{www'} \sim - T_2'^{ww'} \frac{\mu}{\hbar^2 (1 - \zeta\xi) K_f}$$

$$\cdot \int_0^\infty dr \, w(r) \exp\left\{ iK_f (1 - \zeta\xi) r + \frac{i\mu\varepsilon'_f}{\hbar^2 (1 - \zeta\xi) K_f} r \right\} \sin(1 - \zeta\xi) K_f r. \qquad (4.71)$$

For $w(0)$ being finite as in $e^{-\kappa r}$ we obtain asymptotically

$$T_3'^{www'} \sim T_2'^{ww'} \frac{\mu}{2i\hbar^2 (1 - \zeta\xi) K_f} \int_0^\infty dr \, w(r) \qquad (4.71a)$$

which shows that $T_3'^{www'}$ is by $E_i^{-1/2}$ smaller than $T_2'^{ww'}$ and can be dropped asymptotically.

For a screened Coulomb potential $V = A e^{-\kappa r}/r$ (4.71) yields

$$T_3^{\prime www'} \sim T_2^{\prime ww'} \frac{\mu A}{2i\hbar^2(1-\zeta\xi)K_f} \ln \frac{2(1-\zeta\xi)^2 \hbar^2 K_f^2}{\mu\varepsilon_f' + i\kappa K_f \hbar^2 (1-\zeta\xi)} . \qquad (4.71b)$$

For $\kappa \neq 0$ one can drop $\mu\varepsilon_f'$ with respect to $i\kappa K_f \hbar^2 (1-\zeta\xi)$. The Coulomb case corresponds to $\kappa = 0$ in (4.71b). In either case $T_3^{\prime www'}$ is by $E_i^{-1/2} \ln E_i$ smaller than $T_2^{\prime ww'}$ and can be neglected asymptotically.

As in the case of excitation and ionisation the Coulomb potential does not lead to divergencies in the Lippman Schwinger approach for charge exchange. We see in (4.71b) that this is due to ε_f' being finite, which also means that the exchanged mass m is localized near M'. This prevents the great distances in the long range Coulomb force to supply a substantial contribution in the cross section. They are too weak to transfer m from a localized state φ_i around M to a localized state φ_f' around M'.

b) Charge Exchange for Coulomb Interactions

In the following we will discuss briefly the results for Coulomb interaction. Here all the integrations can be carried out analytically because the ground states are known

$$\varphi_{f,i}(r) = \varphi_{f,i}(0)e^{-\kappa_{f,i} r}; \qquad \varphi_{f,i}(0) = \left\{ \frac{\kappa_{f,i}^3}{\pi} \right\}^{1/2}$$

and the interactions and their Fourier transforms are simply

$$w = -\frac{qQ}{r}; \qquad \tilde{w} = -\frac{qQ}{2\pi^2 k^2}; \qquad w = -\frac{qQ'}{r}; \qquad \tilde{w}' = -\frac{qQ'}{2\pi^2 k^2};$$

$$W = \frac{QQ'}{r}; \qquad \tilde{W} = \frac{QQ'}{2\pi^2 k^2}$$

$$2|\varepsilon_i| = qQ\kappa_i = \hbar^2\kappa_i^2/\mu; \qquad 2|\varepsilon_f| = qQ'\kappa_f = \hbar^2\kappa_f^2/\mu'; \qquad a_{f,i} = 1/\kappa_{f,i}$$

where $Q', -q, Q$ are the charges of M', m, M. In an electron exchange, $m \doteq m_{el}$, $q = e$, this corresponds to a transition from a bound state at nucleus M, Q to another M', Q'. In particular

for proton-hydrogen scattering: $M = M' = m_p$, $Q = Q' = e$, (4.70a)

for deuteron-hydrogen scattering: $M = M'/2 = m_p$, $Q = Q' = e$, (4.70b)

for $_2^3$He-tritium scattering: $M = M' = 3m_p$, $Q = Q'/2 = e$. (4.70c)

The scattering of $_2^3$He nuclei has been included as an example for an approximately equal mass case where particles M and M' are distinguishable. In proton-hydrogen scattering, where also $M = M'$, the critical back scattering cannot be observed; further, the Fermi statistics of the protons would have to be included. The deuteron-hydrogen scattering contains two critical angles $\vartheta^{ww'}$ and ϑ^{wW}.

Coulomb scattering has been discussed before by several authors though in less detail and without critical angle scattering. It might, therefore, be useful to discuss briefly the results.

As a reference total cross section we take the cross section originally derived by *Brinkman* and *Kramers* [18] using only $T_B'^w$ as transition amplitude, which does not contain any critical angles,

$$\sigma_{\mathrm{BK}} \sim \frac{vv'}{(2\pi)^2 \hbar^4} \frac{K_f}{K_i} \int 2\pi \, d\alpha |T_B'^w|^2, \qquad \alpha = \cos\vartheta .$$

The calculation is straightforward and gives

$$\sigma_{\mathrm{BK}} \sim \frac{2^8 \pi}{5} a_f a_i \frac{|\varepsilon_f^3| |\varepsilon_i^3|}{E_i^6} \frac{(1-\eta')^4}{\xi} \left\{ \frac{1}{(1-\sqrt{\eta'})^{10}} - \frac{1}{(1+\sqrt{\eta'})^{10}} \right\}. \quad (4.71)$$

For small m one has approximately

$$1 - \eta' \cong \frac{m(M+M')}{MM'}, \qquad 1 - \sqrt{\eta'} \cong \frac{1}{2} \frac{m(M+M')}{MM'}, \qquad \xi \cong 1,$$

and only the first term in the bracket due to the forward peak gives a substantial contribution:

$$\sigma_{\mathrm{BK}} \sim \frac{2^{18} \pi}{5} a_f a_i \frac{|\varepsilon_f^3| |\varepsilon_i^3|}{E_i^6} \left(\frac{MM'}{m(M+M')} \right)^6 = \frac{2^{18} \pi}{5} a_f a_i \left(\frac{E^*}{E_i} \right)^6 \quad (4.72)$$

with

$$E^* = \frac{MM'}{m(M+M')} \sqrt{|\varepsilon_f| |\varepsilon_i|} .$$

This agrees with the expansion of *Brinkmann-Kramers'* result for proton-hydrogen scattering, where $a_f = a_i = a_0$, $|\varepsilon_f| = |\varepsilon_i| = \varepsilon_0$, $E^* = M\varepsilon_0/2m$ and a_0, ε_0 are the Bohr radius and the ground state energy of the hydrogen atom. As has been discussed before, the asymptotic expansion requires $E \gg E^*$.

Next let us consider critical back scattering for $M = M'$, $\zeta = \xi$ where T_B^W alone is predominant. The total cross section for critical back scattering, $\sigma_B'^W$, is given by (4.55). Again the calculation is straight forward and we obtain

$$\sigma_B'^W \sim \frac{2^8 \pi}{3} \frac{\kappa_f \kappa_i}{(\kappa_f + \kappa_i)^4} \frac{(|\varepsilon_i| |\varepsilon_f|)^{3/2}}{E_i^3} \frac{QQ'}{q^2} \frac{1}{(1+\xi)^2 (1-\xi)}$$

and for small mass m where $1 - \xi \cong m/M$ and $1 + \xi \cong 2$

$$
\begin{aligned}
\sigma_B'^{W} &\sim \frac{2^5\pi}{3} \frac{\kappa_f \kappa_i}{(\kappa_f + \kappa_i)^4} \frac{(|\varepsilon_i| |\varepsilon_f|)^{3/2}}{E_i^3} \frac{M}{m} \frac{QQ'}{q^2} \\
&= \frac{2^8\pi}{3} \frac{\kappa_f \kappa_i}{(\kappa_f + \kappa_i)^4} \left(\frac{E^*}{E_i}\right)^3 \frac{m^2}{M^2} \frac{QQ'}{q^2}
\end{aligned}
\tag{4.73}
$$

with the same definition of E^* as in (4.72). Since $\kappa_f \kappa_i/(\kappa_f + \kappa_i)^4$ is of the order $a_f a_i$ and since QQ'/q^2 is of the order 1, one recognizes that indeed $\sigma_B'^{W}$ is asymptotically much larger than σ_{BK}, but the small factor m^2/M^2 shows that for instance using the values (4.70c) E_i has to be at least 3 orders of magnitude larger than E^* for $\sigma_B'^{W}$ to become predominant. This result, of course, depends on the special choice of Coulomb interaction and has no general validity. If all the masses were of comparable magnitude, then already for $E_i \gg E^*$ the term $\sigma_B'^{W}$ would prevail, such as in simple nuclear reactions.

We have seen that near the forward direction one has $T' \sim T_B'^{w} + T_2'^{ww'}$. This expression determines the cross section including critical ww' scattering but excluding critical scattering into angles far from forward direction. For the sake of simplicity we discuss only proton-hydrogen scattering ($\zeta = \xi$). Here one has

$$
T_B'^{w} \sim -2^5 \pi a_0 \left(\frac{\varepsilon_0}{E_i}\right)^3 \frac{\hbar^2}{v} \frac{(1 - \xi^2)^2}{b^3}, \qquad b = (1 + \xi^2 - 2\xi\alpha).
$$

The amplitude $T_2'^{ww'}$ is obtained from (4.61) and (4.64). With

$$
K_0' = k_0' = \frac{\hbar^2}{\mu} \xi K_i \sqrt{b}, \qquad D^{ww'} = E_i\{1 - \xi^2 - \zeta^2 b/(1 - \xi^2)\}
$$

one has

$$
\begin{aligned}
T_2'^{ww'} &\sim 2^5 \pi a_0 \left(\frac{\varepsilon_0}{E_i}\right)^3 \frac{\hbar^2}{v} \frac{(1 - \xi^2)}{b^2} \frac{E_i}{D^{ww'} + 2i\kappa_0 k_0'} \\
&= 2^5 \pi a_0 \left(\frac{\varepsilon_0}{E_i}\right)^3 \frac{\hbar^2}{v} \frac{(1 - \xi^2)^2}{b^2 \xi^2} \frac{1}{b_c - b + i4\sqrt{\dfrac{\varepsilon_0}{E_i}(1 - \xi^2) b/\xi^2}}
\end{aligned}
$$

where $b_c = (1 - \xi^2)^2/\xi^2$ is the critical value of b for $\vartheta = \vartheta_c = \vartheta^{ww'}$ or $\alpha = \alpha_c$. The factor b in the square root of the denominator can be replaced by the critical value b_c because $\sqrt{\varepsilon_0/E_i}$ is small,

$$
4\{\varepsilon_0(1 - \xi^2) b_c/E_i\xi^2\}^{1/2} = 4\{\varepsilon_0(1 - \xi^2)^3/E_i\xi^4\}^{1/2} = \beta,
$$

and we have

$$T' = 2^5 \pi a_0 \left(\frac{\varepsilon_0}{E_i}\right)^3 \frac{\hbar^2}{v} \frac{(1-\xi^2)^2}{\xi^2} f(b),$$

$$f(b) = \frac{1}{b^2}\left\{\frac{1/\xi^2}{b_c - b + i\beta} - \frac{1}{b}\right\}.$$

The cross section, σ_D, for this scattering amplitude is that given by *Drisko*:

$$\sigma_D = \frac{v^2}{(2\pi)^2 \hbar^4} 2\pi \int_{-1}^{1} d\alpha \, |T'|^2 = 2^9 \pi a_0^2 \left(\frac{\varepsilon_0}{E_i}\right)^6 (1-\xi^2)^4 \int_{-1}^{1} d\alpha \, |f(b)|^2$$

$$= 2^8 \pi a_0^2 \left(\frac{\varepsilon_0}{E_i}\right)^6 \frac{(1-\xi^2)^4}{\xi} I$$

where

$$I = \int_{(1-\xi)^2}^{(1+\xi)^2} db \, |f(b)|^2 \cong \int_{(1-\xi)^2}^{\infty} db \, |f(b)|^2.$$

The upper limit in I can be replaced by infinity if $1 - \xi \ll 1$; its contribution corresponds to back scattering and can be dropped for $m \ll M$. Also, the ξ occurring in $f(b)$ can be replaced by 1 and $\beta \cong 4\{\varepsilon_0(1-\xi^2)^3/E_i\}^{1/2}$

$$I = \int_{(1-\xi)^2}^{\infty} db \left\{\frac{1}{b^6} + \frac{1}{b^4} \frac{1}{(b_c-b)^2+\beta^2} - 2\frac{b_c-b}{b^5} \frac{1}{(b_c-b)^2+\beta^2}\right\}.$$

One sees that I will possess an expansion for small* β,

$$I \cong \frac{C_{-1}}{\beta} + C_0 + C_1\beta + \cdots.$$

The coefficient C_{-1} is easily obtained by taking the limit $\beta \to 0$ of βI, because then $\beta/\{(b_c-b)^2+\beta^2\} \to \pi\delta(b_c-b)$. Therefore** $C_{-1} = \pi/b_c^4$, or with $b_c \cong 4(1-\xi)^2$, $C_{-1} = \pi/2^8(1-\xi)^8$. The quantity C_0 can be obtained analogously by straightforward calculation. The first term b^{-6} in the bracket corresponds to the *Brinkman-Kramers* calculation and contributes $1/5(1-\xi)^{-10}$. The final result is

$$I = \frac{\pi/\beta}{2^8(1-\xi)^8} + \frac{1}{(1-\xi)^{10}}\left\{\frac{1}{5} - \frac{55}{3.27} - \frac{7}{2^8}\ln 3\right\}.$$

---

* $\beta/b_c \ll 1$ or $4\sqrt{E^*/E_i} \ll 1$. Since β is the width of critical scattering, this condition means that the relative width of critical scattering is small. Consequently for $E_i \gg E^*$ the critical scattering shows a well defined peak.

** The lower limit is approximately $b_c/4$ and the singularity is contained in the interval of integration.

With

$$\beta = 4\sqrt{\frac{\varepsilon_0}{E_i}}(1-\xi^2)^{3/2} \cdot \beta/(1-\xi)^2 \cong 2^4\sqrt{M\varepsilon_0/2mE_i} = 2^4\sqrt{E^*/E_i}$$

$$I = \frac{1}{(1-\xi)^{10}}\left\{\underbrace{\frac{1}{5} - \frac{55}{3.27} - \frac{7}{2^8}\ln 3}_{=0.35} + \frac{\pi}{2^{12}}\sqrt{E_i/E^*}\right\}$$

we have

$$\sigma_D = \frac{2^{18}\pi}{5}a_0^2\left(\frac{E^*}{E_i}\right)^6\left\{0.3 + \frac{5\pi}{2^{12}}\sqrt{E_i/E^*}\right\} \cong \sigma_{BK}\left\{0.3 + \frac{5\pi}{2^{12}}\sqrt{E_i/E^*}\right\}.$$

$$(4.74)$$

The cross section of *Brinkman-Kramers* is reduced by 0.3 and contains an additive term proportional to $\sigma_{BK}\sqrt{E_i/E^*}$ due to critical scattering. This would be the asymptotically predominant term. The factor in front of the critical scattering term shows that E_i must be many orders of magnitude larger than E^* for predominant critical scattering.

Finally, let us briefly consider the critical scattering off back and forward direction. This would correspond to (4.70b), where the critical angle 9^{wW} is about 97°. To obtain the leading term one replaces the arguments of $\tilde{w}, \tilde{W}$ in (4.65) by their critical values, calculates F^{wW} analogously to $F^{ww'}$, and considers only the leading term proportional to $1/\beta$, in the total cross section σ'^{wW}, which becomes

$$\sigma'^{wW} \sim \sigma_{BK}\frac{5\cdot 3^4}{\pi 11^2 \cdot 2^{10}}\left(\frac{m}{M}\right)^2\sqrt{E_i/E^*} \cong 10^{-3}\sigma_{BK}\left(\frac{m}{M}\right)^2\sqrt{E_i/E^*}.$$

It contains a factor m^2/M^2, similar to critical backscattering, and will, therefore, be unobservably small.

Fig. 7 shows for charge exchange in proton-hydrogen collisions the *Brinkman-Kramers* result (4.72) and the second Born approximation (4.74) compared with experimental results of several authors. The experiments have been performed with molecular hydrogen and the results have been scaled to atomic hydrogen by multiplying with 1/2. *Oppenheimer* [17] showed for the first Born approximation that capture into excited states φ_n' is proportional to n^{-3} and would change $\sigma'(1s-1s)$ only by a small amount. One clearly sees that the *Brinkman-Kramers* calculation provides a too large cross section, while for energies larger than 1 MeV the second Born approximation is within the discrepencies of the different experimental results.

In conclusion one can say the following. The first and second Born approximations provide the correct asymptotic expansion for high incident energies. Scattering into critical angles will be important for masses of comparable magnitude. If the exchanged mass is small, critical scattering is suppressed unless the incident energy is exceedingly large. The asymptotically leading terms, if critical scattering occurs, carry numerically small factors, and the subsequent terms are more important. This fact will also influence an estimate of the errors involved in the asymptotic expansion. The expansion given here applies only to potentials of which the Fourier transforms can be expanded for large k values. For

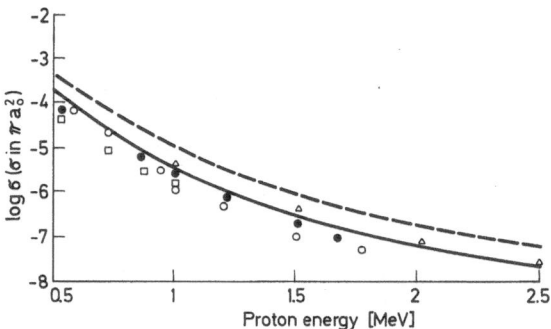

Fig. 7. Cross section for charge exchange in proton-hydrogen collisions. Theoretical: –––– *Brinkman-Kramers*, ——— Second Born approximation. Experimental: ■ *Barnett* and *Reynolds* (1958), ▲ *Schryber* (1968), ● *Toburen et al.* (1968), ○ *Williams* (1967)

other potentials it is still true that the Fourier transforms of $\tilde{\varphi}_{i,f}$ and $\tilde{w}$ are concentrated near $k = 0$. However, it does not hold in general that $\tilde{\varphi}_{i,f}$ vanishes faster than $\tilde{w}$. As a consequence, a discussion of the asymptotic behaviour for more general potentials will be much more involved.

3. Impact Parameter Treatment

We have shown in Chapter III that the impact parameter theory is valid for $m/M, M' \ll 1$ and $|\varepsilon_{i,f}|/E_i \ll 1$. Therefore the high energy limit for atomic collisions can also be investigated in the impact parameter treatment. As an example we show that the high energy cross section for excitation or ionisation is the same in wave- and impact parameter treatment. To this end it is sufficient to consider the first Born approximation.

In the impact parameter theory we obtain from the first term in (3.30) in Fourier representation

$$f_b^B = -\frac{(2\pi)^4 i}{\hbar} \int d\boldsymbol{k}_1 \, d\boldsymbol{k}_2 \, \tilde{\varphi}_f^*(\boldsymbol{k}_1) \, \tilde{w}'(\boldsymbol{k}_2)$$

$$\cdot \tilde{\varphi}_i(\boldsymbol{k}_1 + \boldsymbol{k}_2) \, e^{-i\boldsymbol{k}_2 \boldsymbol{b}} \, \delta\left(\frac{\varepsilon_f}{\hbar} - \frac{\varepsilon_i}{\hbar} - k_2 v\right)$$

$$\varphi_f^*(\boldsymbol{r}) = \int \tilde{\varphi}_f^*(\boldsymbol{k}) \, e^{i\boldsymbol{k}\boldsymbol{r}} \, d\boldsymbol{k} \quad \text{etc.}$$

With (3.9), (3.10) the cross section is in Born approximation

$$\sigma_B = \frac{(2\pi)^{10}}{\hbar^2 v^2} \int d\boldsymbol{k}_1 \, d\boldsymbol{k}_2 \, d\boldsymbol{k}_1' \, \tilde{\varphi}_f^*(\boldsymbol{k}_1) \, \tilde{\varphi}_f(\boldsymbol{k}_1') \, \tilde{w}'^2(\boldsymbol{k}_2)$$

$$\cdot \tilde{\varphi}_i(\boldsymbol{k}_1 + \boldsymbol{k}_2) \, \tilde{\varphi}_i(\boldsymbol{k}_1' + \boldsymbol{k}_2) \, \delta\left(\frac{\varepsilon_f}{\hbar v} - \frac{\varepsilon_i}{\hbar v} - k_2 \frac{v}{v}\right). \tag{4.75}$$

In the wave treatment we obtain from (4.1), (4.3)

$$\sigma_B = \frac{(2\pi)^{10} v^2 K_f}{\hbar^4 K_i} \int d\Omega \, d\boldsymbol{k}_1 \, d\boldsymbol{k}_1' \, \tilde{w}'^2(\boldsymbol{K}_i - \boldsymbol{K}_f) \, \tilde{\varphi}_f^*(\boldsymbol{k}_1) \tag{4.76}$$

$$\tilde{\varphi}_i(\boldsymbol{k}_1 - \zeta(\boldsymbol{K}_i - \boldsymbol{K}_f)) \tilde{\varphi}_f^*(\boldsymbol{k}_1') \, \tilde{\varphi}_i(\boldsymbol{k}_1' - \zeta(\boldsymbol{K}_i - \boldsymbol{K}_f))$$

$$\sigma_B = \frac{(2\pi)^{10} v^2 K_f}{\hbar^4 K_i} \int d\boldsymbol{k}_2 \, d\boldsymbol{k}_1 \, d\boldsymbol{k}_1' \, \tilde{w}'^2(\boldsymbol{k}_2) \, \tilde{\varphi}_f^*(\boldsymbol{k}_1) \, \tilde{\varphi}_i(\boldsymbol{k}_1 + \zeta\boldsymbol{k}_2)$$

$$\cdot \tilde{\varphi}_f(\boldsymbol{k}_1') \, \tilde{\varphi}_i(\boldsymbol{k}_1' + \zeta\boldsymbol{k}_2) \cdot \int d\Omega \, \delta(\boldsymbol{k}_2 - \zeta(\boldsymbol{K}_f - \boldsymbol{K}_i)). \tag{4.76a}$$

With (4.11) is in forward direction $K_i - K_f \cong \dfrac{\varepsilon_f}{\hbar v} - \dfrac{\varepsilon_i}{\hbar v}$ for $|\varepsilon_{i,f}|/E_i \ll 1$, while outside the forward direction $\boldsymbol{K}_i - \boldsymbol{K}_f \to \infty$ for $v \to \infty$. Because of the δ-function is $\boldsymbol{k}_2 = \zeta(\boldsymbol{K}_f - \boldsymbol{K}_i)$ in (4.76a). As $\tilde{w}'^2(\boldsymbol{k}_2)$ is strongly peaked at $\boldsymbol{k} = 0$ only the forward direction contributes in (4.76a). Introducing in (4.75a) for $\boldsymbol{k}_2$ one coordinate k_{2i} in direction $\boldsymbol{K}_i$ and two coordinates k_{2v} vertical on $\boldsymbol{K}_i$ we have with $d\Omega \cong dk_{2v}/K_i^2$

$$\int d\Omega \, \delta(\boldsymbol{k}_2 - \zeta(\boldsymbol{K}_f - \boldsymbol{K}_i)) = \frac{1}{K_i^2} \int dk_{2v} \, \delta(k_{2i} - \zeta(K_{fi} - K_i)) \, \delta(k_{2v} - \zeta K_{fv})$$

$$= \frac{1}{K_i^2} \, \delta(k_{2i} - \zeta(K_{fi} - K_i))$$

where $K_{fi}(K_{fv})$ is the component of $\boldsymbol{K}_f$ in direction $\boldsymbol{K}_i$ (vertical on $\boldsymbol{K}_i$). As only the forward direction contributes, $K_{fi} \cong K_f$ and with (4.11) and

$k_{2i} = k_2 v/v$

$$\int d\Omega \, \delta(k_2 - \zeta(K_f - K_i)) \cong \frac{1}{K_i^2} \delta\left(\frac{\varepsilon_f}{\hbar v} - \frac{\varepsilon_i}{\hbar v} - k_2 \frac{v}{v}\right). \qquad (4.77)$$

From (4.76a) with $\zeta \cong 1$ for $m/v \ll 1$, $K_f \cong K_i$ for $|\varepsilon_{i,f}|/E_i \ll 1$ and $\hbar K_i = vv$ we obtain with (4.77) the impact parameter result (4.75).

In the same way it can be shown for charge exchange that wave- and impact parameter treatment yield identical results in the high energy limit as far as the neighbourhood of the forward direction is concerned. This means that only the ww'-critical scattering is included in the impact parameter theory. On the other hand we have seen that for Coulomb potentials the critical scattering in $\sigma_B'^W$, for $M = M'$, and in σ'^{wW} or $\sigma'^{Ww'}$ is proportional to $(m/M)^2$. So the high energy behaviour for charge exchange with Coulomb interaction is described properly by the impact parameter treatment for large v and infinitely heavy masses M, M', i.e. for $v \to \infty$ after $M, M' \to \infty$. For $W \sim 1/r^2$ on the other hand $\sigma_B'^W$ in (4.55) stays finite for $v \to \infty$ and an impact parameter treatment is only possible for the forward peak.

V. Variational Methods

We saw in Chapter IV that the high energy treatment of atomic collisions based on the first or second Born approximation is confined to energies of 1 MeV or more for a proton as projectile. Often it turns out however, particularly for excitation and ionisation, that the first Born approximation provides results which are in reasonable agreement with experiments down to a few KeV of the incoming proton [16]. This suggests a variational procedure extrapolating the first or second Born approximation to lower energies.

We therefore formulate the *Schwinger* variational principle for inelastic scattering, i.e. direct and rearrangement collisions. With a simple ansatz we then obtain a renormalized Born approximation.

1. Wave Treatment

In order to calculate the differential cross section for direct (1.27) and rearrangement collisions (1.38) we have to calculate T_{if} of (1.20) and T_{if}' of (1.36) on the energy-shell. We consider the transition from a state $\phi_i \to \phi_f'$, where $H_i \phi_i = E_i \phi_i$ according to (1.5) and $H_f \phi_f' = E_f' \varphi_f'$, $E_f' = E_i$. In contrast to (1.2) we do not specify H_f here, so that the direct collisions are treated as a special case with $H_f = H_i$, $\phi_f' = \phi_f$, $T_{if}' = T_{if}$. Though we will use the notation for the three particle system of Chapter I, the following formulas hold for any more complicated system.

We define [2], [10],

$$\hat{T}_{if} = (\phi'_f, V_f \Psi_i^+) + (\Psi'^-_f, V_f[1 + G_f(V_i - V_f)]\phi_i)$$
$$- (\Psi'^-_f, V_f[1 - G_f V_f]\Psi_i^+) \tag{5.1}$$

with arbitrary states Ψ_i^+ and Ψ'^-_f.★

The initial, V_i, and final potential V_f are defined by H_i and H_f and the total Hamiltonian is

$$H = H_i + V_i = H_f + V_f. \tag{5.1a}$$

We shall prove now

a) $$\hat{T}_{if} = T'_{if} \tag{5.2}$$

when

$$\Psi_i^+ = \phi_i + G_i^+ V_i \Psi_i^+; \quad G_i^+ = G_i = \frac{1}{E_i + i\eta - H_i}, \quad \eta \to 0 \tag{5.3a}$$

is the retarted scattering state discussed in (1.15) and

$$\Psi'^-_f = \phi'_f + G_f^- V_f \Psi'^-_f; \quad G_f^- = G_f^\dagger = \frac{1}{E_i - i\eta - H_f}, \quad \eta \to 0 \tag{5.3b}$$

is the advanced scattering state.

b) For variations $\Psi_i^+ + \delta \Psi_i^+$, $\Psi'^-_f + \delta \Psi'^-_f$ $\hat{T}_{if}$ is stationary for Ψ_i^+ and Ψ'^-_f according to (5.3a), (5.3b). Then the stationary value is $\hat{T}_{if} = T'_{if}$.

To prove a) we write in the second term of (5.1) with (5.3b)

$$(\Psi'^-_f, V_f G_f(V_i - V_f)\phi_i) = (G_f^- V_f \Psi'^-_f, (V_i - V_f)\phi_i) = (\Psi'^-_f - \phi'_f, (V_i - V_f)\phi_i).$$

This can be simplified with

$$(\phi'_f, (V_i - V_f)\phi_i) = (\phi'_f, (H - H_i) - (H - H_f)\phi_i) = (E_i - E'_f)(\phi'_f, \phi_i) = 0$$

and we obtain

$$(\Psi'^-_f, V_f G_f(V_i - V_f)\phi_i) = (\Psi'^-_f, (V_i - V_f)\phi_i). \tag{5.4}$$

Introducing (5.4) in (5.1) yields

$$\hat{T}_{if} = (\Psi'^-_f, V_i \phi_i) + (\phi'_f, V_f \Psi_i^+) - (\Psi'^-_f, V_f[1 - G_f V_f]\Psi_i^+). \tag{5.5}$$

The second and third term cancel, as with (5.3b)

$$(\phi'_f - \Psi'^-_f + G_f^- V_f \Psi'^-_f, V_f \Psi_i^+) = 0$$

and the first term can be rewritten by introducing for Ψ'^-_f the solution of the integral equation (5.3b):

$$\Psi'^-_f = \frac{1}{1 - G_f^- V_f} \phi'_f = \phi'_f + \frac{1}{1 - G_f^- V_f} G_f^- V_f \phi'_f \tag{5.6}$$

★ One can also start with a form of $\hat{T}_{if}$ which is symmetrical with respect to (i, f) [10].

or with (1.32b) for the advanced Green's operators

$$\Psi_f'^- = \phi_f' + G^- V_f \phi_f'.\qquad(5.6a)$$

From (5.6a) and (5.5) we obtain

$$\hat{T}_{if} = (\phi_f', V_i \phi_i) + (\phi_f', V_f G V_i \phi_i).\qquad(5.7)$$

In Chapter I, (1.45), we showed that the first Born approximation can be calculated either with V_i or V_f. Using the representation (1.40) for Ψ_i, which corresponds to (5.6a) for $\Psi_f'^-$, (5.7) becomes

$$\hat{T}_{if} = (\phi_f', V_f[1 + GV_i]\phi_i) = (\phi_f', V_f \Psi_i^+).\qquad(5.8)$$

Comparing with (1.36) we see that (5.2) holds.

In order to prove b) we replace Ψ_i^+ and $\Psi_f'^-$ by $\Psi_i^+ + \delta\Psi_i^+$ and $\Psi_f'^- + \delta\Psi_f'^-$. We have to show that the terms proportional to $\delta\psi_i^+$ and $\delta\Psi_f'^-$ vanish. The term proportional to $\delta\Psi_i^+$ is

$$\delta\hat{T}_{if} = (V_f[\phi_f' - \Psi_f'^- + G_f^- V_f \Psi_f'^-], \delta\Psi_i^+)$$

which is zero with (5.3b). For a variation $\delta\Psi_f'^-$ we have

$$\delta\hat{T}_{if} = (\delta\Psi_f'^-, V_f\{\phi_i - \Psi_i^+ - G_f[V_f \Psi_i^+ + (V_i - V_f)\phi_i]\}).\qquad(5.9)$$

With $\phi_i = (1 - G_i V_i)\Psi_i^+$ according to (1.15) we have in (5.9)

$$\begin{aligned}
G_f(V_f \Psi_i^+ + (V_i - V_f)\phi_i) &= G_f[V_f + (V_i - V_f)(1 - G_i V_i)]\Psi_i^+\\
&= G_f[1 - (V_i - V_f)G_i]V_i \Psi_i^+\\
&= G_f[E - H_i + i\eta - V_i + V_f]G_i V_i \Psi_i^+.
\end{aligned}$$

With (5.1a) is $E - H_i - V_i + V_f = E - H_f$, so that

$$G_f(V_f \Psi_i^+ + (V_i - V_f)\phi_i) = G_i V_i \Psi_i^+.\qquad(5.10)$$

Introducing (5.10) in (5.9) one sees with (5.3a) that $\delta\hat{T}_{if} = 0$.

We now use (5.1) to obtain an approximate expression for the T-matrix. Choosing for Ψ_i^+ and $\Psi_f'^-$ the incoming waves ϕ_i and ϕ_f' leads to that second Born approximation which means in (1.40) an expansion of the total Green's operator in powers of V_f. So this Born series is the most favourable one among the possibilities described in Chapter I. It is also interesting that the simplest ansatz for Ψ_i^+ and $\Psi_f'^-$ yields not the first but second Born approximation, which according to Chapter IV is the correct result for an high energetic charge exchange collision. A simple extension of the second Born approximation to lower energies may be achieved for direct collisions ($V_i = V_f$) with the ansatz

$$\Psi_i^+ = a_1 \phi_i, \qquad \Psi_f^- = a_2^* \phi_f'.\qquad(5.11)$$

The constants a_1 and a_2 are variational parameters to be determined from the condition $\delta \hat{T}_{if} = 0$. With (5.11) and (5.1) and the Born approximation $T_B = (\phi_f, V_f \phi_i)$ we obtain

$$\frac{\partial T_{if}}{\partial a_{1,2}} = 0 = T_B - a_{1,2}\{T_B - T_{2i}^{ii}\}, \qquad T_{2i}^{ii} = (\phi_f, V_i G_i V_i \phi_i).$$

With a_1 and a_2 from these equations Ψ_i^+ and Ψ_f^- in (5.11) are determined. Then $\hat{T}_{if}$ of (5.1) can be calculated and we obtain a renormalized Born approximation for T_{if}:

$$T_{if} \cong \frac{T_B}{T_B - T_{2i}^{ii}} T_B. \qquad (5.12)$$

For high energies $|T_{2i}^{ii}| \ll |T_B|$ and by expansion T_{if} is reduced to the second Born approximation.

A renormalized Born approximation for rearrangement collisions results from (5.1) with an ansatz

$$\Psi_i^+ = \phi_i + a_1 G_i V_i \phi_i \qquad \Psi_f'^- = a_2^* \phi_f. \qquad (5.11a)$$

After variation of a_1 and a_2 one obtains

$$T_{if}' = T_B' + \frac{T_{2i}^{fi} \cdot T_{2f}^{fi}}{T_{2i}^{fi} - T_{3fi}^{ffi}}, \qquad \begin{matrix} T_{2i,f}^{fi} = (\phi_f', V_f G_{i,f} V_i \phi_i) \\ T_{3fi}^{ffi} = (\phi_f', V_f G_f V_f G_i V_i \phi_i). \end{matrix} \qquad (5.12a)$$

For high energies $|T_{3fi}| \ll |T_{2i}|$ and (5.12a) yields the second Born approximation.

For potential scattering one has $V_i = V_f = V$ in (5.12) and

$$G_f = G_0 = \frac{1}{E_i + i\eta - H_0} \quad \text{with} \quad H_0 = -\frac{\hbar^2}{2v} \partial_R^2,$$

and the renormalized Born approximation (5.12) is

$$T_{if} \cong \frac{T_B}{T_B - (\phi_f, V G_0 V \phi_i)} T_B. \qquad (5.12b)$$

For a separable potential $V = v(r) \cdot v(r')$ (5.12a) becomes

$$T_{if} = \frac{T_B}{1 - (v, G_0 v)}, \qquad T_B = (\phi_f, v)(v, \phi_i) \qquad (5.12c)$$

which is the exact result for the T-matrix.

2. Impact Parameter Treatment

In the impact parameter treatment we have to calculate the amplitudes (3.16) for direct- and (3.25) for rearrangement collisions. In analogy to the wave treatment we set up a variational principle for the amplitude

$$f' = \int\limits_{-\infty}^{\infty} dt(\varphi_f'(t), V_f \psi_i^+(t)) \tag{5.13}$$

from the initial state $\varphi_i(t)$ to the final state $\varphi_f'(t)$. The hamiltonian is $h = h_i + V_i = h_f + V_f$, and $\varphi_i(t), \varphi_f'(t)$ are solutions of $h_i \varphi_i = i\hbar \dot{\varphi}_i$, $h_f \varphi_f = i\hbar \dot{\varphi}_f$. For a direct collision $V_f = V_i$. The notation is as in Chapter III, though the following formalism is not restricted to a three particle system.

The quantity to be varied is

$$\hat{f} = \int dt(\varphi_f'(t), V_f \psi_i^+(t))$$
$$+ \int dt\left(\psi_f'^-(t), V_f\left[\varphi_i(t) + \frac{1}{i\hbar} \int\limits_{-\infty}^{t} dt' \, U_f(t, t')\,(V_i(t') - V_f)\,\varphi_i(t')\right]\right) \tag{5.14}\star$$
$$- \int dt\left(\psi_f'^-(t), V_f\left[\psi_i^+(t) - \frac{1}{i\hbar} \int\limits_{-\infty}^{t} dt' \, U_f(t, t')\, V_f \psi_i^+(t')\right]\right)$$

with arbitrary states ψ_i^+ and $\psi_f'^-$.

We shall prove

a) $$\hat{f} = f' \tag{5.15}$$

when ψ_i^+, the retarded scattering state $(\psi_i^+(t \to -\infty) = \varphi_i(t))$, is a solution of

$$\psi_i^+(t) = \varphi_i(t) + \frac{1}{i\hbar} \int\limits_{-\infty}^{t} dt' \, U_i(t - t')\, V_i(t')\, \psi_i^+(t'), \tag{5.15a}$$
$$U_i(t) = e^{-\frac{i}{\hbar} h_i t}$$

and $\psi_f'^-$, the advanced scattering state $(\psi_f'^-(t \to \infty) = \varphi_f'(t))$, is a solution of

$$\psi_f'^-(t) = \varphi_f'(t) + \frac{1}{i\hbar} \int\limits_{t}^{\infty} dt' \, U_f(t, t')\, V_f \psi_f'^-(t'), \tag{5.15b}$$
$$U_f(t, t') = T \exp\left\{-\frac{i}{\hbar} \int\limits_{t'}^{t} h_f(t'')\, dt''\right\}.$$

b) $\hat{f}$ is stationary for variations $\psi_i^+ + \delta\psi_i^+$ and $\psi_f'^- + \delta\psi_f'^-$ with ψ_i^+ and $\psi_f'^-$ according (5.15a) and (5.15b). Then the stationary value is $\hat{f} = f'$.

$\star$ Integral signs without limits mean integration over all t: $\int dt = \int\limits_{-\infty}^{\infty} dt$.

To prove a) we write the second term of (5.14)

$$\frac{1}{i\hbar}\int dt\, dt'(\psi_f'^-(t),\, V_f U_f(t,t')\,\Theta(t-t')\,(V_i(t')-V_f)\,\varphi_i(t'))$$

$$= \frac{1}{i\hbar}\int dt\int dt'\,\Theta(t'-t)\, U_f^+(t',t)\, V_f\psi_f'^-(t'),\,(V_i(t)-V_f)\,\varphi_i(t)) \qquad (5.16)$$

$$= \int dt(\psi_f'^-(t)-\varphi_f'(t),\,(V_i(t)-V_f)\,\varphi_i(t))$$

where we have used $U_f^+(t',t)=U_f(t,t')$ ((3.23), (3.24)). In (5.16) is

$$\int dt(\varphi_f'(t),\,(V_i-V_f)\,\varphi_i(t))=\int dt(\varphi_f'(t),\,[h-h_i-(h-h_f)]\,\varphi_i(t))$$

$$=i\hbar\int dt\{(\dot\varphi_f',\varphi_i)-(\varphi_f',\dot\varphi_i)\}$$

$$=i\hbar\{(\varphi_f'(\infty),\varphi_i(\infty))-(\varphi_f'(-\infty),\varphi_i(-\infty))\}$$

by integration by parts. At $t=\pm\infty$ the bound states in a charge exchange process are well separated and the overlap vanishes. Then

$$\int dt(\varphi_f'(t),\,(V_i(t)-V_f)\,\varphi_i(t))=0 \qquad (5.17)$$

which also holds in a direct collision with $V_i=V_f$. From (5.14) we obtain with (5.16), (5.17)

$$\hat f=\int dt(\psi_f'^-,V_i\varphi_i)+\int dt(\varphi_f',V_f\psi_i^+) \qquad (5.18)$$

$$-\int dt\left(\psi_f'^-(t),\, V_f\left[\psi_i^+(t)-\frac{1}{i\hbar}\int_{-\infty}^{\infty}dt'\, U_f(t,t')\,\Theta(t-t')\, V_f\psi_i^+(t')\right]\right).$$

The last term can be rewritten with (5.15b)

$$\frac{1}{i\hbar}\int dt\, dt'(\psi_f'^-(t),\, V_f U_f(t',t)\,\Theta(t'-t)\, V_f\psi_i^+(t))$$

$$=\frac{1}{i\hbar}\int dt\left(\int_t^\infty dt'\, U_f(t,t')\, V_f\psi_f'^-(t'),\, V_f\psi_i^+(t)\right)$$

$$=\int dt(\psi_f'^-(t)-\varphi_f'(t),\, V_f\psi_i^+(t))$$

and we obtain from (5.18)

$$\hat f=\int dt(\psi_f'^-(t),\, V_i\varphi_i(t))\,. \qquad (5.19)$$

Corresponding to (3.26) for $\psi_i^+(t)$ the following integral equation holds for $\psi_f'^-(t)$

$$\psi_f'^-(t)=\varphi_f'(t)+\frac{1}{i\hbar}\int_t^\infty dt'\, U(t,t')\, V_f\varphi_f'(t') \qquad (5.20)$$

with the total time evolution operator U according to (3.29). Introducing (5.20) in (5.19), interchanging the t and t' integration in the second term and replacing V_i by V_f in the first term according to (5.17) results in

$$\hat{f} = \int dt \left(\varphi_f'(t), V_f \left[\varphi_i(t) - \frac{1}{i\hbar} \int_{-\infty}^{t} dt' \, U(t, t') \, V_i(t') \, \varphi_i(t') \right] \right)$$

or with (3.26) and (3.25)

$$\hat{f} = \int_{-\infty}^{\infty} dt (\varphi_f'(t), V_f \psi_i^+(t)) = f'. \tag{5.21}$$

In order to prove b) we have to show that the first variation of (5.14) with respect to ψ_i^+ and $\psi_f'^-$ vanishes. This can be done easily with the term proportional to $\delta\psi_i^+$ using (5.15b). The term proportional to $\delta\psi_f'^-$ is

$$\delta\hat{f} = \int dt \left(\delta\psi_f'^-, V_f \left[\psi_i^+(t) - \varphi_i(t) \right. \right.$$
$$\left. \left. - \frac{1}{i\hbar} \int_{-\infty}^{t} dt' \, U_f(t, t') \{ V_f \psi_i^+(t') + (V_i - V_f) \, \varphi_i(t') \} \right] \right). \tag{5.22}$$

We can rewrite the term in the curly brackets with (5.15a)

$$V_f \psi_i^+(t') + (V_i - V_f) \, \varphi_i(t')$$
$$= V_f \psi_i^+(t') + (V_i - V_f) \left(\psi_i^+(t') - \frac{1}{i\hbar} \int_{-\infty}^{t'} dt_1 \, U_i(t', t_1) \, V_i \psi_i^+(t_1) \right)$$
$$= V_i \psi_i^+(t') - \frac{1}{i\hbar} (V_i - V_f) \int_{-\infty}^{t'} dt_1 \, U_i(t', t_1) \, V_i \psi_i^+(t_1).$$

Then the integral over t' in (5.22) is

$$\int_{-\infty}^{t} dt' \, U_f(t, t') \{ V_f \psi_i^+(t') + (V_i - V_f) \, \varphi_i(t') \}$$
$$= \int_{-\infty}^{t} dt' \, U_f(t, t') \left\{ V_i \psi_i^+(t') - \frac{1}{i\hbar} (V_i - V_f) \int_{-\infty}^{t'} dt_1 \, U_i(t', t_1) \, V_i \psi_i^+(t_1) \right\}.$$

By interchanging the t' and t_1 integration we obtain

$$\int_{-\infty}^{t} dt' \, U_f(t, t') \{ V_f \psi_i^+(t') + (V_i - V_f) \, \varphi_i(t') \}$$
$$= \int_{-\infty}^{t} dt' \left\{ U_f(t, t') - \frac{1}{i\hbar} \int_{t'}^{t} dt_1 \, U_f(t, t_1) \, (V_i(t_1) - V_f) \, U_i(t_1, t') \right\} \tag{5.23}$$
$$\cdot V_i(t') \, \psi_i^+(t')$$
$$= \int_{-\infty}^{t} dt' \, U_i(t, t') \, V_i(t') \, \psi_i^+(t').$$

In (5.23) we have used an integral equation for U_i which can be proved in the following way:

We define $\tilde{U}_i(t, t')$ by

$$U_i(t, t') = U_f(t, t')\, \tilde{U}_i(t, t')\,. \tag{5.24}$$

With (3.13a) and (3.22a) we obtain

$$
\begin{aligned}
i\hbar\, \partial_t U_i &= h_f U_i + i\hbar U_f\, \partial_t \tilde{U}_i = h_i U_i \\
i\hbar\, \partial_t \tilde{U}_i &= U_f^{-1}(h_i - h_f)\, U_i\,.
\end{aligned}
\tag{5.25}
$$

With $h_i - h_f = (h - V_i) - (h - V_f) = V_f - V_i$ and $U_i(t', t') = 1$ (5.25) yields

$$U_i(t, t') = 1 + \frac{1}{i\hbar} \int_{t'}^{t} U_f^{-1}(t_1, t')\,(V_f - V_i)\, U_i(t_1, t')\mathrm{d}t_1\,.$$

By multiplying with $U_f(t, t')$ from the left the desired integral equation for $U_i(t, t')$ results:

$$U_i(t, t') = U_f(t, t') - \frac{1}{i\hbar} \int_{t'}^{t} U_f(t, t_1)\,(V_i(t_1) - V_f)\, U_i(t_1, t')\mathrm{d}t_1\,. \tag{5.26}$$

We have used in (5.26) the characteristic properties of a time evolution operator following in the case of U_f from definition (3.21):

$$U_f^{-1}(t_1, t') = U_f(t', t_1) \quad \text{and} \quad U_f(t, t')\, U_f(t', t_1) = U_f(t, t_1)\,.$$

Introducing (5.23) in (5.22) one sees with (5.15a) that for variations $\psi_f'^- + \delta\psi_f^-$ the quantity $\delta\hat{f}$ vanishes.

In analogy to (5.12) the renormalized Born approximation for direct collisions results with the ansatz

$$\psi_i^+(t) = a_1 \varphi_i(t), \qquad \psi_f^-(t) = a_2^* \varphi_f(t)\,. \tag{5.27}$$

After variation of $\hat{f}$ in (5.14) with respect to a_1 and a_2 we obtain by a straightforward calculation analogous to (5.12)

$$f \cong \frac{f_{\mathrm{B}}}{f_{\mathrm{B}} - f_{2i}^{ii}}\, f_{\mathrm{B}} \tag{5.28}$$

with the Born approximation

$$f_{\mathrm{B}} = -\frac{i}{\hbar} \int \mathrm{d}t(\varphi_f(t), V_i \varphi_i(t))$$

and

$$f_{2i}^{ii} = -\frac{1}{\hbar^2} \int \mathrm{d}t \int_{-\infty}^{t} \mathrm{d}t'(\varphi_f(t), V_i U_i(t, t')\, V_i \varphi_i(t'))\,.$$

For high velocity v is $|f_{2i}^{ii}| \ll |f_B|$ and by expansion f in (5.28) is reduced to the second Born approximation.

As in the wave treatment a renormalized Born approximation for rearrangement collisions results from (5.14) with an ansatz

$$\psi_i^+(t) = \varphi_i(t) + \frac{a_1}{i\hbar} \int_{-\infty}^{t} dt' \, U_i(t, t') \, V_i(t') \, \varphi_i(t')$$

$$\psi_f'^-(t) = a_2^* \, \varphi_f'(t).$$

$$(5.27a)$$

After variation of a_1 and a_2 one obtains

$$f' = f_B' + \frac{f_{2i}^{fi} f_{2f}^{fi}}{f_{2i}^{fi} - f_3 f_i^{fi}},$$

$$f_{2i,f}^{fi} = -\frac{1}{\hbar^2} \int dt \int_{-\infty}^{t} dt' (\varphi_f'(t), V_f U_i(t, t') V_i \varphi_i(t')) \quad \text{etc.}$$

$$(5.28a)$$

For high velocity $|f_{3fi}| \ll |f_{2i}^{fi}|$ and (5.28a) yields the second Born approximation.

3. Examples for Potential Scattering

It is difficult to assess how reliable the results of the renormalized Born approximation are for low energies. To get a rough idea we calculate the differential cross section for potential scattering from a square well for $E_i = 0$ in the renormalized Born approximation and compare with the exact result.

The differential cross section for scattering of a particle with mass v from a potential V is with (1.27)

$$\frac{d\sigma}{d\Omega} = \int d\sigma_{if} K'^2 dK' = \frac{v^2}{(2\pi)^2 \hbar^4} |T(K_f, K_i)|^2, \quad K_f = K_i. \quad (5.29)$$

The T-matrix in the renormalized Born approximation is with (5.12a)

$$T(K_f K_i) \cong \frac{T_B^2(K_f, K_i)}{T_B(K_f, K_i) - T_2(K_f, K_i)};$$

$$T_B = (\phi(K_f), V\phi(K_i)), \quad T_2 = (\phi(K_f), VG_0V\phi(K_i)) \quad (5.30)$$

$$\phi(K) = e^{iKR}, \quad G_0 = \frac{1}{E_i + i\eta - \frac{\hbar^2}{2v} \partial_R^2}.$$

For the square well potential

$$V(R) = V_0 \Theta(R_0 - R), \, \Theta(R_0 - R) = \begin{cases} 1 & R_0 > R \\ 0 & R_0 < R \end{cases} \quad (5.31)$$

the first Born approximation for $K_i = K_f = 0$ is

$$T_B = V_0 \int \Theta(R_0 - R)\mathrm{d}R = \frac{4\pi}{3} R_0^3 V_0 . \qquad (5.32)$$

The Green's function G_0 is in space respresentation [5]

$$G_0(R - R') = - \frac{v}{2\pi\hbar^2} \frac{e^{iK_i|R-R'|}}{|R-R'|} \qquad (5.33)$$

and the second Born approximation for $K_i = K_f = 0$ with the square well potential (5.31) is

$$T_2 = -\frac{v}{2\pi\hbar^2} V_0^2 \int_0^{R_0} \mathrm{d}R \int_0^{R_0} \frac{1}{|R-R'|} \mathrm{d}R' = -\frac{16\pi}{15\hbar^2} R_0^5 V_0^2 . \quad (5.34)$$

The T-matrix in the renormalized Born approximation is then with (5.30), (5.32), (5.34)

$$T(K_f, K_i) \cong \frac{4\pi R_0^3 V_0}{3\left(1 + \dfrac{4}{5}\dfrac{vV_0 R_0^2}{\hbar^2}\right)} \qquad (5.35)$$

and the differential cross section for $E_i = 0$ with (5.29)

$$\frac{\mathrm{d}\sigma(E_i = 0)}{\mathrm{d}\Omega} = R_0^2 \frac{\alpha^4 R_0^4}{9\left(1 - \dfrac{4}{10}\alpha^2 R_0^2\right)^2} ; \qquad \alpha^2 = -\frac{2vV_0}{\hbar^2} . \qquad (5.36)$$

The differential cross section (5.36) is spherically symmetric so that the total cross section for $E_i = 0$ is

$$\sigma(E_i = 0) = 4\pi \frac{\mathrm{d}\sigma(E_i = 0)}{\mathrm{d}\Omega} = 4\pi R_0^2 \frac{\alpha^4 R_0^4}{9\left(1 - \dfrac{4}{10}\alpha_2 R_0^2\right)^2} . \qquad (5.37)$$

For $V_0 \to \infty$ we obtain the hard sphere scattering and (5.37) yields

$$\sigma(E_i = 0) = 4\pi R_0^2 \cdot 0.7 . \qquad (5.37a)$$

The exact value is $\sigma(E_i = 0) = 4\pi R_0^2$. The agreement is reasonable bearing in mind that $E_i = 0$, $V_0 \to \infty$ is the most unfavourable case for a high energy approach. For $V_0 < 0$ resonances and zeros occur in the cross section. Fig. 8 shows that the first zero and resonance is described very well, while for V getting deeper or wider the renormalized Born approximation provides a constant, representing an average value of the variation of the true cross section. Though the agreement for $\alpha R_0 > 1$ is poor it means some progress as the first Born approximation varies like $\alpha^4 R_0^4$ for $\alpha R_0 > 1$.

We can improve the agreement for $\alpha R_0 > 1$ and produce higher resonances by a better guess of the trial functions $\Psi^+(K_i)$ and $\Psi^-(K_f)$ in (5.1). Since the wave length of $\Psi^+(K_i)$ and $\Psi^-(K_f)$ in (5.1) is changed within the range of the potential it is plausible to try an ansatz

$$\Psi(K_i) = a_1 e^{ik_i(R)R} \quad \text{and} \quad \Psi^-(K_f) = a_2^* e^{ik_i(R)R} \tag{5.38}$$

with

$$k_i(R) = \sqrt{K_i^2 - \frac{2v}{\hbar^2} V(R)} \frac{K_i}{K_i}$$

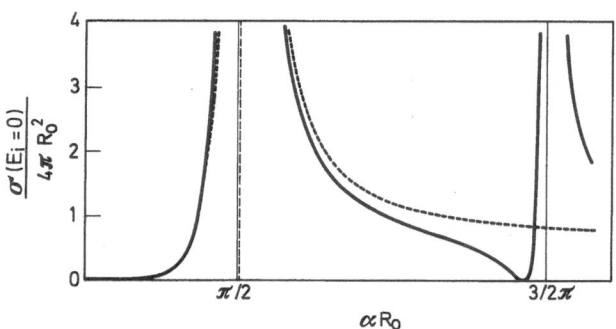

Fig. 8. Total cross section at $E_i = 0$ for scattering from a square well potential
$V = V_0 \Theta(R_0 - R)$, $\alpha^2 = -\dfrac{2v}{\hbar^2} V_0$. ——— exact, - - - - Variation (5.37)

instead of (5.11) in terms of plane waves with the incoming wave vectors K_i, K_f. The calculation of $T(K_f, K_i)$ is the same as before and the result is analogous to (5.30)

$$T(K_f, K_i) \cong \frac{T_B(k_i, K_i)}{T_B(k_i, k_i) - T_2(k_i, k_i)} T_B(K_f, k_i). \tag{5.39}$$

Since $k_i(R)$ occurs in T_B, T_2 only within the potential, we can evaluate T_B, T_2 for the square well potential (5.31) with $k_i = \sqrt{K_i^2 + \alpha^2}$. For $K_i = K_f = 0$ we obtain with $e_i = K_i/K_i$

$$T_B(k_i, k_i) = \int V(R) dR = V_0 \frac{4\pi}{3} R_0^3$$

$$T_B(K_f = 0, k_i) = T_B(k_i, K_i = 0)$$

$$= \int V(R) e^{i\alpha e_i R} dR = \frac{4\pi V_0}{\alpha} \left\{ \frac{1}{\alpha^2} \sin\alpha R_0 - \frac{R}{\alpha} \cos\alpha R_0 \right\} \tag{5.40}$$

$$T_2(k_i, k_i) = -\frac{v V_0^2}{2\pi\hbar^2} \int_0^{R_0} dR\, e^{-i\alpha e_i R} \int_0^{R_0} dR' \frac{1}{|R - R'|} e^{i\alpha e_i R'}$$

$$= -\frac{2\pi v V_0^2}{\hbar^2 \alpha^2} \left\{ \frac{4}{3} R_0^3 - \frac{\sin 2\alpha R_0}{2\alpha^3} + \frac{R_0 \cos 2\alpha R_0}{\alpha^2} \right\}.$$

The integral T_2 is evaluated in Appendix 5. Introducing (5.40) in (5.39) yields

$$T(K_i = K_f = 0) \cong \frac{8\pi\hbar^2 R_0}{v} \frac{\left(\cos\alpha R_0 - \dfrac{\sin\alpha R_0}{\alpha R_0}\right)^2}{\cos 2\alpha R_0 - \dfrac{\sin 2\alpha R_0}{2\alpha R_0}}. \qquad (5.41)$$

The differential cross section for $E_i = 0$ is spherically symmetric and given by (5.29) and (5.41). We then obtain the total cross section for $E_i = 0$

$$\sigma(E_i = 0) = 4\pi \frac{d\sigma(E_i = 0)}{d\Omega} \cong \frac{64\pi R_0^2 \left(\cos\alpha R_0 - \dfrac{\sin\alpha R_0}{\alpha R_0}\right)^4}{\left(\cos 2\alpha R_0 - \dfrac{\sin 2\alpha R_0}{2\alpha R_0}\right)^2}. \qquad (5.42)$$

For $V_0 < 0$ α is real and (5.42) yields resonances when the denominator vanishes. There are also zeros of the cross section for vanishing numerator. In Fig. 9 $\sigma(E_i = 0)$ is plotted according (5.42) and compared with the exact s-scattering cross section. One recognizes that (5.42) produces the

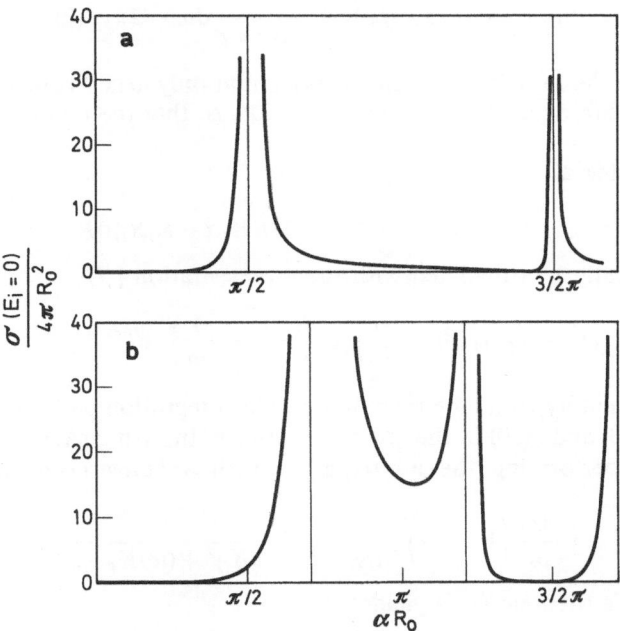

Fig. 9. Total cross section at $E_i = 0$ for scattering from a square well potential $V = V_0 \Theta(R_0 - R)$, $\alpha^2 = -\dfrac{2v}{\hbar^2} V_0$. a exact, b Variation (5.42)

resonances better than (5.37). The resonance near $\alpha R_0 = \pi$ corresponds to the p-wave resonance at $\alpha R_0 = \pi$ in the low energy cross section. The zeros of the cross section are in agreement with the exact result.

In conclusion of the variational methods one can say that the results are exact for high energies and rather good for low energies at least as far as potential scattering is concerned. For inelastic collisions there are not yet any applications of (5.12) or (5.28), but we feel that here is a starting point to extend high energy methods to intermediate and possibly low energies.

Acknowledgements. I would like to thank *P. H. Dederichs* and *G. Leibfried* for many helpful and stimulating discussions.

Appendix

1. We have to prove for $t \to \infty$

$$\frac{e^{-i\omega t}}{\omega + i\eta} \cong 2\pi i \delta(\omega), \quad \eta \to 0. \tag{A.1}$$

For any analytic function $f(\omega)$ is for great t

$$\int_{-\infty}^{\infty} f(\omega) \frac{e^{-i\omega t}}{\omega + i\eta} d\omega \cong f(0) \int_{-\infty}^{\infty} \frac{e^{-i\omega t}}{\omega + i\eta} d\omega = 2\pi i f(0) \tag{A.2}$$

because with the rapidly oscillating exponential only $\omega \cong 0$ contributes. (A.2) is the defining property of the δ-function, so that the relation (A.1) holds.

2. The integral

$$I = \int e^{-iKR} \frac{1}{R} e^{iK_iR} {}_1F_1(-i\alpha, 1, i(K_iR - K_iR)) dR \tag{A.3}$$

can conveniently be calculated with the representation [5]

$$ {}_1F_1(-i\alpha, 1, i\beta) = \frac{1}{2\pi i} \oint \frac{d\omega}{\omega} \left(1 - \frac{1}{\omega}\right)^{i\alpha} e^{i\beta\omega} \tag{A.4}$$

for the confluent hypergeometric function. The integration path encircles the point $(0, 0)$ and $(1, 0)$ in the complex ω-plane. Inserting (A.4) in (A.3) yields after performing the r-integration with a convergence factor $e^{-\varepsilon R}$, $\varepsilon \to 0$

$$I = -2i \int \frac{d\omega}{\omega} \left(1 - \frac{1}{\omega}\right)^{i\alpha} \frac{1}{(K_i - K - \omega K_i)^2 + (i\omega K_i - \varepsilon)^2}. \tag{A.5}$$

With *Cauchy*'s theorem (A.5) yields

$$I = \frac{-4\pi}{(K_i - K)^2 + \varepsilon^2} \left\{ \frac{K^2 - K_i^2 - 2i\varepsilon K_i}{(K_i - K)^2 + \varepsilon^2} \right\}^{i\alpha} \varepsilon \to 0. \tag{A.6}$$

With (A.4) and a convergence factor $e^{-\varepsilon R}$, $\varepsilon \to 0$ also the Fourier transform of $\Psi(R)$ in (2.5) can easily be calculated. The result is

$$\tilde{\Psi}(K) = \frac{8\pi\alpha K_i e^{-\frac{\pi\alpha}{2}}\Gamma(1+i\alpha)}{[(K_i-K)^2+\varepsilon^2][K^2-K_i^2-2i\varepsilon K_i]}\left\{\frac{K^2-K_i^2-2i\varepsilon K_i}{(K_i-K)^2+\varepsilon^2}\right\}^{i\alpha}. \quad \text{(A.6a)}$$

Comparing (2.7) with (A.6a) yields $\tilde{\psi}(K) = G_0(K)\,T_{KK_i}$ or $\Psi = G_0 V \Psi$ with $V = A/R$ and $G_0(K) = \dfrac{2v/\hbar^2}{K_i^2-K^2+i\varepsilon}$.

3. The integral

$$\Theta(t) = -\frac{1}{2\pi i}\int d\xi \frac{e^{-i\xi t}}{\xi+i\varepsilon}, \qquad \varepsilon \to 0 \quad \text{(A.7)}$$

vanishes with *Cauchy*'s theorem after extending the integration path by a large semicircle in the upper half plane for $t < 0$ and is 1 for $t > 0$, which is the definition for the step function $\Theta(t)$.

4. For $t_i < 0$ is

$$I = \int_{-\infty}^{\infty} d\xi\,\xi^{i\alpha}\frac{e^{i\xi t_i}}{\xi+i\varepsilon} = \frac{1}{(-t_i)^{i\alpha}}\int_{-\infty}^{\infty} d\xi\,\xi^{i\alpha}\frac{e^{-i\xi}}{\xi+i\varepsilon}, \qquad \varepsilon \to 0 \quad \text{(A.8)}$$

and after integration by parts

$$I = \frac{1}{\alpha}\frac{1}{(-t_i)^{i\alpha}}\int_{-\infty}^{\infty} d\xi\,\xi^{i\alpha}e^{-i\xi}.$$

We can deform the integration path by encircling a branch cut from 0 to ∞ along the real axis. As $\xi^{i\alpha} = |\xi|^{i\alpha}$ in the upper half plane, we have

$$I = \frac{1-e^{-2\pi\alpha}}{\alpha}\frac{e^{2\pi\alpha}}{(-t_i)^{i\alpha}}\int_{0-i\varepsilon}^{\infty-i\varepsilon} d\xi\,\xi^{i\alpha}e^{-i\xi}$$

or integrating along the imaginary axis

$$I = \frac{1-e^{-2\pi\alpha}}{\alpha}\frac{-ie^{\pi\alpha/2}}{(-t_i)^{i\alpha}}\int_{0}^{\infty} d\xi\,|\xi|^{i\alpha}e^{-\xi}. \quad \text{(A.9)}$$

The integral in (A.9) is the Γ-function [39] and we have

$$I = -\frac{i}{(-t_i)^{i\alpha}}\frac{1-e^{-2\pi\alpha}}{\alpha}e^{\frac{\pi\alpha}{2}}\Gamma(1+i\alpha). \quad \text{(A.10)}$$

5. The integral

$$I = \int_{0}^{R_0} d\boldsymbol{R}\,e^{-i\boldsymbol{KR}}\int_{0}^{R_0} d\boldsymbol{R}'\,\frac{e^{i\boldsymbol{KR}'}}{|\boldsymbol{R}-\boldsymbol{R}'|} \quad \text{(A.11)}$$

can be written with the step function

$$I = \int d\boldsymbol{R}\,d\boldsymbol{R}'\,\Theta(R_0-R)\,\Theta(R_0-R')\,\frac{e^{i\boldsymbol{K}(\boldsymbol{R}-\boldsymbol{R}')}}{|\boldsymbol{R}-\boldsymbol{R}'|}$$

and after substituting $R - R'$ for R'

$$I = \int d\boldsymbol{R} \, d\boldsymbol{R}' \, \Theta(R_0 - R) \, \Theta(R_0 - |\boldsymbol{R} - \boldsymbol{R}'|) \frac{e^{-i\boldsymbol{K}\boldsymbol{R}'}}{R'}.$$

The integral over $\boldsymbol{R}$ is the overlap volume $O(R')$ of two spheres with radius R_0 in distance R'. With

$$O(R') = \frac{\pi}{3} \left(R_0 - \frac{R'}{2} \right)^2 (4R_0 + R')$$

the R'-integration can easily be performed and the result is

$$I = \frac{4\pi^2}{3K} \left\{ \frac{4R_0^3}{K} - \frac{3}{2K^4} \sin 2KR_0 + \frac{3R_0}{K^3} \cos 2KR_0 \right\}. \quad (A.12)$$

6. For $w' = \frac{A}{r} e^{-\varkappa r}$ is

$$I = \int d\boldsymbol{K} \, \frac{\tilde{w}'(\boldsymbol{K} - \boldsymbol{K}_i) \, \tilde{w}'(\boldsymbol{K} - \boldsymbol{K}_f)}{E_i + i\eta - \dfrac{\hbar^2 K^2}{2v}}$$

$$= \frac{A^2}{4\pi^4} \int d\boldsymbol{K} \, \frac{1}{(\boldsymbol{K} - \boldsymbol{K}_i)^2 + \varkappa^2} \frac{1}{(\boldsymbol{K} - \boldsymbol{K}_f)^2 + \varkappa^2} \frac{1}{E_i + i\eta - \dfrac{\hbar^2 K^2}{2v}}. \quad (A.13)$$

With the Feyman integral

$$\frac{1}{ab} = \int_0^1 \frac{dx}{[ax + b(1-x)]^2}$$

the two centre integral (A.13) is

$$I = \frac{A^2}{4\pi^4} \int_0^1 dx \int d\boldsymbol{K}$$

$$\cdot \frac{\left(E_i + i\eta - \dfrac{\hbar^2 K^2}{2v} \right)^{-1}}{\{K^2 - 2K[(\boldsymbol{K}_i - \boldsymbol{K}_f)x + \boldsymbol{K}_f] + x(K_i^2 - K_f^2) + K_f^2 + \varkappa^2\}^2}. \quad (A.14)$$

According to (4.11) is $\boldsymbol{K}_i - \boldsymbol{K}_f \sim 1/\sqrt{E_i}$ for $\boldsymbol{K}_f$ parallel to $\boldsymbol{K}_i$ and can be neglected in the denominator of (A.14). Performing the $\boldsymbol{K}$ integration in (A.14) yields with $K_i^2 - K_f^2 = 2v(\varepsilon_f - \varepsilon_i)/\hbar^2$

$$I = \frac{vA^2 i}{2\hbar^2 \pi^2 K_f} \left\{ \int_0^1 \frac{\dfrac{2v}{\hbar^2} dx}{\left(\sqrt{\dfrac{2v}{\hbar^2} E_i} - K_f \right)^2 + \dfrac{2vx}{\hbar^2}(\varepsilon_f - \varepsilon_i) + \varkappa^2} \right.$$

$$\left. - \int_0^1 \frac{iK_f \, dx}{\sqrt{\dfrac{2v}{\hbar^2} x(\varepsilon_f - \varepsilon_i) + \varkappa^2} \left[E_i - \dfrac{\hbar^2}{2v} \left(K_f + i\sqrt{\dfrac{2vx}{\hbar^2}(\varepsilon_f - \varepsilon_i) + \varkappa^2} \right)^2 \right]} \right\}. \quad (A.15)$$

Asymptotically for great E_i is with

$$E_i = \frac{\hbar^2 K_i^2}{2v} + \varepsilon_i = \frac{\hbar^2 K_f^2}{2v} + \varepsilon_f$$

and (4.11)

$$\sqrt{\frac{2v}{\hbar^2} E_i} - K_f \cong \frac{v\varepsilon_f}{\hbar^2 K_i}$$

and

$$E_i - \frac{\hbar^2}{2v}\left(K_f + i\sqrt{\frac{2vx}{\hbar^2}(\varepsilon_f - \varepsilon_i) + \kappa^2}\right)^2$$

$$\cong \varepsilon_f - \frac{\hbar^2 i}{v} K_f \sqrt{\frac{2vx}{\hbar^2}(\varepsilon_f - \varepsilon_i) + \kappa^2} \, .$$

For $\kappa^2 \neq 0$ (A.15) yields

$$I = \frac{-ivA^2}{4\pi^2 \hbar^2 K_f} \int_0^1 \frac{\mathrm{d}x}{x(\varepsilon_f - \varepsilon_i) + \frac{\hbar^2 \kappa^2}{2v}}$$

$$= \frac{-ivA^2}{4\pi^2 \hbar^2 K_f(\varepsilon_f - \varepsilon_i)} \ln\left(1 + \frac{(\varepsilon_f - \varepsilon_i)2v}{\hbar^2 \kappa^2}\right) \qquad (A.16a)$$

and for $\kappa = 0$ with the substitution $\sqrt{x} = y$ in the second integral of (A.15)

$$I = \frac{ivA^2}{8\pi^2 \hbar^2 K_f(\varepsilon_f - \varepsilon_i)} \ln \frac{2v^3 \varepsilon_f^6}{\hbar^6 K_i^6(\varepsilon_f - \varepsilon_i)^3} \, . \qquad (A.16b)$$

References

1. *Glauber, R. J.:* High-energy collision theory, Lectures in Theoretical Physics, Vol. I, Boulder 1958. New York: Interscience 1959.
2. *Lippmann, B. A., Schwinger, J.:* Phys. Rev. **79**, 469 (1950).
3. *Faddeev, L. D.:* JETP **12**, 1014 (1961).
4. *Goldberger, M. L., Watson, K. M.:* Collision theory. New York: John Wiley & Sons 1964.
5. *McDowell, M. R. C., Coleman, J. P.:* Introduction to the theory of ion-atom collisions. Amsterdam-London: North Holland Publishing Company 1970.
6. *Mott, N. F., Massey, H. S. W.:* The theory of atomic collisions. Oxford: Clarendon Press 1965.
7. Atomic and molecular processes, ed. by *Bates, D. R.* New York: Academic Press 1962.
8. *Rodberg, L. S., Thaler, R. M.:* Introduction to the quantum theory of scattering. New York-London: Academic Press 1967.
9. *Newton, R. G.:* Scattering theory of waves and particles. New York, San Francisco, St. Louis, Toronto, London, Sydney: Mc Graw-Hill Book Comp. 1966.
10. *Lippmann, B. A.:* Phys. Rev. **102**, 264 (1956).

206   *K. Dettmann:* High Energy Treatment of Atomic Collisions

11. *Gell-Mann, M., Goldberger, M. L.:* Phys. Rev. **91**, 398 (1953).
12. *Bethe, H.:* Quantenmechanik der Ein- und Zwei-Elektronenprobleme. Handbuch der Physik, Band 24/I. Berlin: Springer 1933.
13. *Holt, A. R., Moiseiwitsch, B. L.:* Born expansions, advances in atomic and molecular physics; Vol. 4, ed. by *Bates, D. R., Estermann, I.:* New York, London: Academic Press 1968.
14. *Bates, D. R., McCarroll, R.:* Charge transfer. Advan. Phys. **11**, 39 (1962).
15. *Bransden, B. H.:* Rearrangement collisions, lectures in theoretical physics, Vol. XI–C, Boulder 1968. New York: Gordon and Breach 1969.
16. *Bates, D. R., Griffing, G.:* Proc. Phys. Soc. (London) A **66**, 961 (1953).
17. *Oppenheimer, J. R.:* Phys. Rev. **31**, 349 (1928).
18. *Brinkman, H. C., Kramers, H. A.:* Proc. Acad. Sci. Amsterdam **33**, 973 (1930).
19. *Drisko, R. M.:* Carnegie Inst. of Techn., Thesis (1955).
20. *Jackson, J. D., Schiff, H.:* Phys. Rev. **89**, 359 (1953).
21. *Mapleton, R. A.:* Proc. Phys. Soc. (London) **83**, 895 (1964).
22. *McCarroll, R., Salin, A.:* Proc. Phys. Soc. (London) **90**, 63 (1967).
23. — — Proc. Roy. Soc. (London). Ser. A, **300**, 202 (1967).
24. *Dettmann, K., Leibfried, G.:* Z. Physik **210**, 43 (1968).
25. — — Z. Physik **218**, 1 (1969).
26. *Coleman, J., McDowell, M. R. C.:* Proc. Phys. Soc. (London) **83**, (907) 1964.
27. *Bassel, R. H., Gerjoy, E.:* Phys. Rev. **117**, 749 (1960).
28. *Aaron, R., Amado, R. D, Lee, B. W.:* Phys. Rev. **121**, 319 (1961).
29. *Greider, K. R., Dodd, L. R.:* Phys. Rev. **146**, 671 (1966).
30. *Dettmann, K., Leibfried, G.:* Phys. Rev. **148**, 1271 (1966).
31. *Corbett, J. V.:* J. Math. Phys. **9**, 891 (1968).
32. *McGuire, J. B.:* J. Math. Phys. **5**, 439 (1964).
33. *Moiseiwitsch, B. L.:* Proc. Phys. Soc. (London) **87**, 885 (1966).
34. *Mittleman, M. H.:* Phys. Rev. **122**, 499 (1961).
35. *West, G. B.:* J. Math. Phys. **8**, 942 (1967).
36. *Dalitz, R. H.:* Proc. Roy. Soc. (London), Ser. A **206**, 509 (1951).
37. *Schwinger, J.:* J. Math. Phys. **5**, 1606 (1964).
38. *Nutt, G. L.:* J. Math. Phys. **9**, 796 (1968).
39. *Magnus, W., Oberhettinger, F.:* Formeln und Sätze für die speziellen Funktionen der mathematischen Physik. Berlin-Göttingen-Heidelberg: Springer 1948.
40. *Gilbody, H. B., Ireland, J. V.:* Proc. Roy. Soc. (London), Ser. A **277**, 137 (1963).
41. *Bethe, H.:* Ann. Physik **5**, 325 (1930).
42. *Thomas, L. H.:* Proc. Roy. Soc. (London) **114**, 561 (1927).
43. *Barnett, C. F., Reynolds, H. K.:* Phys. Rev. **109**, 355 (1958).
44. *Schryber, U.:* Helv. Phys. Acta **40**, 1023 (1968).
45. *Williams, J. F.:* Phys. Rev. **157**, 97 (1967).
46. *Toburen, L. H., Nakai, M. Y., Langley, R. A.:* Phys. Rev. **171**, 114 (1968).

Dr. *Klaus Dettmann*
Institut für Theoretische Physik
der Technischen Hochschule
D-5100 Aachen
Germany

Dynamisches Verhalten von Metallen unter Stoßwellenbelastung

K. H. SCHRAMM

Inhaltsverzeichnis

Zusammenfassung

Das dynamische Verhalten von Metallen bzw. Metallplatten bei der Beschleunigung durch Stoßwellenbelastung, speziell durch detonierende Sprengstoffe wird diskutiert. Die Untersuchungen gehen von speziellen Ansätzen für den thermodynamischen Zustand komprimierter, fester

Metalle unter Drücken bis zu etwa 1 Mbar aus. Damit werden die ablaufenden Bewegungsvorgänge nach bekannten Stoßwellengleichungen berechnet. Besondere Aufmerksamkeit wird den mehrfachen Stoß- und Entlastungswellenreflexionen an den Plattengrenzflächen geschenkt. Die Gleichungen werden für 22 reine Metalle numerisch ausgewertet, wobei der Rechnung die Daten von Composition B (60:40) als des in Experimenten meist verwendeten Sprengstoffs zugrunde gelegt werden. Die erhaltenen Ergebnisse werden im Hinblick auf ihre allgemeine physikalische Bedeutung diskutiert.

1. Einleitung

Das dynamische Verhalten von Metallen unter Stoßwellenbelastung hat in den vergangenen 15 Jahren zunehmend an physikalischem Interesse und an technischer Bedeutung gewonnen. Das physikalische Interesse konzentriert sich dabei vornehmlich auf das Zustandsverhalten fester Metalle unter hohen Drücken, während die technische Bedeutung in der Entwicklung neuartiger Verfahren zur Verformung und Bearbeitung metallischer Werkstoffe liegt.

Die ersten Untersuchungen zum Verhalten der Materie unter hohen Drücken gehen auf *Bridgman* [1, 2] zurück. Dabei wird der Druck in der zu untersuchenden Probe statisch durch äußeres Pressen aufgebracht. Die Weiterentwicklung dieser statischen Untersuchungsmethoden im Hinblick auf eine Ausweitung des erfaßbaren Druck- und Temperaturbereichs dürfte inzwischen im wesentlichen abgeschlossen sein. Einen Überblick über die angewandten experimentellen Methoden und die gewonnenen Meßergebnisse gibt *Swenson* [3]. Die meisten Messungen wurden im Druckbereich unterhalb 100 kbar durchgeführt. Höhere Drücke können zwar erreicht werden, erfordern dann aber einen sehr großen experimentellen Aufwand. Der Grund hierfür liegt hauptsächlich darin, daß bei hohem Druck das die Probe einschließende Material selbst zu zerfließen beginnt.

Für Messungen bei noch höheren Drücken empfiehlt sich daher die Anwendung dynamischer Verfahren, bei denen die Probe nur kurzzeitig einer sehr hohen Druckbelastung ausgesetzt wird. Hierzu bietet sich in erster Linie die Verwendung von Sprengstoffen an, mit denen Druckbelastungen von mehreren hundert Kilobar bis zu einigen Megabar erzeugt werden konnten. Zusammenfassende Darstellungen durchgeführter Arbeiten geben *Rice et al.* [4] und *Alt'Shuler* [5]. Unter Ausnutzung der in Abschnitt 3.7 beschriebenen irregulären Stoßwellenreflexion konnten in neuester Zeit sogar Druckbelastungen bis zu 15 Mbar mittels Sprengstoffen erzeugt werden [6].

Durch eine starke dynamische Druckbelastung wird in dem Probe-körper eine Stoßwelle erzeugt, die durch ihn hindurch läuft und zu mannigfaltigen mechanischen, elektrischen, optischen und strukturellen Erscheinungen Anlaß gibt. Einen Überblick über die beobachteten Phänomene und ein ausführliches Literaturverzeichnis geben *Doran* und *Linde* [7].

Bei technischen Fertigungsverfahren lassen sich Sprengstoffe zur Verformung sowie insbesondere zum Schneiden, Schweißen und Plattieren metallischer Werkstoffe verwenden. Wegen der auftretenden hohen Drücke und der großen Geschwindigkeit, mit der diese Vorgänge ablaufen, werden dabei metallurgische und technische Probleme auf-geworfen, die sich grundlegend von denjenigen der konventionellen Ver-fahrenstechnik unterscheiden. Eine ausführliche Darstellung findet sich in einer Monographie von *Rinehart* und *Pearson* [8].

Die vorliegende Arbeit behandelt die Kompression und den Material-fluß in Metallen bzw. in Metallplatten, die durch detonierende Spreng-stoffe beschleunigt werden. Dabei werden außer Problemen der Zustands-gleichung Erscheinungen diskutiert, die mit Reflexion und Brechung von Stoß- und Entlastungswellen zusammenhängen. Numerische Daten werden für eine größere Anzahl reiner Metalle angegeben.

2. Zustand der Metalle unter hohen Drücken

2.1 Zustandsgleichung fester Metalle

Über Zustandsgleichungen fester Metalle liegen bereits mehrere zu-sammenfassende Arbeiten vor [9, 10], so daß es im folgenden genügt, sich auf eine kurze, skizzenhafte Darstellung zu beschränken.

Die innere Energie U eines Metallvolumens V hängt außer von V selbst noch von der Temperatur T ab und ist als Funktion der beiden Zustandsgrößen V und T bis auf eine beliebige Normierungskonstante eindeutig bestimmt. Um die kalorische Zustandsgleichung $U = U(V, T)$ explizit angeben zu können, muß man von der atomaren Struktur des betrachteten Mediums ausgehen. Metalle im festen Zustand bestehen aus gitterförmig angeordneten Metallionen und mehr oder weniger frei beweglichen Leitungselektronen. Die Energie U setzt sich dem-entsprechend aus den potentiellen Wechselwirkungsenergien zwischen diesen Teilchen und deren thermischer Bewegungsenergie zusammen [11].

Die temperaturabhängige Komponente des von den Elektronen her-rührenden Anteils zur Energie U kann in dem hier zur Diskussion stehenden Zustandsbereich gegenüber der thermischen Schwingungs-energie des Ionengitters vernachlässigt werden. Sie spielt nur bei hochionisierten Hochtemperaturplasmen eine Rolle [12] sowie in der Nähe des absoluten Nullpunkts, da dort die thermische Energie des Ionengitters mit $T \to 0$ sehr viel stärker verschwindet als diejenige der Elektronen.

Der Anteil U_g der thermischen Gitterschwingungen zur Energie U wird meist nach der Theorie von *Debye* berechnet [11]. Nach der Quantentheorie kann ein Oszillator der Eigenfrequenz v nur die diskreten Energiewerte u_v

$$u_v = \frac{h v}{e^{\frac{h v}{k T}} - 1} + \frac{1}{2} h v \qquad (1)$$

annehmen (h = Plancksches Wirkungsquantum; k = Boltzmann-Konstante). Faßt man daher das Ionengitter als ein System gekoppelter Oszillatoren auf, so erhält man U_g durch Integration von u_v über alle in V auftretenden Eigenschwingungen. Deren Gesamtzahl ist wegen der drei Schwingungsfreiheitsgrade eines jeden Oszillators gleich $3N$ (N = Anzahl der Metallionen in V). Macht man weiterhin nach *Debye* die Annahme, daß die auftretenden Eigenfrequenzen kontinuierlich von Null bis zu einem oberen Grenzwert v_m verteilt sind, so erhält man in Verbindung mit der Theorie der Eigenschwingungen eines elastischen Körpers [13]

$$U_g = \frac{9N}{v_m^3} \cdot \int_0^{v_m} u_v \cdot v^2 \cdot dv. \qquad (2)$$

Aus den Gl. (1) und (2) folgt, wenn man noch durch Hinzufügen der temperaturabhängigen potentiellen Energieterme wieder von U_g zu U übergeht:

$$U = \Phi(V) + 3NkT \cdot D(x) \qquad (3)$$

mit

a) $D(x) = \frac{3}{x^3} \cdot \int_0^x \frac{\xi^3 \cdot d\xi}{e^\xi - 1}$;   b) $x = \frac{\Theta_D(V)}{T}$;   c) $\Theta_D = \frac{h \cdot v_m}{k}$. $\qquad (4)$

Dabei wird die durch Gl. (4c) definierte charakteristische Temperatur Θ_D als nur von V abhängig vorausgesetzt. Die Funktion $D(x)$ ist tabelliert, z. B. in [10]. Spezielle Werte sind:

$$D(0) = 1 \quad \text{und} \quad D(\infty) \doteq \frac{\pi^4}{5x^3}.$$

Ausgehend von speziellen Ansätzen für die intermolekularen Wechselwirkungspotentiale im Gitter lassen sich auch für den in Gl. (3) noch unbestimmt gebliebenen Term $\Phi(V)$ explizite Ausdrücke angeben [10,14]. Diese Vorgehensweise ist jedoch für die Stoßwellenberechnungen in der vorliegenden Arbeit zu ungenau. Im folgenden wird daher stattdessen die

Volumensabhängigkeit aller auftretenden Größen durch geeignete Anpassung an die Meßergebnisse von Stoßwellenexperimenten erfaßt. Hierzu sind einige zusätzliche theoretische Betrachtungen erforderlich. Zwischen dem Kompressionsmodul K

$$K = -V \cdot \left(\frac{\partial p}{\partial V}\right)_T \tag{5}$$

(p = Druck) und dem thermischen Volumenausdehnungskoeffizienten α

$$\alpha = \frac{1}{V} \cdot \left(\frac{\partial V}{\partial T}\right)_p \tag{6}$$

besteht der allgemeingültige Zusammenhang [15]

$$\alpha \cdot K = \left(\frac{\partial p}{\partial T}\right)_V = \left(\frac{\partial S}{\partial V}\right)_T \tag{7}$$

(S = Entropieinhalt von V).

Die Funktion $S = S(V, T)$ läßt sich aufgrund der Beziehung

$$T \cdot \left(\frac{\partial S}{\partial T}\right)_V = \left(\frac{\partial U}{\partial T}\right)_V = C_v$$

(C_v = spezifische Wärme), aus Gl. (3) ableiten. Man erhält [15]

a) $C_v = 3Nk \cdot [D(x) - x \cdot D'(x)]$;

b) $S = -3Nk \cdot \int_\infty^x \left[\frac{D(\xi)}{\xi} - D'(\xi)\right] d\xi$, $\qquad$ (8)

wenn man berücksichtigt, daß nach dem Nernstschen Wärmesatz $S(V, T = 0) = 0$ ist.

Einsetzen der Gl. (8) in Gl. (7) liefert in Verbindung mit Gl. (4b):

$$\frac{\alpha \cdot K \cdot V}{C_v} = -\frac{d \ln \Theta_D(V)}{d \ln V} \equiv \gamma(V) . \tag{9}$$

Dabei wurde mit: $C_v(V, T = 0) = 0$ nochmals der Nernstsche Wärmesatz angewandt. Die durch Gl. (9) definierte Größe γ heißt Grüneisen-Konstante; sie hängt nur von V ab, da voraussetzungsgemäß auch Θ_D nur von V abhängt.

Bei dieser Herleitung [15] der Gl. (9) wurde von der speziellen analytischen Form der Funktion $D(x)$ nach Gl. (4a) überhaupt kein Gebrauch gemacht; es wurde lediglich die Aussage benutzt, daß $D(x)$ dem Nernstschen Wärmesatz gehorcht. Dies bedeutet, daß Gl. (9) unter viel allgemeineren Voraussetzungen gültig ist als unter denen, die der Debyeschen Theorie zugrunde liegen.

14*

Aus der kalorischen Zustandsgleichung Gl. (3) läßt sich jetzt leicht ein entsprechender Ausdruck für die thermische Zustandsgleichung $p = p(V, T)$ herleiten. Da nach Gl. (8b) $S = $ const. die Aussage $x = $ const. beinhaltet, so besteht der thermodynamische Zusammenhang

$$p = -\left(\frac{\partial U}{\partial V}\right)_x,$$

was nach den Gl. (3) und (4b) auf

$$p = -\Phi'(V) + \gamma(V) \cdot 3\,\frac{N}{V}\,kT \cdot D(x) \qquad (10)$$

führt.

2.2 Nullpunktisothermen

Hinsichtlich der Volumenabhängigkeit der einzelnen Zustandsfunktionen ist es zweckmäßig, deren Werte am absoluten Nullpunkt der Temperatur als Bezugsgrößen einzuführen. Bezeichnet man mit $p_T(V) = p(V, T = 0)$ und $U_T(V) = U(V, T = 0)$ die Nullpunktisothermen im $p - V$- bzw. $U - V$-Diagramm, so erhält man aus den Gl. (3) und (10)

$$p(V, T) - p_T(V) = \gamma(V) \cdot \frac{U(V, T) - U_T(V)}{V}. \qquad (11)$$

Molekulartheoretische Überlegungen führen auf einen Zusammenhang zwischen den beiden in Gl. (11) auftretenden Funktionen $p_T(V)$ und $\gamma(V)$

$$\gamma = -\frac{V}{2} \cdot \frac{\mathrm{d}^2(p_T\,V^{2/3})/\mathrm{d}V^2}{\mathrm{d}(p_T\,V^{2/3})/\mathrm{d}V} - \frac{1}{3}. \qquad (12)$$

Diese Gleichung geht auf *Dugdale* und *Mac Donald* [16] zurück und hat sich als gut brauchbare Näherung bewährt [4]. Ihre ursprüngliche Begründung [16] ist umstritten; es läßt sich aber folgende Plausibilitätsbetrachtung durchführen, die an eine stark vereinfachende Darstellung in [11] anknüpft. Dabei werden die in Gl. (9) eingehenden Größen α, K und C_v mittels eines molekulartheoretischen Modells berechnet.

Der Vorgang einer Volumenänderung läßt sich am Verhalten eines einzelnen Gitterions im Potentialfeld seines nächsten Nachbarn studieren. Fig. 1a zeigt den Potentialverlauf $\varphi = \varphi(r)$ als Funktion des Teilchenabstands r. Die übrigen Gitterionen und die Leitungselektronen denke man sich bei dieser nur orientierenden Betrachtungsweise über den ganzen Raum verschmiert, so daß sich ihr Einfluß summarisch durch geeignete Definition der Funktion $\varphi(r)$ näherungsweise miterfassen läßt.

Wenn keine äußeren Kräfte auf das Gesamtsystem wirken ($p_T = 0$) und außerdem die Gitterionen keine thermische Bewegung ausführen ($T = 0$), so stellt sich im Gleichgewicht der Abstand $r = r_0$ ein, bei dem die potentielle Wechselwirkungsenergie $\varphi(r)$ zum Minimum, d.h. $\varphi'(r_0) = 0$ wird. Man hat r_0 als eine Art mittleren Teilchenabstand im Gitter anzusehen und kann dementsprechend $V_0 = N \cdot a \cdot r_0^3$ setzen. Dabei ist a ein im einzelnen nicht interessierender Formfaktor der Größenordnung Eins.

Unter dem Einfluß eines äußeren Drucks $p_T \neq 0$ (bei $T = 0$) verringert sich der Teilchenabstand von r_0 auf r_p (vgl. Fig. 1a), das Gesamtvolumen also von V_0 auf $V_p = N \cdot a \cdot r_p^3$. Bei dieser Kompression wird an dem System die Arbeit A

$$A = - \int_{V_0}^{V_p} p_T \cdot \mathrm{d}V = -3aN \int_{r_0}^{r_p} p_T \cdot r^2 \cdot \mathrm{d}r$$

geleistet, die gleich dem Zuwachs $E_p = N \cdot \{\varphi(r_p) - \varphi(r_0)\}$ an potentieller Wechselwirkungsenergie der Teilchen sein muß. Gleichsetzen von A und E_p und Differenzieren nach r_p ergibt

a) $\displaystyle p_T(V_p) = -\frac{1}{3a}\frac{\varphi'(r_p)}{r_p^2}$;   b) $V_p = N \cdot a \cdot r_p^3$.   (13)

Hieraus folgt nach Gl. (5) für den Kompressionsmodul K

$$K(V_p) = \frac{1}{9a}\left\{\frac{\varphi''(r_p)}{r_p} - 2\frac{\varphi'(r_p)}{r_p^2}\right\}.$$ (14)

Bei nicht verschwindender Temperatur $T > 0$ (bei $p_T = 0$) führt das betrachtete Gitterion thermische Schwingungen um seine Nullpunktslage r_0 aus. Wie in Fig. 1b angedeutet, ist diese Schwingung zwischen

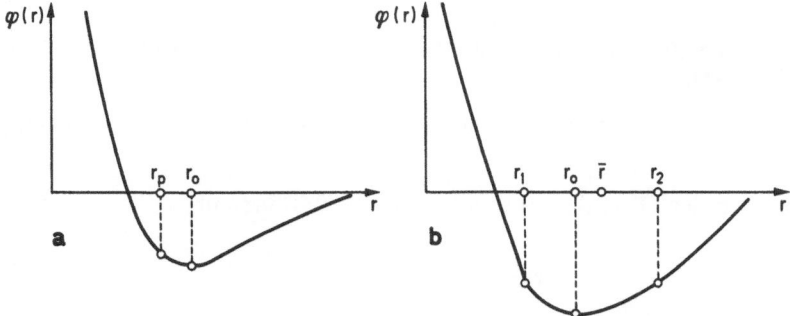

Fig. 1a u. b. Intermolekulares Wechselwirkungspotential $\varphi(r)$. a Kompression bei $T = 0$; b Erwärmung bei $p = 0$

r_1 und r_2 nicht symmetrisch zu r_0; für den zeitlichen Mittelwert $\bar{r}$ des Teilchenabstands gilt vielmehr $\bar{r} > r_0$, was einer thermischen Volumenvergrößerung von V_0 auf $V_T = N \cdot a \cdot \bar{r}^3$ entspricht. Der Wert $\bar{r}$ wird durch die Forderung bestimmt, daß beim Fehlen eines äußeren Drucks p_T im zeitlichen Mittel keine Kraft auf das schwingende Ion wirkt, d. h. daß $\overline{\varphi'(r)} = 0$ sein muß. Setzt man die Teilchenbewegung nach Fig. 1 b als gestörte harmonische Schwingung an

$$\text{a) } r(t) = r_0 + \xi(t); \quad \text{b) } \xi(t) = \xi_0 \cdot \cos\omega t + \varepsilon(t); \quad |\varepsilon| \ll \xi_0 \quad (15)$$

so gilt $\varphi'(r) = \varphi'(r_0) + \xi(t) \cdot \varphi''(r_0) + \frac{1}{2}\xi^2(t) \cdot \varphi'''(r_0) \dots$ und wegen $\overline{\varphi'(r)} = 0$ weiter $0 = \bar{\xi} \cdot \varphi''(r_0) + \frac{1}{2}\overline{\xi^2} \cdot \varphi'''(r_0)$. Nach Gl. (15b) ist $\bar{\xi} = \bar{\varepsilon}$ und $\overline{\xi^2} = \frac{1}{2}\xi_0^2 + O(\varepsilon^2)$, so daß man

$$\bar{\xi} = -\frac{1}{4}\xi_0^2 \frac{\varphi'''(r_0)}{\varphi''(r_0)} \quad (16)$$

erhält.

An dieser Stelle läßt sich der Volumenausdehnungskoeffizient α einführen. Nach Gl. (6) und der Definition von $\bar{r}$ bzw. $\bar{\xi}$ gilt

$$\int_0^T \alpha \cdot dT\big|_{p=0} = \frac{V_T - V_0}{V_0} = \frac{\bar{r}^3 - r_0^3}{r_0^3} = 3 \cdot \frac{\bar{\xi}}{r_0}$$

und daher weiter nach Gl. (16)

$$\alpha = -\frac{3}{4} \frac{\varphi'''(r_0)}{r_0 \cdot \varphi''(r_0)} \left(\frac{\partial \xi_0^2}{\partial T}\right)_{p=0}. \quad (17)$$

Bei Temperaturanstieg wird dem System die Wärme $Q = \int_0^T C_v \cdot dT\big|_{p=0}$ zugeführt; diese muß gleich dem Zuwachs E_T

$$E_T = N\{\varphi(r_2) - \varphi(r_0)\} = \frac{1}{2} N \cdot \xi_0^2 \cdot \varphi''(r_0)$$

an Teilchenenergie sein. Gleichsetzen von Q und E_T und Differenzieren nach T ergibt

$$C_v = \frac{1}{2} N \varphi''(r_0) \cdot \left(\frac{\partial \xi_0^2}{\partial T}\right)_{p=0}. \quad (18)$$

Einsetzen der Gl. (14), (17) und (18) in Gl. (9) führt auf

$$\gamma(V_0) = -\frac{1}{6} \frac{r_0 \cdot \varphi'''(r_0)}{\varphi''(r_0)}. \quad (19)$$

Aus Gl. (19) läßt sich die Grüneisen-Konstante γ für spezielle Wechselwirkungspotentiale $\varphi(r)$ berechnen. Man erhält beispielsweise für den

Potenzansatz nach *Mie*

$$\gamma = \frac{m+n+3}{6} \quad \text{für} \quad \varphi(r) = \frac{a}{r^m} - \frac{b}{r^n}. \tag{20}$$

Setzt man p_T nach Gl. (13) in Gl. (12) ein, so erhält man mit $r_p \to r_0$ gerade den rechts in Gl. (19) stehenden Ausdruck für γ. Damit ist die Dugdale-MacDonald-Formel Gl. (12) für $p_T \to 0$ unter stark vereinfachenden Voraussetzungen hergeleitet. Die Berechtigung ihrer Anwendung auch bei höheren Drücken ergibt sich aber im Grunde genommen nur durch den Erfolg bei der Interpretation experimenteller Daten [4]. Hierauf wird weiter unten bei den numerischen Ergebnissen nochmals eingegangen.

2.3 Hugoniot-Kurven

Ähnlich wie in Gasen und Flüssigkeiten lassen sich auch in Festkörpern durch starke äußere Druckeinwirkung Stoßwellen erzeugen. Dabei bildet sich eine Stoßfront aus, die im Sinne der Kontinuumsphysik eine Diskontinuitätsfläche darstellt, an der Druck, Temperatur und Dichte sprunghaft auf höhere Werte ansteigen. Bezüglich der experimentellen Vorgehensweise zur Erzeugung der Stoßwellen und zur meßtechnischen Erfassung der ablaufenden Vorgänge muß auf die Literatur [4, 5, 6, 17, 18, 19] verwiesen werden.

Die folgende Diskussion der Vorgänge an der Stoßfront knüpft an bekannte Darstellungen an [4, 20, 21].

Fig. 2a zeigt die Verhältnisse schematisch. Die Stoßfront bewegt sich mit der Geschwindigkeit v von rechts nach links in das ruhende Metall hinein; hinter ihr fließt das verdichtete Material mit der Geschwindigkeit u nach. Die Zustandsgrößen vor dem Stoß sind mit dem Index 0, diejenigen danach mit dem Index H bezeichnet.

Die vorliegende Arbeit bezieht sich auf den Druckbereich oberhalb 100 kbar. Gegenüber derartig hohen Drücken spielt die Festigkeit des Materials, die bei den im folgenden betrachteten reinen Metallen unterhalb $50 \, \text{kp} \cdot \text{mm}^{-2} = 5 \, \text{kbar}$ liegt, im allgemeinen nur eine untergeordnete Rolle und kann daher in den Bewegungsgleichungen in guter Näherung vernachlässigt werden. Dies bedeutet, daß man ein festes Metall bei diesen Drücken kontinuums-mechanisch als ideal fluides Medium auffassen und die Vorgänge an der Stoßfront nach einfachen, bereits aus der Gasdynamik her bekannten Gleichungen berechnen darf.

Hierzu ist es zweckmäßig, vom Laborsystem (Fig. 2a) mittels der Transformation

$$\text{a)} \; w_0 + v = 0; \quad \text{b)} \; w_H + v = u \tag{21}$$

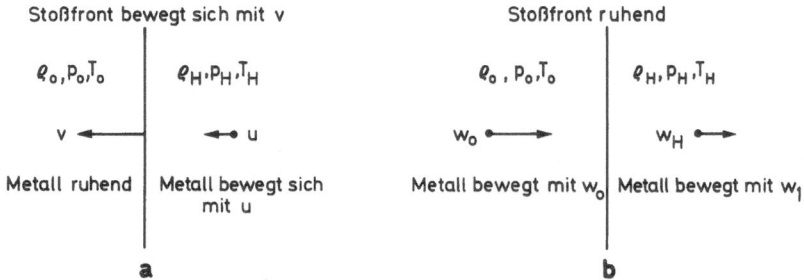

Fig. 2a u. b. Verhältnisse an der Stoßfront. a Laborsystem; b Mitbewegtes System

zu dem mit der Stoßfront mitbewegten System (Fig. 2b) überzugehen. In diesem System ist der Bewegungsablauf stationär, so daß man die Strömungsdifferentialgleichungen [22] leicht integrieren kann. Man erhält an der Stoßfront [22] (ϱ = Massendichte)

a) für die Massenerhaltung

$$\varrho_H \cdot w_H = \varrho_0 \cdot w_0, \tag{22}$$

b) für die Impulserhaltung

$$\varrho_H \cdot w_H^2 + p_H = \varrho_0 w_0^2 + p_0, \tag{23}$$

c) für die Energieerhaltung

$$\frac{w_H^2}{2} + i_H = \frac{w_0^2}{2} + i_0 \tag{24}$$

(i = Enthalpie pro Masseneinheit).

Aus den Gl. (21), (22) und (23) folgt

$$a)\ u = \left(1 - \frac{\varrho_0}{\varrho_H}\right) \cdot v; \quad b)\ p_H = \varrho_0 \cdot u \cdot v + p_0. \tag{25}$$

Die Gl. (24) und (25) lassen sich auswerten, wenn man die Zustandsgleichung des betreffenden Materials kennt. Schließlich führt dies auf die Frage nach dem funktionalen Zusammenhang zwischen den beiden Geschwindigkeiten u und v. Hierüber liegen für eine große Anzahl reiner Metalle experimentelle Daten vor [4, 17, 23]. Diese zeigen, daß sich in allen Fällen die Funktionen $v = v(u)$ über einen großen Druckbereich überraschend exakt durch eine lineare Beziehung der Form

$$v = C + S \cdot u \tag{26}$$

darstellen lassen, wobei C und S Materialkonstanten sind. Die Ursache dieser Linearität konnte bisher nicht befriedigend geklärt werden.

Aus den Gl. (25) und (26) erhält man die sogenannte Hugoniot-Kurve $p_H = p_H(\varrho_H)$. Fig. 3 zeigt als Beispiel deren Verlauf für Kupfer [17].

Die experimentell bestimmten Hugoniot-Kurven liefern einen weiteren Zusammenhang zwischen den in Gl. (11) auftretenden Funktionen $\gamma(V)$, $p_T(V)$ und $U_T(V)$, die durch Hinzunahme der Gl. (12) dann vollständig bestimmt sind:

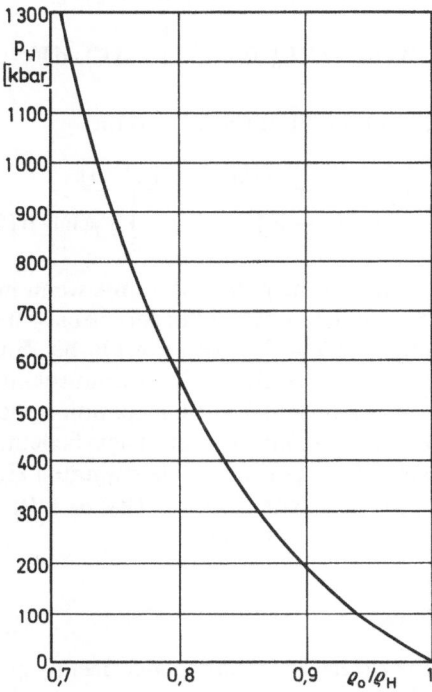

Fig. 3. Hugoniot-Kurve für Cu (nach [17])

Es sei V_0' ein Metallvolumen am absoluten Nullpunkt $T = 0$ bei verschwindendem äußeren Druck ($p = 0$). Ferner seien unter $p_H = p_H(V)$ und $U_H = U_H(V)$ die vom Zustand ($V_0'; p_0 = 0; T_0 = 0$) ausgehenden Hugoniot-Kurven im $p - V$- bzw. $U - V$-Diagramm verstanden. Dann gilt nach Gl. (11)

$$\gamma(V) = V \frac{p_H(V) - p_T(V)}{U_H(V) - U_T(V)}; \quad p_T(V_0') = 0 = p_H(V_0'). \qquad (27)$$

Der Nenner in Gl. (27) läßt sich aufspalten in

$$U_H(V) - U_T(V) = [U_H(V) - U(V_0', 0)] + [U(V_0', 0) - U_T(V)].$$

Für den ersten Term erhält man nach Gl. (24), wenn man die Energie U mittels der Definitionsgleichung $I \equiv i\varrho V = U + pV$ auf die Enthalpie I zurückführt und $p_0 = 0$ sowie $U_0(V_0', 0) = U_H(V_0')$ berücksichtigt:

$$U_H(V) - U(V_0', 0) = \frac{1}{2}(V_0' - V) \cdot p_H(V) \qquad (28)$$

und für den zweiten Term wegen $U(V_0', 0) = U_T(V_0')$

$$U(V, 0) - U(V_0', 0) = - \int_{V_0'}^{V} p_T(V) \cdot dV. \qquad (29)$$

Einsetzen der Gl. (28) und (29) in Gl. (27) ergibt:

$$\gamma(V) = \frac{V \cdot [p_H(V) - p_T(V)]}{\frac{1}{2}[V_0' - V] \cdot p_H(V) + \int_{V_0'}^{V} p_T(V) \cdot dV}; \qquad (30)$$

Gl. (30) ist auch dann noch eine gute Näherung, wenn man für das Ausgangsvolumen des betrachteten Metallkörpers bei $p = 0$ statt des Wertes V_0' am absoluten Nullpunkt $T = 0°$ K den Wert V_0 bei Zimmertemperatur $T = 293°$ K einsetzt. Die thermische Volumenausdehnung $V_0 - V_0'$ ist nämlich klein gegenüber den weiter zur Diskussion stehenden Volumenänderungen $V_0 - V$ durch Kompression. Dies bedeutet, daß man in Gl. (30) näherungsweise die experimentell bestimmten Hugoniot-Kurven $p_H(V)$ einsetzen darf, die nicht vom Zustand $(V_0'; p_0 = 0; T_0 = 0)$, sondern von $(V_0; p_0 = 0; T_0 = 293°$ K$)$ ausgehen.

2.4 Reihenentwicklungen

Für die spätere numerische Berechnung der Isentropen ist es zweckmäßig, die Hilfsgröße η

$$\eta = \frac{V_0}{V} - 1 = \frac{\varrho}{\varrho_0} - 1 \qquad (31)$$

einzuführen und alle auftretenden Funktionen von V bzw. ϱ nach Potenzen von η zu entwickeln. Da im folgenden stets $|\eta| \ll 1$ ist, so kann man sich dabei in guter Näherung jeweils auf einige wenige Reihenglieder beschränken.

Nach Voraussetzung ist $p_0 \equiv p_H(V_0) = 0$; man hat dementsprechend nach Gl. (31)

$$p_H(V) = A_H \cdot \eta + B_H \cdot \eta^2 + C_H \cdot \eta^3 + \cdots \qquad (32)$$

für die durch den Anfangspunkt $(p_0 = 0; T_0 = 293°$ K$)$ gehende Hugoniot-Kurve anzusetzen. Weiterhin gilt $p_T(V_0') = 0$ und daher, wenn man, wie oben dargelegt, V_0' mit V_0 identifiziert:

$$p_T(V) = A_T \cdot \eta + B_T \cdot \eta^2 + C_T \cdot \eta^3 + \cdots. \tag{33}$$

Bei der Funktion $\gamma(V)$ ist es sinnlos, mehr als zwei Reihenglieder zu berücksichtigen, da auch die experimentell bestimmten Daten für γ keine größere Genauigkeit aufweisen:

$$\gamma(V) = \gamma_0 + \gamma_1 \cdot \eta + \cdots. \tag{34}$$

Aus der Volumenabhängigkeit der Grüneisen-Konstante γ läßt sich diejenige der charakteristischen Temperatur Θ_D herleiten. Mit dem Reihenansatz

$$\Theta_D(V) = \Theta_{D0} + \Theta_{D1} \cdot \eta + \Theta_{D2} \cdot \eta^2 + \cdots \tag{35}$$

erhält man nach Gl. (9), (31) und (34)

a) $\Theta_{D1} = \Theta_{D0} \cdot \gamma_0;$ b) $\Theta_{D2} = \Theta_{D0} \dfrac{\gamma_0(\gamma_0 - 1) + \gamma_1}{2}.$ (36)

Gl. (12) liefert einen Zusammenhang zwischen den Koeffizienten der Gl. (33) und (34)

a) $\gamma_0 = \dfrac{B_T}{A_T};$

b) $\gamma_1 = 3\dfrac{C_T}{A_T} + 2\dfrac{B_T}{A_T} - \left(\dfrac{B_T}{A_T} + \dfrac{1}{3}\right) \cdot \left(2\dfrac{B_T}{A_T} - \dfrac{1}{3}\right)$ (37)

und Gl. (30) einen weiteren Zusammenhang zwischen denjenigen der Gl. (32), (33) und (34)

a) $A_H - A_T = 0;$ b) $B_H - B_T = 0;$ c) $C_H - C_T = \dfrac{A_H + B_H}{6} \cdot \gamma_0.$ (38)

Aus den Gl. (25) und (26) folgt in Verbindung mit Gl. (31)

$$p_H = \varrho_0 \cdot C^2 \dfrac{\eta(\eta + 1)}{[1 - \eta(S - 1)]^2}. \tag{39}$$

Daraus ergibt sich durch Reihenentwicklung für die Koeffizienten in Gl. (32)

a) $A_H = \varrho_0 \cdot C^2;$ b) $B_H = \varrho_0 \cdot C^2(2S - 1);$

c) $C_H = \varrho_0 \cdot C^2 \cdot (S - 1) \cdot (3S - 1).$ (40)

Die Koeffizienten in den Gl. (33), (34) und (35) können damit nach Gl. (36), (37) und (38) ebenfalls auf die gemessenen Daten für ϱ_0, C und S zurückgeführt werden. Man erhält speziell für γ_0 und γ_1 nach Gl. (37) und (38)

a) $\gamma_0 = 2S - 1;$ b) $\gamma_1 = -S^2 + \dfrac{1}{3}S - \dfrac{5}{9}.$ (41)

2.5 Phasenübergänge

Die bisherigen Betrachtungen setzten implizit voraus, daß die behandelten Zustandsänderungen stetig verlaufen. Tatsächlich können jedoch bei Stoßwellenexperimenten auch unstetige Phasenübergänge auftreten. Dabei kann es sich sowohl um fest-fest-Übergänge als auch um Schmelzvorgänge im Material handeln.

Die bekannteste bei Stoßwellenexperimenten beobachtete fest-fest-Umwandlung ist diejenige des Eisens bei 132 kbar. Dieser von *Bancroft*, *Peterson* und *Minshall* [24] entdeckte Phasenübergang wurde anfangs als $\alpha \to \gamma$ Umwandlung des kubisch raumzentrierten Gitters in ein kubisch flächenzentriertes gedeutet. Umfangreiche Stoßwellenexperimente [25] bei unterschiedlichen Ausgangstemperaturen haben jedoch später gezeigt, daß es sich bei der beobachteten Hochdruckmodifikation nicht um γ-Fe, sondern um eine neue Phase handelt, die einem hexagonalen Gittertyp (ε-Fe) entspricht [26, 27].

Fig. 4 zeigt die gemessene Hugoniot-Kurve (ausgezogene Linie) für Eisen (Meßwerte für $p \leq 200$ kbar nach [24]; oberer Kurventeil für $p > 200$ kbar nach [18]). Die gestrichelte Kurve erhält man, wenn man auch für $p < 200$ kbar formal nach Gl. (39) mit den in [18] im oberen Druckbereich bestimmten Konstanten C und S rechnet. Sie ist an Stelle des tatsächlichen Kurvenverlaufs so lange als hinreichende Näherung verwendbar, wie es sich nur um Zusammenhänge zwischen rein mechanischen Größen wie Druck, Dichte und Geschwindigkeit handelt. Bei der Diskussion thermischer Zustandsgrößen muß man dagegen den tatsächlichen Kurvenverlauf und die Umwandlungswärme beim Phasenübergang berücksichtigen.

Hinsichtlich des Schmelzens unter Druck finden sich in der Literatur zahlreiche theoretische und experimentelle Arbeiten, die jedoch teilweise einander widersprechen [10, 28]. Nach *Kraut* und *Kennedy* [28] sollte die Schmelzkurve dann eine besonders einfache Gestalt annehmen, wenn man die Schmelztemperatur T_s nicht über p, sondern über V aufträgt. Die genannten Autoren fanden empirisch den linearen Zusammenhang

$$T_s(V) = T_s(V_0) \cdot \left\{ 1 + c_s \frac{V_0 - V}{V_0} \right\} \qquad (42)$$

(c_s = Materialkonstante), den sie an Hand zahlreicher Meßdaten verschiedener Autoren bestätigen konnten.

Im Gegensatz zu einem fest-fest-Übergang bleibt die Phasenumwandlung beim Schmelzen praktisch ohne Einfluß auf den Verlauf der Hugoniot-Kurve im $p - V$-Diagramm; insbesondere tritt kein Knick im Kurvenverlauf wie in Fig. 4 auf. Diese Aussage ergibt sich aus einer experimentellen Arbeit von *Kormer et al.* [29] und aus theoretischen Überlegungen von *Horie* [30]. Im p-T-Diagramm treten wegen der Schmelzwärme natürlich Unstetigkeiten in der Hugoniot-Kurve auf, die auch experimentell verifiziert werden konnten [29].

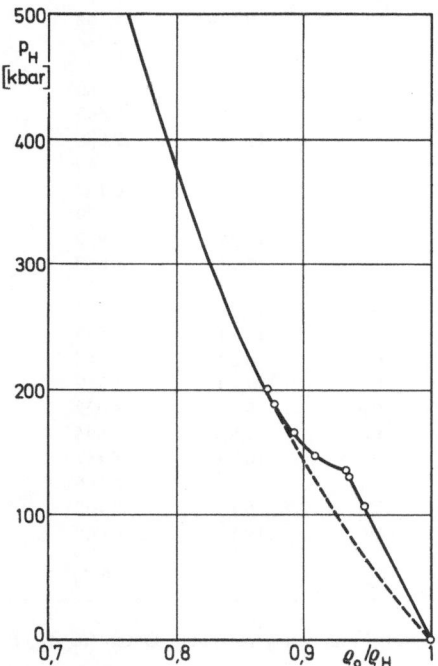

Fig. 4. Hugoniot-Kurve für Fe (nach [18] und [24])

Aus dem Gesagten folgt, daß der Einfluß eventueller Phasenumwandlungen für die Materialbewegung unter Stoßwellenbelastung nur von untergeordneter Bedeutung ist und lediglich bei der Temperaturermittlung berücksichtigt werden muß.

2.6 Numerische Ergebnisse

In Tab. 1 sind die in Stoßwellenexperimenten gemessenen Daten ϱ_0, C und S für eine größere Anzahl von Metallen zusammengestellt. Diese Metalle lassen sich hinsichtlich ihres Verhaltens bei Stoßwellenbelastung mittels Sprengstoffen in vier Gruppen einteilen, wie weiter unten gezeigt wird. Tab. 1 ist bereits entsprechend gegliedert.

Tabelle 1. *Metalldaten aus Stoßwellenexperimenten*

Gruppe	Metall	ϱ_0 [g · cm^3]	C [mm/μs]	S	Ref.
I) Alkaligruppe	K	0,862	2,088	1,154	[23]
	Li	0,534	4,142	1,186	[23]
	Na	0,971	2,446	1,283	[23]
II) Aluminiumgruppe	Al	2,785	5,460	1,318	[18]
	Cd	8,64	2,443	1,671	[17]
	Pb	11,34	2,028	1,517	[17]
	Sn	7,28	2,640	1,476	[17]
	Th	11,68	2,132	1,278	[17]
	Ti	4,51	4,779	1,089	[17]
	Tl	11,84	1,859	1,515	[17]
	Zn	7,14	3,050	1,559	[17]
III) Kupfergruppe	Ag	10,49	3,243	1,586	[17]
	Au	19,24	3,075	1,560	[17]
	Co	8,82	4,748	1,330	[17]
	Cr	7,10	5,217	1,465	[17]
	Cu	8,93	4,022	1,480	[18]
	Fe	7,85	3,694	1,848	[18]
	Mo	10,20	5,157	1,238	[17]
	Ni	8,86	4,646	1,445	[17]
	V	6,1	5,108	1,210	[17]
	W	19,17	4,005	1,268	[17]
IV) Beryllium	Be	1,86	(7,972)	(1,085)	[4]

(Für Be wurden C und S aus den gemessenen Daten für A_H, B_H und C_H ermittelt).

In Tab. 2 finden sich einige weitere Metalldaten, die im folgenden benötigt werden.

Die Gl. (12) und (30) für γ wurden aus vereinfachenden Annahmen hergeleitet. Einen Eindruck von der Güte ihrer Näherung gewinnt man, wenn man die Zahlenwerte für γ_0 nach der aus den Gl. (12) und (30) abgeleiteten Gl. (41a) mit denjenigen nach Gl. (9) vergleicht. Tab. 3 gibt dazu einen Überblick. Beim Vergleich der Zahlen muß man beachten, daß die Datenangaben der Literatur zum Volumenausdehnungskoeffizienten α teilweise um mehr als 20 % voneinander abweichen, was eine entsprechende Unsicherheit im Zahlenwert für γ nach Gl. (12) bedingt. Unter Beachtung dieser Unsicherheit erscheint die Übereinstimmung zwischen den Gl. (9) und (41) befriedigend. Wegen der erheblich größeren Meßgenauigkeit für S gegenüber derjenigen für α wird daher im folgenden γ stets nach Gl. (41) berechnet.

Tabelle 2. *Weitere Metalldaten*

Gr.	Metall	K [dyn/cm^2]	μ	α [grd^{-1}]	c_v [erg/g]	$T_s(V_0)$ [°C]	Θ_{D0} [°K]
I	K	$0{,}399 \cdot 10^{11}$	0,352	$252 \cdot 10^{-6}$	$7{,}547 \cdot 10^6$	63,2	90
	Li	$1{,}362 \cdot 10^{11}$	0,359	$168 \cdot 10^{-6}$	$34{,}063 \cdot 10^6$	180,5	370
	Na	$0{,}719 \cdot 10^{11}$	0,342	$213 \cdot 10^{-6}$	$12{,}274 \cdot 10^6$	97,8	158
II	Al	$7{,}312 \cdot 10^{11}$	0,339	$70{,}8 \cdot 10^{-6}$	$9{,}021 \cdot 10^6$	659	428
	Cd	$4{,}359 \cdot 10^{11}$	0,262	$77{,}0 \cdot 10^{-6}$	$2{,}316 \cdot 10^6$	321	188
	Pb	$4{,}036 \cdot 10^{11}$	0,434	$79{,}1 \cdot 10^{-6}$	$1{,}294 \cdot 10^6$	327,4	110
	Sn	$5{,}244 \cdot 10^{11}$	0,332	$62{,}5 \cdot 10^{-6}$	$2{,}221 \cdot 10^6$	231,9	199
	Th	$5{,}40 \cdot 10^{11}$	0,258	$40{,}5 \cdot 10^{-6}$	$1{,}178 \cdot 10^6$	1695	170
	Ti	$12{,}351 \cdot 10^{11}$	0,359	$26{,}0 \cdot 10^{-6}$	$5{,}219 \cdot 10^6$	1668	420
	Tl	$2{,}818 \cdot 10^{11}$	0,454	$86{,}45 \cdot 10^{-6}$	$1{,}292 \cdot 10^6$	303,5	87
	Zn	$5{,}838 \cdot 10^{11}$	0,235	$89{,}0 \cdot 10^{-6}$	$3{,}885 \cdot 10^6$	419,5	310
III	Ag	$9{,}967 \cdot 10^{11}$	0,363	$58{,}6 \cdot 10^{-6}$	$2{,}364 \cdot 10^6$	961,3	226
	Au	$17{,}085 \cdot 10^{11}$	0,424	$42{,}70 \cdot 10^{-6}$	$1{,}288 \cdot 10^6$	1064,8	164
	Co	$18{,}30 \cdot 10^{11}$	0,310	$45{,}8 \cdot 10^{-6}$	$4{,}184 \cdot 10^6$	1493	445
	Cr	$18{,}505 \cdot 10^{11}$	0,245	$19{,}29 \cdot 10^{-6}$	$4{,}490 \cdot 10^6$	1903	630
	Cu	$13{,}70 \cdot 10^{11}$	0,343	$50{,}7 \cdot 10^{-6}$	$3{,}859 \cdot 10^6$	1083	343
	Fe	$16{,}793 \cdot 10^{11}$	0,291	$36{,}8 \cdot 10^{-6}$	$4{,}491 \cdot 10^6$	1536	467
	Mo	$28{,}51 \cdot 10^{11}$	0,307	$14{,}6 \cdot 10^{-6}$	$2{,}478 \cdot 10^6$	2620	450
	Ni	$18{,}64 \cdot 10^{11}$	0,304	$37{,}5 \cdot 10^{-6}$	$4{,}437 \cdot 10^6$	1455	450
	V	$15{,}586 \cdot 10^{11}$	0,363	$24{,}9 \cdot 10^{-6}$	$4{,}805 \cdot 10^6$	1890	360
	W	$33{,}81 \cdot 10^{11}$	0,299	$12{,}96 \cdot 10^{-6}$	$1{,}309 \cdot 10^6$	3390	400
IV	Be	$12{,}557 \cdot 10^{11}$	0,118	$36{,}9 \cdot 10^{-6}$	$18{,}240 \cdot 10^6$	1283	1160

K = Kompressionsmodul
μ = Poissonsche Querkontraktionszahl $\bigg\}$ nach [33].
α = Volumenausdehnungskoeffizient, nach [34], K, Li, Na, V, Be nach [31].
c_v = spezifische Wärme, nach [31]; $T_s(V_0)$ = Schmelztemperatur nach [31].
Θ_{D0} = charakteristische Temperatur, nach [32].

In Tab. 4 sind die aus den Daten in Tab. 1 nach Gl. (39), (40) und (41a) folgenden Zahlenwerte für A_H, B_H, C_H und A_T, B_T, C_T zusammengestellt.

2.7 Plastische und elastische Schallgeschwindigkeit

Beim Grenzübergang zu beliebig kleinen Drucksprüngen $p_H - p_0 \to 0$ auf der Hugoniot-Kurve geht nach Gl. (25b): $u \to 0$ und daher nach Gl. (26): $v \to C$. Der Grenzwert $v = C$ für die Fortpflanzungsgeschwindigkeit kleinster Druckstörungen ist die plastische Schallgeschwindigkeit des

Tabelle 3. *Zahlenwerte zur Grüneisen-Konstante (berechnet nach Tab. 1 und 2)*

Gruppe	Metall	$\gamma = \dfrac{\alpha \cdot K \cdot V}{C_v}$	$\gamma_0 = 2S - 1$	$\gamma_1 = -S^2 + \frac{1}{3}S - \frac{5}{9}$
I	K	1,546	1,308	−1,503
	Li	1,258	1,372	−1,567
	Na	1,285	1,566	−1,774
II	Al	2,061	1,636	−1,853
	Cd	1,677	2,342	−2,791
	Pb	2,176	2,034	−2,351
	Sn	2,027	1,952	−2,242
	Th	1,590	1,556	−1,763
	Ti	1,364	1,178	−1,378
	Tl	1,593	2,030	−2,346
	Zn	1,873	2,118	−2,466
III	Ag	2,355	2,172	−2,542
	Au	2,944	2,120	−2,469
	Co	2,271	1,660	−1,881
	Cr	1,120	1,930	−2,213
	Cu	2,016	1,960	−2,252
	Fe	1,753	2,696	−3,355
	Mo	1,647	1,476	−1,676
	Ni	1,778	1,890	−2,162
	V	1,324	1,420	−1,616
	W	1,746	1,536	−1,741
IV	Be	1,366	1,170	−1,371

betreffenden Mediums; für sie gilt [22]

$$C = \sqrt{\left(\frac{\partial p}{\partial \varrho}\right)_s} \quad \text{für} \quad p_H \to p_0 \, . \tag{43}$$

Sie ist stets kleiner als die Ausbreitungsgeschwindigkeit c_{long} elastischer Kompressionswellen

$$c_{\text{long}} = \sqrt{\frac{K}{\varrho_0} \, 3 \, \frac{1-\mu}{1+\mu}} \tag{44}$$

(μ = Poissonsche Zahl der Querkontraktion), wie aus Tab. 5 hervorgeht. (Bei Tl zeigt sich eine Abweichung, die möglicherweise auf einem Meßfehler beruht.) Dies bedeutet, daß jede plastische Welle nicht zu großer Amplitude einen elastischen Vorläufer hat.

Die plastische Schallgeschwindigkeit C läßt sich auf andere Materialkonstanten zurückführen. Aus den Gl. (43) und (5) folgt

$$\varrho_0 C^2 = K + V_0 \cdot \left\{\left(\frac{\partial p}{\partial V}\right)_T - \left(\frac{\partial p}{\partial V}\right)_s\right\} \tag{45}$$

Tabelle 5. Elastische und plastische Schallgeschwindigkeiten

Gruppe	Metall	c_{long} [mm/µs] nach [33]	C [mm/µs] nach Tab. 1	C [mm/µs] nach Gl. (47)
I	K	2,597	2,088	2,274
	Li	6,033	4,142	5,208
	Na	3,305	2,446	2,831
II	Al	6,355	5,460	5,235
	Cd	2,980	2,443	2,290
	Pb	2,047	2,028	1,933
	Sn	3,300	2,640	2,731
	Th	2,849	2,132	2,169
	Ti	6,263	4,779	5,267
	Tl	1,628	1,859	1,574
	Zn	3,894	3,050	2,933
III	Ag	3,636	3,243	3,146
	Au	3,281	3,075	3,031
	Co	5,732	4,748	4,617
	Cr	6,845	5,217	5,089
	Cu	4,756	4,022	3,974
	Fe	5,952	3,694	4,664
	Mo	6,649	5,157	5,306
	Ni	5,806	4,646	4,621
	V	6,000	5,108	5,109
	W	5,319	4,005	4,194
IV	Be	12,719	7,972	8,278

Tabelle 4. Zahlenwerte für Hugoniot-Kurve und Nullpunktisotherme

Gruppe	Metall	$A_H = A_T$ [kbar]	$B_H = B_T$ [kbar]	C_H [kbar]	C_T [kbar]
I	K	37,6	49,2	14,2	−4,66
	Li	91,6	125,7	43,6	−6,10
	Na	58,1	91,0	46,8	7,93
II	Al	830	1358	780	183
	Cd	516	1208	1389	716
	Pb	466	949	856	377
	Sn	507	990	828	341
	Th	531	826	418	66
	Ti	1030	1213	208	−233
	Tl	409	831	747	328
	Zn	664	1407	1365	634
III	Ag	1103	2396	2430	1163
	Au	1819	3857	3749	1744
	Co	1988	3301	1962	499
	Cr	1932	3730	3051	1229
	Cu	1445	2831	2385	988
	Fe	1071	2888	4128	2349
	Mo	2713	4004	1752	100
	Ni	1913	3615	2838	1097
	V	1592	2260	879	−33
	W	3075	4723	2311	314
IV	Be	1182	1383	227	−274

und weiter durch einfache thermodynamische Umformung:

$$\varrho_0 C^2 = K - V_0 \cdot \left(\frac{\partial p}{\partial T}\right)_V \cdot \left(\frac{\partial T}{\partial V}\right)_S. \tag{46}$$

Nach Gl. (8b) ist $S = \text{const}$ gleichbedeutend mit

$$x = \frac{\Theta_D(V)}{T} = \text{const}$$

was in Verbindung mit Gl. (9) auf:

$$\left(\frac{\partial T}{\partial V}\right)_S = -\frac{T \cdot \gamma}{V}$$

führt. Einsetzen in Gl. (46) ergibt zusammen mit Gl. (7)

$$C = \sqrt{\frac{K}{\varrho_0}(1 + \gamma_0 \cdot \alpha \cdot T_0)}. \tag{47}$$

Die nach Gl. (47) mit den Daten der Tab. 1 und 2 errechneten Zahlenwerte für C finden sich in Tab. 5. Dabei wurde $T_0 = 300°$ K gesetzt und γ_0 nach Gl. (9) berechnet. Da in allen Fällen $\gamma_0 \alpha T_0 \ll 1$ ist, so brauchen keine hohen Genauigkeitsansprüche an die einzusetzenden Zahlenwerte für γ_0, α und T_0 gestellt zu werden.

Man erkennt, daß die nach Gl. (47) berechneten Werte für C recht gut mit den gemessenen nach Tab. 1 übereinstimmen. Die verbleibenden geringen Abweichungen lassen sich im wesentlichen damit erklären, daß für K in Gl. (47) der unter Normalbedingungen gültige Wert eingesetzt wurde. Streng genommen muß man jedoch berücksichtigen, daß der Übergang vom elastischen zum plastischen Bereich nicht bei Normaldruck, sondern erst bei einem kritischen Druck p_K erfolgt, der von der Festigkeit σ_F des Materials abhängt. In erster Näherung gilt [20]

$$p_K = \sigma_F \frac{1-\mu}{1-2\mu}. \tag{48}$$

Für K ist dementsprechend in Gl. (47) der Wert des Kompressionsmoduls beim Druck $p = p_K$ einzusetzen. Die Druckabhängigkeit von K ist von *Bridgman* [1, 2] gemessen worden. Mit dessen Daten konnte abgeschätzt werden, daß die genannte Korrektur in Gl. (47) einige Prozent ausmacht und größenordnungsmäßig dem Unterschied zu den Meßwerten nach Tab. 1 entspricht.

3. Stoßwellenausbreitung in Metallen bei äußerer Belastung

3.1 Bewegung der Metalloberfläche bei Stoßbelastung

Wird ein Metallkörper starken äußeren Stoßbelastungen ausgesetzt, so entstehen an seiner Oberfläche Stoßwellen, die in das Innere hinein-laufen. Dadurch wird das Material verdichtet und zum Fließen gebracht. Die damit verbundenen Vorgänge sollen im folgenden näher untersucht werden.

Wie eingangs bereits erwähnt, beziehen sich die Berechnungen auf den Fall, daß die äußere Stoßbelastung durch detonierende Sprengstoffe aufgebracht wird. Fig. 5 zeigt die zugrunde gelegte Anordnung schema-

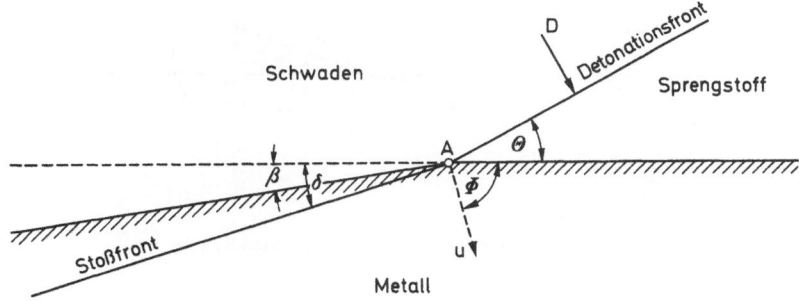

Fig. 5. Detonationsfront und Metalloberfläche

tisch. Sprengstoff- und Metallkörper seien beide als unendlich ausgedehnt gedacht. Bei der Detonation des Sprengstoffs bildet sich eine Reaktions-zone aus, die mit der Detonationsgeschwindigkeit D in den Sprengstoff hineinläuft. Sie ist erfahrungsgemäß so dünn (einige Zehntel Millimeter), daß man sie als Diskontinuitätsfläche, als „Detonationsfront" beschrei-ben kann [20], an der die Zustandsgrößen des Sprengstoffs sprunghaft in diejenigen der Schwaden übergehen. Hinter der Detonationsfront herrscht bei hochbrisanten Sprengstoffen in den Schwaden ein Druck von mehreren hundert Kilobar.

Der Winkel θ, unter dem die Detonationsfront auf die Metallober-fläche auftrifft, kann durch geeignete Anordnung der Zündung beliebig zwischen 0° und 90° vorgegeben werden. Die im Metall erzeugte Stoß-front ist für $\theta > 0$ um den Winkel $\delta > 0$ gegenüber dem ursprünglichen Verlauf der Metalloberfläche geneigt. Hinter ihr fließt das Material, wie nachstehend gezeigt wird, senkrecht zum Frontverlauf mit einer Ge-schwindigkeit u, die sich aus den Gleichungen in Abschnitt 2.3 ergibt. Dieser Materialfluß erfolgt demnach für $\theta > 0$ nicht senkrecht zum

15*

ursprünglichen Verlauf der Metalloberfläche. Das bedeutet, daß diese im jeweiligen Auftreffpunkt A der Detonationsfront um einen Winkel β abgeknickt wird.

Zur Diskussion der Winkelverhältnisse nach Abb. 5 für $\theta > 0$ ist es zweckmäßig, die Hugoniot-Kurve für die eingezeichnete Stoßfront nicht im $p - u$-Diagramm, sondern in einem $p - \beta$-Diagramm darzustellen. Die Funktion $p_H = p_H(\beta)$ hängt dabei natürlich noch vom jeweiligen Auftreffwinkel θ und von der Detonationsgeschwindigkeit D ab.

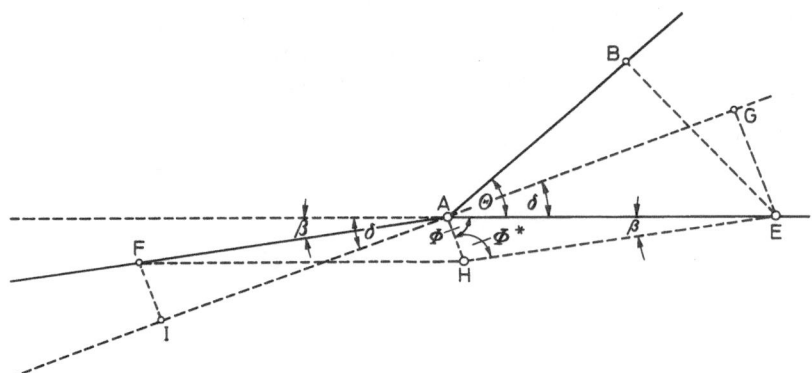

Fig. 6. Geschwindigkeiten im mitbewegten System

Fig. 6 zeigt die Geschwindigkeitsverhältnisse in einem Koordinatensystem, das mit dem Auftreffpunkt A der Detonationsfront mitbewegt wird, in dem die Vorgänge also stationär ablaufen. Die eingetragenen Strecken geben Geschwindigkeiten an, wobei der Maßstab so gewählt wurde, daß die Strecke $\overline{BE}$ gerade die Detonationsgeschwindigkeit D darstellt. Dann muß $\overline{EG}$ die Stoßwellengeschwindigkeit v im Metall sein, so daß ersichtlich

$$\frac{v}{D} = \frac{\sin \delta}{\sin \theta} \tag{49}$$

ist.

Ein Beobachter in A sieht das Material mit der Geschwindigkeit $w_0 = \overline{AE}$ von E her auf sich zukommen und mit $w_H = \overline{AF} = \overline{EH}$ nach F abfließen. Die Anwendung des Sinussatzes auf das Dreieck $\triangle AEH$ liefert

$$\frac{\sin(\Phi + \beta)}{\sin \Phi} = \frac{w_0}{w_H}. \tag{50}$$

Ein Zusammenhang zwischen w_0 und w_H ergibt sich aus den Stoßgleichungen, die hier jedoch etwas anders lauten als in Abschnitt 2.3, da die beiden Geschwindigkeiten w_0 und w_H nicht mehr senkrecht zur Stoßfront gerichtet sind.

Sind w_{0N} und w_{HN} die Normalkomponenten von w_0 bzw. w_H relativ zur Stoßfront und w_{0T} bzw. w_{HT} die Tangentialkomponenten, so lautet die Massenerhaltung analog zu Gl. (22) [22]

$$\varrho_H \cdot w_{HN} = \varrho_0 \cdot w_{0N}. \tag{51}$$

Die Impulserhaltung ist eine Vektorgleichung, die den beiden skalaren Gleichungen [22]

$$\begin{aligned} &\text{a)} \quad \varrho_H \cdot w_{HN}^2 + p_H = \varrho_0 \cdot w_{0N}^2 + p_0; \\ &\text{b)} \quad \varrho_H \cdot w_{HN} \cdot w_{HT} = \varrho_0 \cdot w_{0N} \cdot w_{0T} \end{aligned} \tag{52}$$

entspricht. Die Energieerhaltung ist identisch mit Gl. (24).

Aus den Gl. (51) und (52b) folgt: $w_{HT} = w_{0T}$ und damit weiter nach Fig. 6: $w_0 \cdot \cos\delta = w_H \cdot \cos(\delta - \beta)$. Kombination mit Gl. (50) ergibt

$$\theta + \delta = 90°. \tag{53}$$

Damit bestätigt sich, daß im Laborsystem die Materialflußgeschwindigkeit u, die in Fig. 6 der Strecke $\overline{AH}$ entspricht, senkrecht zur Stoßfront gerichtet ist. Dies bedeutet, daß die Zustandsänderungen an der Stoßfront unabhängig vom Winkel θ den in Abschnitt 2.3 ff. abgeleiteten Beziehungen gehorchen.

Aus den Gl. (49), (50) und (53) erhält man, wenn man noch $w_0 = \overline{AE} = D/\sin\theta$ berücksichtigt

$$\cos\beta = \frac{w_0^2 - v^2 \pm v \cdot \sqrt{v^2 + w_H^2 - w_0^2}}{w_0 \cdot w_H}. \tag{54}$$

Wegen $w_{HT} = w_{0T}$ ist: $w_H^2 = w_{HN}^2 + w_{0T}^2$. Für w_{0T}^2 gilt nach Fig. 6 ersichtlich: $w_{0T}^2 = \overline{AG}^2 = \overline{AE}^2 - \overline{EG}^2 = w_0^2 - v^2$ und für w_{HN}^2 nach Gl. (52): $w_{HN} = w_{0N} - \dfrac{p_H - p_0}{\varrho_0 \cdot w_{0N}}$. Mit $p_0 = 0$ und $w_{0N} = \overline{EG} = v$ ergibt sich somit

$$w_H^2 = w_0^2 + \left(\frac{p_H}{\varrho_0 \cdot v}\right)^2 - 2\frac{p_H}{\varrho_0}. \tag{55}$$

Im Grenzfall $p_H = 0$ muß natürlich $\beta = 0$, d. h. $\cos\beta = 1$ sein. Man erkennt leicht, daß dies nur möglich ist, wenn in Gl. (54) das positive Wurzelzeichen genommen wird. Aus den Gl. (54) und (55) folgt dann als Gleichung der Hugoniot-Kurve $p_H = p_H(\beta)$ bzw. $\beta = \beta(p_H)$ im $p - \beta$-Diagramm

$$\sin\beta = \frac{p_H \cdot \sin\theta}{\varrho_0 \cdot D} \sqrt{\frac{\dfrac{1}{v^2} - \dfrac{\sin^2\theta}{D^2}}{1 - 2\dfrac{p_H \cdot \sin^2\theta}{\varrho_0 \cdot D^2} + \left(\dfrac{p_H \cdot \sin\theta}{\varrho_0 \cdot v \cdot D}\right)^2}}. \tag{56}$$

Dabei wurde wieder $w_0 = D/\sin\theta$ eingesetzt. Die Stoßwellengeschwindigkeit v ist nach Gl. (25b) und (26) eine Funktion von p_H.

3.2 Sprengstoffdaten

Die Zustandsänderungen bei der Detonation von Sprengstoffen lassen sich ähnlich denjenigen beim Stoßwellendurchgang durch Metalle in enger Anlehnung an die entsprechenden Gleichungen der Gasdynamik [22] berechnen. Analog zur Vorgehensweise in Abschnitt 2.3 seien die Verhältnisse in einem mit der Detonationsfront mitbewegten Koordinatensystem betrachtet. Die Gl. (21), (22), (23) und (24) lassen sich dann sinngemäß übertragen. Entsprechend Gl. (25) gilt dann

$$\text{a)} \; u^* = \left(1 - \frac{\varrho_0^*}{\varrho^*}\right) \cdot D; \quad \text{b)} \; p^* = \varrho_0^* \cdot u^* \cdot D + p_0^*. \tag{57}$$

Dabei sind die auf den Sprengstoff bezogenen Zustandsgrößen zur Unterscheidung von den entsprechenden der Metalle mit * bezeichnet; im übrigen ist die Bezeichnungsweise dieselbe wie dort. Größen mit dem Index 0 beziehen sich auf den (nicht reagierten) Sprengstoff, solche ohne Index auf die Schwaden hinter der Detonationsfront.

Weiterhin folgt sinngemäß aus den Gl. (22), (23) und (24)

$$i^* - i_0^* = \frac{1}{2}\left(\frac{1}{\varrho^*} + \frac{1}{\varrho_0^*}\right)(p^* - p_0^*). \tag{58}$$

Die Enthalpiedifferenz $i^* - i_0^*$ in Gl. (58) enthält auch die bei der Detonation frei werdende Reaktionswärme.

Die Gl. (57) und (58) bestimmen die Hugoniot-Kurve $p^* = p^*(\varrho^*)$ der Detonation. Die Werte für Druck und Dichte in den Schwaden liegen auf der Hugoniot-Kurve und werden durch die Zusatzbedingung bestimmt, daß dort der Entropiezuwachs gegenüber dem Ausgangszustand (p_0^*, ϱ_0^*) zum Minimum wird [22]. Diese Bedingung geht auf *Chapman* und *Jouguet* zurück. Mit der zu Gl. (43) analogen Gleichung für die Schallgeschwindigkeit c^* in den Schwaden führt sie nach Gl. (58) auf

$$-\varrho_{\text{CJ}}^{*2} \cdot c_{\text{CJ}}^{*2} = \frac{p_{\text{CJ}}^* - p_0^*}{\dfrac{1}{\varrho_{\text{CJ}}^*} - \dfrac{1}{\varrho_0^*}}. \tag{59}$$

Der Index CJ deutet an, daß es sich um denjenigen Zustandspunkt (p^*, ϱ^*) auf der Hugoniot-Kurve handelt, der der Chapman-Jouguet-Bedingung gehorcht.

Aus den Gl. (57) und (59) folgt

$$\varrho_{\text{CJ}}^* \cdot c_{\text{CJ}}^* = \varrho_0^* \cdot D \tag{60}$$

was nach Gl. (57 a) auf

$$D = u_{CJ}^* + c_{CJ}^* \tag{61}$$

führt.

In den folgenden Abschnitten hat man es mit Stoß- und Verdünnungswellen in den Sprengstoffschwaden zu tun. Die damit verbundenen Zustandsänderungen innerhalb der Schwaden sind praktisch isentrop und lassen sich näherungsweise gut durch die Polytropengleichung

$$p^* \sim \varrho^{*\gamma^*} \tag{62}$$

Tabelle 6. *Meßwerte für Sprengstoffgemische*

Mischungsverhältnis Hexogen:TNT	ϱ_0^* [g·cm^{-3}]	D [mm·μs^{-1}]	p_{CJ}^* [kbar]
100 : 0	1,767	8,639	337,9
77 : 23	1,743	8,252	312,5
64 : 36	1,713	8,018	292,2
0 :100	1,637	6,942	189,1

darstellen [20]. Aus dieser folgt mit der zu Gl. (43) analogen Gleichung für die Schallgeschwindigkeit c^*

$$c^* = \sqrt{\gamma^* \frac{p^*}{\varrho^*}}. \tag{63}$$

Den folgenden numerischen Berechnungen wird Composition B (60:40) als Sprengstoff zugrunde gelegt; das ist ein Gemisch aus 60% Hexogen und 40% Trinitrotoluol (TNT), dem aus technischen Gründen häufig noch ein geringer Prozentsatz inerter Stoffe beigefügt ist.

Die Eigenschaften verschiedener Hexogen-TNT-Gemische sind von *Deal* [35] gemessen worden; die von ihm ermittelten Daten für ϱ_0^*, D und p_{CJ}^* zeigt Tab. 6. Trägt man sie über dem Mischungsverhältnis Hexogen:TNT auf, so lassen sich jeweils glatte Kurvenzüge mit minimaler Streuung hindurch legen, auf denen sich durch Interpolation die Werte für das Mischungsverhältnis 60:40 ablesen lassen. Aus diesen erhält man die übrigen Sprengstoffdaten mittels vorstehender Gleichungen. Sukzessive ergibt sich so

a) u_{CJ}^* nach Gl. (57 b) mit $p_0^* = 0$,
b) c_{CJ}^* nach Gl. (61),
c) ϱ_{CJ}^* nach Gl. (60),
d) γ^* nach Gl. (63).

Tab. 7 zeigt die so ermittelten Daten für das Mischungsverhältnis
60:40, mit denen weiterhin gerechnet wird. Sie können als repräsentativ
für den Sprengstoff Composition B (60:40) angesehen werden. Beim
Vergleich mit experimentellen Untersuchungen muß jedoch berücksichtigt werden, daß kleinere Abweichungen auftreten können, die durch
die Porosität des Sprengstoffs und seiner inerten Zusätze bedingt sind
und von Fall zu Fall variieren.

Tabelle 7. *Sprengstoffdaten für Composition B (60:40)*

$$\varrho_0^* = 1{,}715 \text{ g} \cdot \text{cm}^{-3}; \qquad \varrho_{CJ}^* = 2{,}333 \text{ g} \cdot \text{cm}^{-3}$$
$$p_{CJ}^* = 287{,}5 \text{ kbar}; \qquad D = 7{,}956 \text{ mm} \cdot \mu\text{s}^{-1}$$
$$u_{CJ}^* = 2{,}107 \text{ mm} \cdot \mu\text{s}^{-1}; \qquad c_{CJ}^* = 5{,}849 \text{ mm} \cdot \mu\text{s}^{-1}$$
$$\gamma^* = 2{,}776$$

3.3 Senkrecht auftreffende Detonationsfront

Die Verhältnisse sind besonders einfach, wenn sich die Detonationsfront
senkrecht auf die Metalloberfläche zu bewegt, d. h. wenn $\theta = 0°$ ist (vgl.
Fig. 5). In diesem Fall ist auch $\beta = 0°$ und $\delta = 0°$.

Im Metall herrschen hinter der Stoßfront der Druck p_H und die
senkrecht zur Oberfläche gerichtete Materialflußgeschwindigkeit u_H. Der
Zustandspunkt (p_H, u_H) liegt auf der Hugoniot-Kurve nach Gl. (25)
und (26).

In den Sprengstoffschwaden herrschen hinter der Detonationsfront
der Druck p_{CJ}^* und die Materialflußgeschwindigkeit u_{CJ}^*. Beim Auftreffen
der Detonationsfront auf die Metalloberfläche wird dort je nach den
Eigenschaften des Metalls, entweder eine Stoßwelle oder aber eine
Verdünnungswelle in die Schwaden zurück reflektiert. Durch eine reflektierte Stoßwelle werden die Schwaden weiter verdichtet; infolge-
dessen herrscht hinter der Stoßfront, d. h. in dem der Metalloberfläche
zugekehrten Schwadenbereich, ein Druck $p_s^* > p_{CJ}^*$; die Materialflußgeschwindigkeit u_s^* in Richtung auf das Metall zu wird dadurch gegenüber dem Wert u_{CJ}^* vor dem Stoß erniedrigt. Im Fall der reflektierten
Verdünnungswelle gilt umgekehrt $p_s^* < p_{CJ}^*$ und $u_s^* > u_{CJ}^*$.

Es ist leicht einzusehen, daß im Gleichgewicht Geschwindigkeit u
und Druck p in beiden Medien jeweils gleich, d. h. $u_s^* = u_H$ und $p_s^* = p_H$
sein müssen. Die Bedingung $u_s^* = u_H$ ist trivial, da sich zwischen Metall
und Schwaden kein leerer Raum ausbilden kann. Andererseits würde
ein Druck $p_s^* \neq p_H$ so lange die Stoßwelle verstärken bzw. abschwächen,
bis Kräftegleichgewicht $p_s^* = p_H$ herrscht. Der sich einstellende Gleich-
gewichtszustand $(p_\perp, u_\perp)$ ergibt sich damit einfach als Schnittpunkt der

Hugoniot-Kurve $p_H = p_H(u_H)$ des jeweiligen Metalls mit der „Reflexionskurve" $p_s^* = p_s^*(u_s^*)$ der Sprengstoffschwaden.

Im Hinblick auf die spätere Diskussion von Mehrfachreflexionen wird die Reflexionskurve $p_s^* = p_s^*(u_s^*)$ gleich unter allgemeineren Voraussetzungen hergeleitet. Die reflektierte Welle bewege sich in einem Schwadenbereich beliebig vorgebbaren Zustands $(p_1^*, \varrho_1^*, u_1^*)$ hinein; der Fall $p_1^* = p_{CJ}^*, \varrho_1^* = \varrho_{CJ}^*, u_1^* = u_{CJ}^*$ ist darin als Spezialfall enthalten. Hinter der reflektierten Welle herrsche der Zustand $(p_s^*, \varrho_s^*, u_s^*)$.

Es muß zwischen den beiden Kurvenästen $p_s^* \geqq p_1^*$ und $p_s^* \leqq p_1^*$ unterschieden werden. Im Fall $p_s^* \geqq p_1^*$ hat man es mit einer reflektierten Stoßwelle zu tun, auf die sich die Gl. (22) und (23) sinngemäß anwenden lassen. Versteht man unter v_1^* die Geschwindigkeit der Stoßfront im Laborsystem, so sind $w_s^* = u_s^* + v_1^*$ und $w_1^* = u_1^* + v_1^*$ die Materialflußgeschwindigkeiten in dem mit der Stoßfront mitbewegten System. In diesem gilt nach Gl. (22) und (23)

$$ w_1^* - w_s^* = \sqrt{(p_s^* - p_1^*)\left(\frac{1}{\varrho_1^*} - \frac{1}{\varrho_s^*}\right)}. \tag{64}$$

In Verbindung mit der Polytropengleichung (62) und wegen $w_1^* - w_s^* = u_1^* - u_s^*$ folgt aus Gl. (64)

$$ (u_1^* - u_s^*)^2 = \frac{p_1^*}{\varrho_1^*}\left[\frac{p_s^*}{p_1^*} - 1\right]\left[1 - \left(\frac{p_1^*}{p_s^*}\right)^{\frac{1}{\gamma^*}}\right]. \tag{65}$$

Im Fall $p_s^* \leqq p_1^*$ hat man es mit einer reflektierten Verdünnungswelle zu tun. Bei dieser erfolgt die Zustandsänderung nicht sprunghaft wie bei einer Stoßfront, sondern kontinuierlich über eine längere Wegstrecke. Ist x die Raumrichtung, in die sich die Welle bewegt, so lauten die differentiellen Bewegungsgleichungen im mitbewegten System [22]

$$ \text{a) } \frac{d(\varrho^* w^*)}{dx} = 0 \quad \text{b) } \varrho^* w^* \frac{dw^*}{dx} + \frac{dp^*}{dx} = 0. \tag{66}$$

Die Gl. (66) sind insofern nicht ganz korrekt, als es streng genommen gar kein mitbewegtes Bezugssystem gibt, in dem die Bewegung völlig stationär verläuft. Die Schallgeschwindigkeit als Relativgeschwindigkeit zum Laborsystem hängt nämlich selbst noch von Druck und Dichte ab und ist daher nicht räumlich konstant. Dieser Umstand, der sich in einer Abflachung der Verdünnungswelle mit der Zeit äußert, spielt jedoch für die Verhältnisse an der Grenzfläche Schwaden: Metall nur eine untergeordnete Rolle und soll daher hier ignoriert werden.

Aus den Gl. (66) folgt

$$ w^* \frac{d\varrho^*}{dp^*} = -\varrho^* \frac{dw^*}{dp^*}. \tag{67}$$

In der Verdünnungswelle erfolgen die Zustandsänderungen isentrop, so daß man Gl. (43) für die Schallgeschwindigkeit c^* anwenden kann.

Wegen der vorausgesetzten Stationarität im mitbewegten System ist außerdem $dw^* = du^*$. Multipliziert man Gl. (67) mit dw^*/dp^* und eliminiert anschließend $w^* \, dw^*/dp^*$ mittels Gl. (66b), so erhält man

$$u_s^* = u_1^* \pm \int\limits_{p_1^*}^{p_s^*} \frac{dp^*}{\varrho^* \cdot c^*}. \tag{68}$$

Die auf das Metall hin gerichtete Schwadengeschwindigkeit u_s^* ist um so größer, je kleiner p_s^* ist; es muß also das Minuszeichen vor dem Integral in Gl. (68) gelten.

Mittels der Gl. (60), (62) und (63) läßt sich Gl. (68) integrieren und umformen zu

$$u_s^* = u_1^* + \frac{2\gamma^*}{\gamma^{*2}-1} D \left(\frac{p_1^*}{p_{CJ}^*}\right)^{\frac{\gamma^*-1}{2\gamma^*}} \left[1 - \left(\frac{p_s^*}{p_1^*}\right)^{\frac{\gamma^*-1}{2\gamma^*}}\right]. \tag{69}$$

Mit $p_1^* = p_{CJ}^*$ erhält man nach Gl. (65) und (69) die beiden Äste der in Fig. 7 für den Sprengstoff Comp. B (60:40) dargestellten Reflexions-

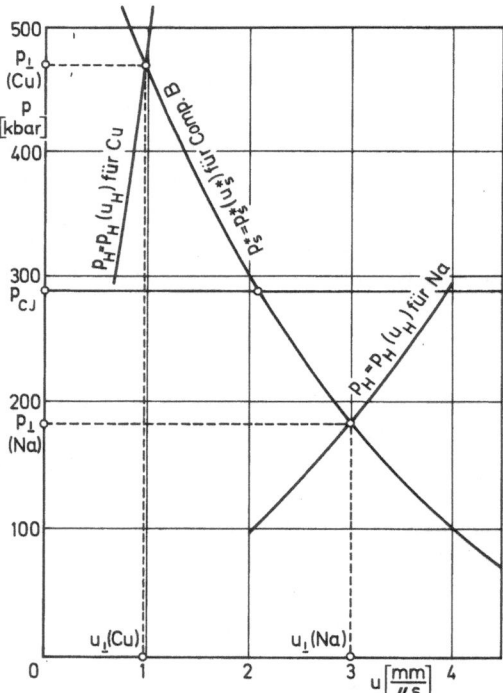

Fig. 7. Bestimmung von $(p_\perp, u_\perp)$

kurve $p_s^* = p_s^*(u_s^*)$. Diese wurde zur Ermittlung der Gleichgewichtswerte $(p_\perp, u_\perp)$ mit den Hugoniot-Kurven $p_H = p_H(u_H)$ der einzelnen Metalle zum Schnitt gebracht. Fig. 7 zeigt als Beispiele Kupfer und Natrium.

Die insgesamt für Composition B (60:40) ermittelten Zahlenwerte $p_\perp, u_\perp, \eta_\perp$ und $T_\perp$ sind in Tab. 8 zusammengestellt. Dabei wurde $\eta_\perp$ nach Gl. (39) aus $p_\perp$ errechnet. Zur Bestimmung von $T_\perp$ wurde von Gl. (11) in Verbindung mit Gl. (3) ausgegangen, wobei $p_T(\eta_\perp)$ und $\gamma(\eta_\perp)$ nach Gl. (33) bzw. (34) gebildet wurde.

Wie man erkennt, zeichnen sich die Metalle der Gruppe I dadurch aus, daß bei ihnen eine Verdünnungswelle in die Schwaden reflektiert wird ($p_\perp < p_{CJ}^* = 287,5$ kbar); bei allen anderen Gruppen wird eine Stoßwelle reflektiert ($p_\perp > p_{CJ}^*$).

Die Zahl $\eta_\perp$ erweist sich für K, Li und Na als so groß, daß die Näherungsdarstellungen für $p_T(\eta_\perp)$ und $\gamma(\eta_\perp)$ nach Gl. (33) bzw. (34) nicht mehr gut genug sind, um als Basis für weitere Berechnungen dienen zu können. Insbesondere ergeben sich keine sinnvollen Werte mehr für $T_\perp$.

Die Genauigkeit der Angaben für $T_\perp$ ist insgesamt gesehen schlecht, wie beispielsweise der sinnlose Wert für Be zeigt. Abgesehen von den Unsicherheiten in der stark eingehenden Grüneisen-Konstante γ rührt

Tabelle 8. *Numerische Ergebnisse im Fall senkrecht auftreffender Detonationsfront*

Gruppe	Metall	$p_\perp$[kbar]	$u_\perp$[mm/µs]	$\eta_\perp$	$T_\perp$[°K]
I	K	161,5	3,233	1,247	
	Li	149,5	3,380	0,709	
	Na	182,5	2,995	0,909	
II	Al	353,2	1,659	0,277	608
	Cd	437,5	1,157	0,359	(1939)
	Pb	450,9	1,083	0,419	(3139)
	Sn	417,5	1,270	0,391	(2176)
	Th	449,4	1,091	0,448	2973
	Ti	396,1	1,395	0,284	617
	Tl	450,4	1,086	0,449	(3481)
	Zn	427,6	1,212	0,325	1047
III	Ag	475,2	0,953	0,251	766
	Au	530,7	0,670	0,194	495
	Co	482,2	0,916	0,181	338
	Cr	468,3	0,989	0,174	426
	Cu	472,1	0,969	0,216	446
	Fe	458,8	1,041	0,227	(706)
	Mo	504,0	0,803	0,150	319
	Ni	482,7	0,913	0,181	365
	V	443,5	1,124	0,210	377
	W	545,5	0,597	0,143	310
IV	Be	330,1	1,811	0,223	237

diese schlechte Genauigkeit daher, daß man zur Berechnung von $T_\perp$ von der kleinen Differenz zweier großer Zahlen $p_\perp$ und $p_T(\eta_\perp)$ ausgeht, die ihrerseits mit Näherungsfehlern behaftet sind. Für diejenigen Metalle, die beim Stoßwellendurchgang eine Phasenumwandlung erleiden, wurde zudem der Einfluß der Umwandlungswärmen auf $T_\perp$ nicht berücksichtigt; ihre $T_\perp$-Werte sind in Tab. 8 eingeklammert.

3.4 Tangential entlanglaufende Detonationsfront

Bei tangential zur Metalloberfläche laufender Detonationsfront, d. h. für $\theta = 90°$ (vgl. Fig. 5), erfolgt stets eine seitliche Expansion der Sprengstoffschwaden, verbunden mit einem Abknicken der Metalloberfläche ($\beta > 0$). Der Gleichgewichtsdruck $p_{||}$ in Metall und Schwaden ist hier immer kleiner als p_{CJ}^*.

Metalle der Gruppe IV (bei Verwendung von Composition B ist dies nur Be) zeichnen sich dadurch aus, daß ihre plastische Schallgeschwindigkeit C größer ist als die Detonationsgeschwindigkeit D des verwendeten Sprengstoffs. Daher läuft parallel zur Metalloberfläche eine Druckwelle der Detonationsfront voraus, so daß man es mit einem Unterschallvorgang im Metall zu tun hat. Die folgenden Betrachtungen beschränken sich daher auf die Gruppen I bis III.

Im Prinzip ist die Vorgehensweise ähnlich wie in Abschnitt 3.3 für $\theta = 0°$. Der Gleichgewichtsdruck $p_{||}$ bestimmt sich dadurch, daß die Reflexionskurve der Sprengstoffschwaden mit der Hugoniot-Kurve des betreffenden Metalls zum Schnitt gebracht wird. Eine zusätzliche Schwierigkeit tritt hier jedoch dadurch auf, daß Gleichheit der Materialflußgeschwindigkeiten in Metall und Schwaden nur für deren Komponenten senkrecht zur abgeknickten Metalloberfläche gefordert werden kann. Wegen des Fehlens einer genügend starken Haftreibung an der Grenzfläche können nämlich beide Medien aneinander entlang gleiten. Man müßte daher die Kurven $p_s^* = p_s^*(u_s^*)$ und $p_H = p_H(u_H)$ zuvor auf die entsprechenden Normalkomponenten von u_s^* bzw. u_H umrechnen. Anstelle dieser etwas umständlichen Vorgehensweise werden im folgenden die beiden Kurven statt im p-u-Diagramm, im p-β-Diagramm zum Schnitt gebracht. Da zwischen Metall und Schwaden kein leerer Raum entstehen kann, müssen nämlich auch die Abknickwinkel β beider Medien einander gleich sein.

Die Hugoniot-Kurve $p_H = p_H(\beta)$ der Metalle wird durch Gl. (56) gegeben. Zur Berechnung der Reflexionskurve $p_s^* = p_s^*(\beta)$ der Schwaden betrachte man den Materialfluß in einem mit der Abknickstelle A mitbewegten Bezugssystem. Fig. 8 zeigt die Verhältnisse schematisch. Ein Beobachter in A sieht den unverbrannten Sprengstoff mit der Geschwindigkeit $D = \overline{AE}$ von E her auf sich zukommen und die verdünnten Schwaden mit $w_s^* = \overline{AC}$ nach C abfließen.

Die Strömung der Schwaden um die bei A liegende Ecke herum kann als Prandtl-Meyer-Strömung [22] behandelt werden. Hierzu führt man Zylinderkoordinaten (r, φ, z) ein und unterteilt den Schwadenbereich in zwei Teilbereiche II und III. In III sind Druck und Geschwindigkeit konstant $(p^* = p_s^*; w^* = w_s^*)$. Der Druckabfall vom Wert p_{CJ}^* unmittelbar hinter der Detonationsfront auf den Wert p_s^* erfolgt kontinuierlich in II,

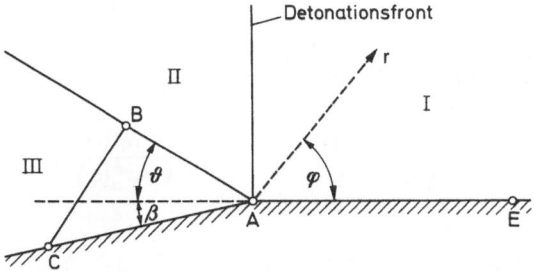

Fig. 8. Winkelverhältnisse für $\theta = 90°$. Bereich I: Sprengstoff; Bereich II: Verdünnungs-fächer; Bereich III: Schwaden konstanter Dichte

dem sogenannten Verdünnungsfächer. Man kann dort voraussetzen, daß in der Nähe von A alle Größen praktisch nur von φ, nicht aber auch von r und z abhängen und daß keine Geschwindigkeitskomponenten in z-Richtung auftreten. Aus den Vektorgleichungen für Massen- und Impulserhaltung [22]

$$\text{a) } \operatorname{div}(\varrho^* w^*) = 0,$$
$$\text{b) } \varrho^* \{\tfrac{1}{2} \operatorname{grad} w^{*2} - w^* \times \operatorname{rot} w^*\} + \operatorname{grad} p^* = 0 \tag{70}$$

$(w^* = \text{Geschwindigkeitsvektor})$ folgt dann

$$\text{a) } \varrho^* w_r^* + \frac{\mathrm{d}(\varrho^* w_\varphi^*)}{\mathrm{d}\varphi} = 0; \quad \text{b) } w_\varphi^* = \frac{\mathrm{d}w_r^*}{\mathrm{d}\varphi};$$
$$\text{c) } \varrho^* w_\varphi^* \left(w_r^* + \frac{\mathrm{d}w_\varphi^*}{\mathrm{d}\varphi} \right) + \frac{\mathrm{d}p^*}{\mathrm{d}\varphi} = 0 \tag{71}$$

$(w_r^*, w_\varphi^* = \text{Komponenten von } w^*)$.

Unter Verwendung der Polytropengleichung (62) lassen sich die Funktionen $p^*(\varphi)$, $w_r^*(\varphi)$ und $w_\varphi^*(\varphi)$ aus den Gl. (71) eliminieren, so daß man eine Differentialgleichung für $\varrho^*(\varphi)$ erhält

$$\frac{\gamma^* + 1}{2} \frac{1}{\varrho^*} \frac{\mathrm{d}^2 \varrho^*}{\mathrm{d}\varphi^2} + \frac{(\gamma^* - 3)(\gamma^* + 1)}{4} \frac{1}{\varrho^{*2}} \left(\frac{\mathrm{d}\varrho^*}{\mathrm{d}\varphi} \right)^2 + 1 = 0. \tag{72}$$

Diese läßt sich einmal integrieren zu

$$\gamma^* \frac{1}{\varrho^*} \frac{d\varphi^*}{d\varphi} \equiv \frac{1}{p^*} \frac{dp^*}{d\varphi} = -\frac{2\gamma^*}{\sqrt{\gamma^{*2}-1}} \tan\left\{\left|\sqrt{\frac{\gamma^*-1}{\gamma^*+1}} (\varphi - \bar{\varphi})\right\} \right. \tag{73}$$

und ein zweites Mal zu

$$p^*(\varphi) = \overline{p^*} \left|\cos\left[\left|\sqrt{\frac{\gamma^*-1}{\gamma^*+1}} (\varphi - \bar{\varphi})\right]\right|^{\frac{2\gamma^*}{\gamma^*-1}} \tag{74}$$

$(\bar{\varphi}, \overline{p^*} = \text{Integrationskonstanten}).$

Weiterhin ergibt sich der Zusammenhang

$$\frac{1}{p^*(\varphi)} \frac{dp^*(\varphi)}{d\varphi} = -\frac{2\gamma^*}{\gamma^*+1} \frac{w_r^*(\varphi)}{w_\varphi^*(\varphi)} \tag{75}$$

sowie

a) $w_\varphi^*(\varphi) = \dfrac{\varrho_0^*}{\varrho_{CJ}^*} D \left(\dfrac{\overline{p^*}}{p_{CJ}^*}\right)^{\frac{\gamma^*-1}{2\gamma^*}} \cos\left[\left|\sqrt{\dfrac{\gamma^*-1}{\gamma^*+1}} (\varphi - \bar{\varphi})\right]\right.$

$\hspace{11cm}$ (76)

b) $w_r^*(\varphi) = \dfrac{\varrho_0^*}{\varrho_{CJ}^*} \sqrt{\dfrac{\gamma^*+1}{\gamma^*-1}} D \left(\dfrac{\overline{p^*}}{p_{CJ}^*}\right)^{\frac{\gamma^*-1}{2\gamma^*}} \sin\left[\left|\sqrt{\dfrac{\gamma^*-1}{\gamma^*+1}} (\varphi - \bar{\varphi})\right].\right.$

Randbedingungen für $p^*(\varphi)$, $w_r^*(\varphi)$ und $w_\varphi^*(\varphi)$ liefert die Forderung, daß unmittelbar hinter der Detonationsfront, d. h. für $\varphi = 90°$ die Chapman-Jouguet-Bedingungen gelten müssen.

a) $p^*\left(\dfrac{\pi}{2}\right) = p_{CJ}^*;$ b) $w_r^*\left(\dfrac{\pi}{2}\right) = 0;$ c) $w_\varphi^*\left(\dfrac{\pi}{2}\right) = D - u_{CJ}^*.$ (77)

Mit diesen ergeben sich die Integrationskonstanten nach Gl. (74) und (76) zu

a) $\overline{p^*} = p_{CJ}^*;$ b) $\bar{\varphi} = \dfrac{\pi}{2}.$ (78)

Die Grenze zwischen den Teilbereichen II und III ist durch $\varphi = \pi - \vartheta$ definiert. Dort muß der Druck p^* stetig in den im Teilbereich III herr-schenden Druck p_s^* übergehen, so daß nach Gl. (74) und (78)

$$p_s^* = p_{CJ}^* \left|\cos\left[\left|\sqrt{\frac{\gamma^*-1}{\gamma^*+1}} \left(\frac{\pi}{2} - \vartheta\right)\right]\right|^{\frac{2\gamma^*}{\gamma^*-1}} \tag{79}$$

ist.

Andererseits ist nach Fig. 8 ersichtlich: $\tan(\beta + \vartheta) = \dfrac{\overline{BC}}{\overline{AB}} = \dfrac{w_\varphi^*}{w_r^*}$ und damit nach Gl. (73) und (75)

$$\tan(\beta + \vartheta) = \sqrt{\frac{\gamma^* - 1}{\gamma^* + 1}} \left| \cot\left[\sqrt{\frac{\gamma^* - 1}{\gamma^* + 1}} \left(\frac{\pi}{2} - \vartheta \right) \right] \right|. \tag{80}$$

Elimination von ϑ aus den beiden Gl. (79) und (80) führt auf die gesuchte Reflexionskurve $p_s^* = p_s^*(\beta)$ der Sprengstoffschwaden.

Der Schnittpunkt dieser Reflexionskurve mit den jeweiligen Hugoniot-Kurven $p_H = p_H(\beta)$ der Metalle liefert die Gleichgewichtswerte $p_{\parallel}$ und $\beta_{\parallel}$. Aus $p_{\parallel}$ erhält man dann $u_{\parallel}$ nach Gl. (25) und (26) und $\eta_{\parallel}$ nach Gl. (18). Die so ermittelten Daten sind in Tab. 9 zusammengestellt.

Tabelle 9. *Numerische Ergebnisse im Fall tangential entlanglaufender Detonationsfront*

Gruppe	Metall	$p_{\parallel}$ [kbar]	$\beta_{\parallel}$ [°]	$u_{\parallel}$ [mm/µs]	$\eta_{\parallel}$
II	Al	176,3	4,1	0,944	0,164
	Cd	177,4	4,0	0,597	0,210
	Pb	180,4	3,8	0,554	0,240
	Sn	171,3	4,3	0,653	0,221
	Th	181,2	3,8	0,548	0,240
	Ti	178,8	3,9	0,714	0,147
	Tl	179,9	3,9	0,561	0,261
	Zn	177,4	4,0	0,619	0,182
III	Ag	195,6	3,0	0,468	0,133
	Au	215,3	2,1	0,314	0,097
	Co	206,8	2,4	0,440	0,090
	Cr	206,3	2,5	0,490	0,090
	Cu	198,9	2,8	0,472	0,111
	Fe	192,1	3,2	0,525	0,127
	Mo	216,7	2,0	0,378	0,072
	Ni	206,3	2,5	0,441	0,091
	V	197,4	2,9	0,559	0,107
	W	224,8	1,7	0,270	0,066

3.5 Schräg auftreffende Detonationsfront im Winkelbereich der Verdünnungswellenreflexion

Geht man von $\theta = 90°$ aus zu kleineren Auftreffwinkeln θ (vgl. Fig. 5) über, so steigt der Gleichgewichtsdruck p_g in Metall und Schwaden von $p_g = p_{\parallel}$ aus an. Er bleibt dabei zunächst unterhalb p_{CJ}^*, so daß man es weiterhin mit einer reflektierten Verdünnungswelle in den Schwaden zu tun hat, die sich nach den Gleichungen des Verdünnungsfächers berechnen läßt.

Ein mitbewegter Beobachter in A (vgl. Fig. 6) sieht den Sprengstoff mit der Geschwindigkeit $w_\mathrm{I} = \overline{\mathrm{AE}} = \dfrac{D}{\sin\theta}$ von E her auf sich zukommen.

Für die Komponenten von w_I normal und tangential zur Detonationsfront gilt $w_\mathrm{IN} = \overline{\mathrm{BE}} = D$ und $w_\mathrm{IT} = \overline{\mathrm{AB}} = \dfrac{D\cos\theta}{\sin\theta}$. Wendet man die Gl. (51) und (52) sinngemäß auf die Verhältnisse an der Detonationsfront an, so erhält man für die entsprechenden Komponenten in den Schwaden $w_\mathrm{IIN} = \varrho_0^*/\varrho_\mathrm{CJ}^* \cdot w_\mathrm{IN}$ und $w_\mathrm{IIT} = w_\mathrm{IT}$. In Verbindung mit Gl. (60) und (61) folgen daraus die Randbedingungen für $\varphi = \theta$:

$$\text{a) } p^*(\theta) = p_\mathrm{CJ}^*; \quad \text{b) } w_r^*(\theta) = -D\,\frac{\cos\theta}{\sin\theta}; \quad\quad (81)$$
$$\text{c) } w_\varphi^*(\theta) = D - u_\mathrm{CJ}^*.$$

Es ist nun aber nicht möglich, weiterhin analog wie in Abschnitt 3.4 vorzugehen und dabei lediglich die Randbedingungen Gl. (77) durch Gl. (81) zu ersetzen. Auf diese Weise erhielte man nämlich ein sinnloses Ergebnis. Man erkennt das sehr schnell, wenn man Gl. (81) in Gl. (75) einsetzt. Es wäre dann $\dfrac{\mathrm{d}p^*}{\mathrm{d}\varphi} > 0$ für $\varphi = \theta$; ein solcher Anstieg des Schwadendrucks im Verdünnungsfächer hinter der Detonationsfront ist aber physikalisch nicht verständlich.

Diese Anomalie tritt nicht auf, wenn der Schwadenbereich, wie in Fig. 9 dargestellt, nicht in zwei, sondern in drei Teilbereiche II, III und IV unterteilt wird. Druck und Geschwindigkeit sind in II und IV jeweils konstant; es ist $p^* = p_s^*$ in IV und $p^* = p_\mathrm{CJ}^*$ in II. Der Verdünnungsfächer

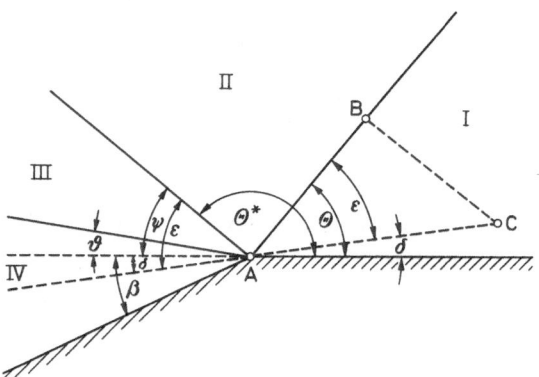

Fig. 9. Winkelverhältnisse im Fall des Verdünnungsfächers. Bereich I: Sprengstoff; Bereich II: Schwaden konstanter Dichte; Bereich III: Verdünnungsfächer; Bereich IV: Schwaden konstanter Dichte

umfaßt nur noch den Teilbereich III, wobei sich der Winkel θ^* aus den Randbedingungen bestimmt.

Aus den Gl. (74) und (76a) folgt

$$w_\varphi^*(\varphi) = \frac{\varrho_0^*}{\varrho_{CJ}^*} D \left[\frac{p^*(\varphi)}{p_{CJ}^*} \right]^{\frac{\gamma^*-1}{2\gamma^*}} . \tag{82}$$

Für $\varphi = \theta^*$ muß $p^* = p_{CJ}^*$ und damit nach Gl. (82) in Verbindung mit Gl. (60) und (61): $w_\varphi^* = D - u_{CJ}^*$, d. h. nach Gl. (81c): $w_\varphi^*(\theta^*) = w_\varphi^*(\theta)$ sein. Wegen $w^{*2} = w_\varphi^{*2} + w_r^{*2} = \text{const}$ in III folgt daraus: $w_r^*(\theta^*) = \pm w_r^*(\theta)$, wobei jedoch nur das Minuszeichen einem Winkel $\theta^* \neq \theta$ entspricht und daher hier zu nehmen ist. Damit lauten die Randbedingungen des Verdünnungs-fächers III für $\varphi = \theta^*$

a) $p^*(\theta^*) = p_{CJ}^*$;   b) $w_r^*(\theta^*) = D \dfrac{\cos\theta}{\sin\theta}$;

c) $w_\varphi^*(\theta^*) = D - u_{CJ}^*$.
$\tag{83}$

Durch die Symmetriebedingungen $w_r^*(\theta^*) = -w_r^*(\theta)$ und $w_\varphi^*(\theta^*) = w_\varphi^*(\theta)$ werden die in Fig. 9 dargestellten Winkelbeziehungen festgelegt. Dort entspricht der Vektor $\overrightarrow{CA}$ der Strömungsgeschwindigkeit w im Teil-bereich II. Es gilt ersichtlich $\varepsilon - \psi = \theta - \varepsilon = \delta$ mit $\psi = 180° - \theta^*$ sowie $\tan\varepsilon = \overline{BC}/\overline{AB} = -w_\varphi^*(\theta)/w_r^*(\theta)$.

Damit ergibt sich in Verbindung mit den Gl. (81), (60) und (61) als Be-stimmungsgleichung für θ^*:

a) $\theta^* = 180° + \theta - 2\varepsilon$;   b) $\tan\varepsilon = \dfrac{\varrho_0^*}{\varrho_{CJ}^*} \tan\theta$.
$\tag{84}$

Im Grenzfall $\theta = 90°$ ist speziell $\varepsilon = 90°$ und somit $\theta^* = \theta = 90°$.

Nach Gl. (83) errechnen sich die Integrationskonstanten $\overline{p^*}$ und $\overline{\varphi}$ in den Gl. (74) und (76) aus

a) $\tan\left\{ \sqrt{\dfrac{\gamma^*-1}{\gamma^*+1}} \, (\theta^* - \overline{\varphi}) \right\} = \dfrac{\varrho_{CJ}^*}{\varrho_0^*} \sqrt{\dfrac{\gamma^*-1}{\gamma+1}} \, \dfrac{1}{\tan\theta}$,

b) $\dfrac{1}{\overline{p^*}} = \dfrac{1}{p_{CJ}^*} \left\{ \cos\left[\sqrt{\dfrac{\gamma^*-1}{\gamma^*+1}} \, (\theta^* - \overline{\varphi}) \right] \right\}^{\frac{2\gamma^*}{\gamma^*-1}}$.
$\tag{85}$

Die Grenze zwischen den Teilbereichen III und IV ist durch $\varphi = \pi - \vartheta$ gegeben. Dort ist $p^* = p_s^*$, und es gilt daher nach Gl. (74) und (85)

$$p_s^* = p_{CJ}^* \left\{ \frac{\cos\left[\sqrt{\dfrac{\gamma^*-1}{\gamma+1}} \, (\pi - \overline{\varphi} - \vartheta) \right]}{\cos\left[\sqrt{\dfrac{\gamma^*-1}{\gamma^*+1}} \, (\theta^* - \overline{\varphi}) \right]} \right\}^{\frac{2\gamma^*}{\gamma^*-1}} . \tag{86}$$

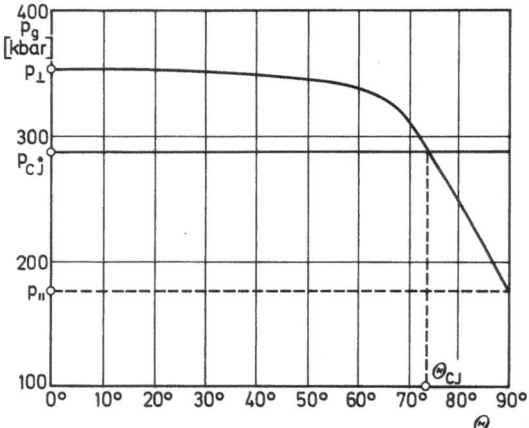

Fig. 10. $p_g = f(\theta)$ für Aluminium bei Verwendung von Composition B (60:40)

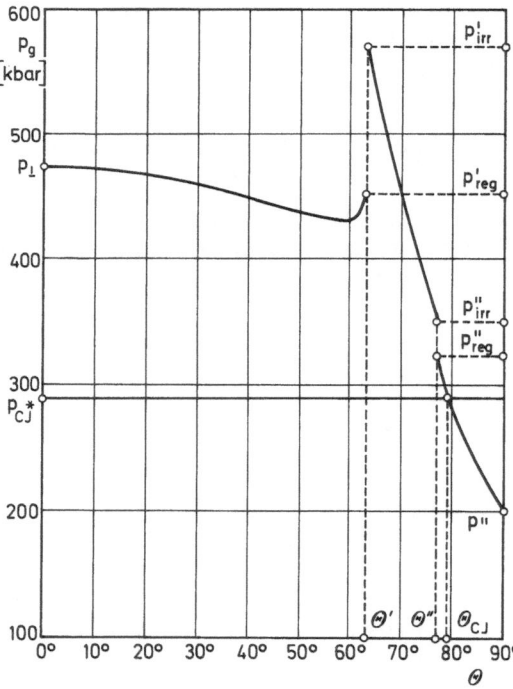

Fig. 11. $p_g = f(\theta)$ für Kupfer bei Verwendung von Composition B (60:40)

Elimination von $\overline{\varphi}$ und ϑ aus den Gl. (85a), (86) und (80) führt auf die gesuchte Reflexionskurve $p_s^* = p_s^*(\beta)$ der Sprengstoffschwaden. Dabei ergibt sich für $\theta \to 0$ als Grenzfall wieder Gl. (69), wie sich zeigen läßt.

Der Schnittpunkt der Kurven $p_s^* = p_s^*(\beta)$ und $p_H = p_H(\beta)$ liefert die Gleichgewichtswerte p_g und β_g in Metall und Schwaden. In Fig. 10 ist der Gleichgewichtsdruck p_g für Aluminium und in Fig. 11 für Kupfer in Abhängigkeit vom Auftreffwinkel θ aufgetragen. Man erkennt, daß p_g im Winkelbereich des Verdünnungsfächers mit kleiner werdendem θ sehr rasch ansteigt und bei einem Grenzwinkel θ_{CJ} den Wert p_{CJ}^* erreicht.

Für $\theta < \theta_{CJ}$ wird keine Verdünnungswelle mehr von der Metalloberfläche aus in die Schwaden reflektiert, sondern eine Stoßwelle. In Tab. 10 sind die θ_{CJ}-Werte der einzelnen Metalle zusammengestellt.

Tabelle 10. *Grenzwinkel* θ_{CJ}

Gruppe II		Gruppe III	
Metall	θ_{CJ} [°]	Metall	θ_{CJ} [°]
Al	73,8	Ag	78,3
Cd	73,4	Au	82,6
Pb	74,4	Co	80,8
Sn	71,0	Cr	80,9
Th	74,4	Cu	79,1
Ti	72,7	Fe	77,6
Tl	74,1	Mo	82,8
Zn	73,1	Ni	80,8
		V	78,6
		W	84,2

3.6 Schräg auftreffende Detonationsfront im Winkelbereich der regulären Stoßwellenreflexion

Im Winkelbereich $0 \leq \theta \leq \theta_{CJ}$ muß man verschiedene Teilbereiche unterscheiden, da die Stoßwellenreflexion in den Schwaden auf zweierlei Weise, regulär oder irregulär erfolgen kann.

Eine reguläre Reflexion entspricht den in Fig. 12 dargestellten Verhältnissen. Detonationsfront und reflektierte Stoßfront laufen in einem Punkt A an der Metalloberfläche zusammen. Ein in A mitbewegter Beobachter sieht den unverbrannten Sprengstoff mit der Geschwindigkeit $w_I = \overline{AE} = D/\sin\theta$ von E her auf sich zukommen und die Schwaden im Bereich II mit $w_{II} = \overline{AF}$ nach F hin abfließen. Zwischen den Komponenten von w_I und w_{II} normal und tangential zur Detonationsfront

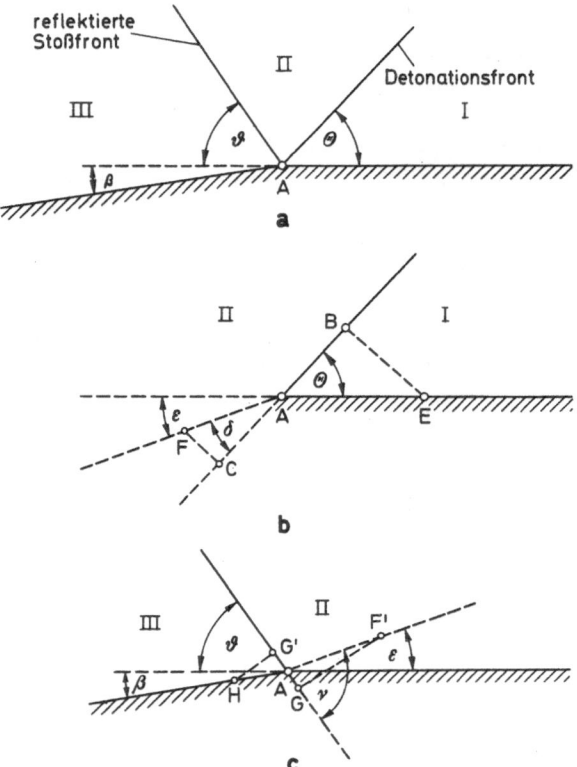

Fig. 12a–c. Winkelverhältnisse im Fall der regulären Stoßwellenreflexion. a Lage der Stoßfront; b Schwadengeschwindigkeit in II; c Schwadengeschwindigkeit in III

bestehen wiederum die Beziehungen

$$w_{IIN} = \frac{\varrho_0^*}{\varrho_{CJ}^*} \, w_{IN} \quad \text{und} \quad w_{IIT} = w_{IT}$$

entsprechend Gl. (51) und (52). Andererseits gilt ersichtlich $\tan\theta = w_{IN}/w_{IT}$ und $\tan\delta \equiv \tan(\theta - \varepsilon) = w_{IIN}/w_{IIT}$, so daß man

$$\tan(\theta - \varepsilon) = \frac{\varrho_0^*}{\varrho_{CJ}^*} \tan\theta \tag{87}$$

als Bestimmungsgleichung für den Hilfswinkel ε erhält. Für den Geschwindigkeitsbetrag w_{II} ergibt sich

$$w_{II} = \frac{c_{CJ}^*}{\sin(\theta - \varepsilon)} . \tag{88}$$

Hinsichtlich des Übergangs von II nach III sieht der in A mitbewegte Beobachter die Schwaden in II mit der Geschwindigkeit $w_{II} = \overline{AF} = \overline{AF'}$ von F' her auf sich zukommen und in III mit $w_{III} = \overline{AH}$ nach H hin abfließen. Durch sinngemäße Anwendung der Stoßgleichungen (51) und (52) erhält man in Verbindung mit der Polytropengleichung (62) für die Komponente w_{IIn} von w_{II} normal zur reflektierten Stoßfront

$$w_{IIn} = \sqrt{\frac{p_{II}^*}{\varrho_{II}^*} \frac{\dfrac{p_{III}^*}{p_{II}^*} - 1}{1 - \left(\dfrac{p_{II}^*}{p_{III}^*}\right)^{1/\gamma^*}}} \, . \tag{89}$$

Andererseits gilt ersichtlich mit $\overline{GF'} = w_{IIn}$ und $\overline{AF'} = w_{II}$

$$\sin v = \frac{w_{IIn}}{w_{II}} \, . \tag{90}$$

Kombination der Gl. (88), (89) und (90) ergibt mit $\varrho_{II}^* = \varrho_{CJ}^*$, $p_{II}^* = p_{CJ}^*$ und $p_{III}^* = p_s^*$ zusammen mit Gl. (63)

$$a) \; \sin v = A^* \sin(\theta - \varepsilon); \quad b) \; A^* = \sqrt{\frac{1}{\gamma^*} \frac{\dfrac{p_s^*}{p_{CJ}^*} - 1}{1 - \left(\dfrac{p_{CJ}^*}{p_s^*}\right)^{1/\gamma^*}}} \, . \tag{91}$$

Weiterhin ist $\tan(\beta + \vartheta) = w_{IIIn}/w_{IIIt}$ und $\tan v = w_{IIn}/w_{IIt}$, wobei w_{IIt} und w_{IIIt} die Tangentialkomponenten von w_{II} bzw. w_{III} bezüglich der reflektierten Stoßfront sind. Durch sinngemäße Anwendung der Stoßgleichungen (51) und (52) sowie der Polytropengleichung (62) erhält man daher

$$\tan(\beta + v - \varepsilon) = \left(\frac{p_{CJ}^*}{p_s^*}\right)^{1/\gamma^*} \tan \varepsilon \, . \tag{92}$$

Elimination der Hilfswinkel ε und v aus den Gl. (87), (91) und (92) führt auf die gesuchte Reflexionskurve $p_s^* = p_s^*(\beta)$ der Schwaden.

Bringt man diese mit der jeweiligen Hugoniot-Kurve $p_H = p_H(\beta)$ des Metalls zum Schnitt, so findet man zu jedem Auftreffwinkel θ in der Regel zwei verschiedene Schnittpunkte (p_g, β_g). Davon entspricht der eine einem stabilen, der andere einem labilen Gleichgewicht zwischen Metall und Schwaden. Da stets $dp_s^*/d\beta < 0$ ist, so läßt sich leicht einsehen, daß stabiles Gleichgewicht dann herrscht, wenn $dp_H/d\beta > 0$ in $\beta = \beta_g$ ist. Wenn sich nämlich beispielsweise infolge einer momentanen Störung der Klappwinkel β vergrößert, so sinkt wegen $dp_s^*/d\beta < 0$ der wirksame Schwadendruck p_s^* ab. Dadurch wiederum sinkt der Druck p_H im Metall, wodurch wegen $dp_H/d\beta > 0$ der Winkel β wieder verkleinert, die Störung also rückgängig gemacht wird.

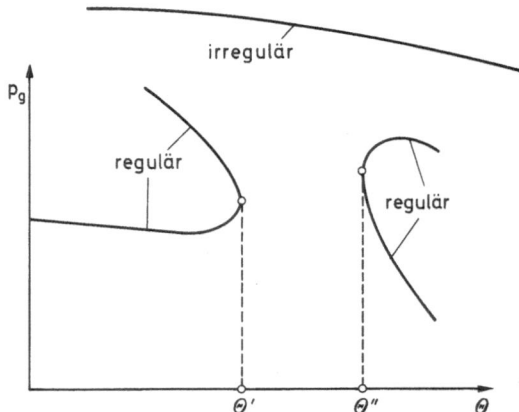

Fig. 13. Regulärer und irregulärer Kurvenverlauf $p_g = f(\theta)$

Die Stabilitätsbedingung $dp_H/d\beta > 0$ ist, wie sich abschätzen läßt, bei allen diskutierten Metallen für den kleineren der beiden Schnittwinkel β_g erfüllt. Dem kleineren Winkel β_g entspricht auch der kleinere der beiden Druckwerte p_g.

Die in Fig. 10 dargestellte Funktion $p_g = f(\theta)$ für Aluminium bezieht sich auf den stabilen Gleichgewichtswert p_g. Für $p_g = p^*_{CJ}$ geht der Kurvenanteil der regulären Stoßwellenreflexion stetig in denjenigen der Verdünnungswellenreflexion über. Im Grenzfall $\theta \to 0$ ist $p_g \to p_\perp$.

Geschlossene Kurvenzüge, wie in Fig. 10, ergeben sich für alle Metalle der Gruppe II. Bei den Metallen der Gruppe III sind die Verhältnisse verwickelter, wie Fig. 13 schematisch zeigt. Dort zerfällt der Winkelbereich regulärer Stoßwellenreflexion in zwei voneinander getrennte Teilbereiche $0 \leq \theta \leq \theta'$ und $\theta'' \leq \theta \leq \theta_{CJ}$. Für $\theta = \theta'$ und $\theta = \theta''$ fallen jeweils stabiles und labiles Gleichgewicht zusammen; für $\theta' < \theta < \theta''$ ergibt sich überhaupt keine Lösung (p_g, β_g) mehr. Da andererseits aber jeder beliebige Auftreffwinkel θ, also auch $\theta' < \theta < \theta''$ vorgegeben kann, so muß in diesem Winkelbereich eine andersartige, im folgenden als irregulär bezeichnete Stoßwellenreflexion stattfinden.

3.7 Schräg auftreffende Detonationsfront im Winkelbereich der irregulären Stoßwellenreflexion

Es konnte gezeigt werden [36], daß es sich bei der im Winkelbereich $\theta' < \theta < \theta''$ auftretenden irregulären Stoßwellenreflexion um eine MACH-Reflexion handelt. Dort bildet sich die Stoßfront als „Gabelstoß" aus, wie Fig. 14 schematisch zeigt. Detonationsfront und Stoßfront laufen in einem Tripelpunkt A' oberhalb der Metalloberfläche zusammen. Dort gabelt sich die Detonationsfront in zwei Äste I und II. Der Ast II ist gekrümmt, da der Schwadendruck von A' nach A nicht konstant ist.

Interessanterweise geht die Zustandsänderung an der Stoßfront in die nachfolgende Berechnung der Reflexionskurve $p_s^* = p_s^*(\beta)$ überhaupt nicht explizit ein. Es wird lediglich die Aussage verwertet, daß ein gekrümmter Kurvenzug AA' existiert, auf dem Detonationsfront und re-

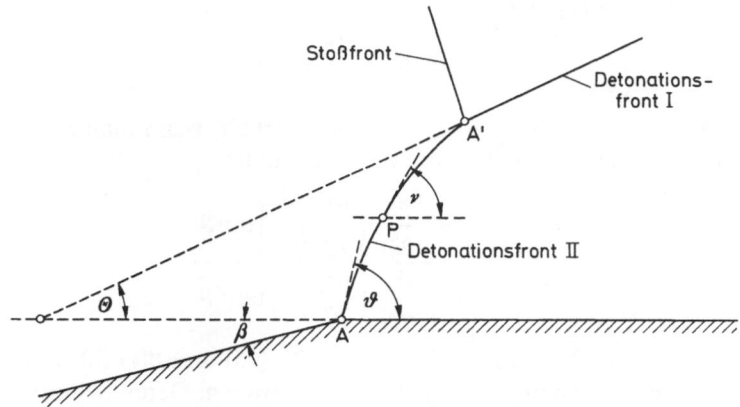

Fig. 14. Winkelverhältnisse im Fall der irregulären Stoßwellenreflexion

flektierte Stoßfront gewissermaßen zusammenfallen. Die Bedingung dieses Zusammenfallens liefert eine Aussage über den Winkel ϑ. Der Winkel θ tritt in den Gleichungen auf, weil sich im Laborsystem der Punkt A in Fig. 14 mit der gleichförmigen Geschwindigkeit $D/\sin\theta$ längs der Metalloberfläche bewegt.

Es seien die Verhältnisse in einem beliebigen Punkt P der Detonationsfront II betrachtet. Dort herrsche in den Schwaden der Druck p^* und die Dichte ϱ^*. Rechts von der Detonationsfront II liegt der Sprengstoff im Ausgangszustand vor. Betrachtet man wiederum die Strömung im mitbewegten Bezugssystem, so gilt für die Komponente w_{0N} der Strömungsgeschwindigkeit w_0 des Sprengstoffs normal zur Stoßfront in P

$$w_{0N} = D\,\frac{\sin\nu}{\sin\theta}\,. \tag{93}$$

Sinngemäße Anwendung der Stoßgleichungen (51) und (52) liefert mit $p^* \gg p_0^*$:

$$p^* = \varrho_0^*\,w_{0N}^2\left(1 - \frac{\varrho_0^*}{\varrho^*}\right). \tag{94}$$

Aus den Gl. (93) und (94) erhält man in Verbindung mit der Polytropengleichung (62) einen funktionalen Zusammenhang zwischen p^* und v
für jeden beliebigen Punkt P der Detonationsfront II. Speziell für P = A
ist $p^* = p_s^*$ und $v = \vartheta$; man erhält dort

$$p_s^* = \varrho_0^* D^2 \left[1 - \frac{\varrho_0^*}{\varrho_{CJ}^*} \left(\frac{p_{CJ}^*}{p_s^*} \right)^{1/\gamma^*} \right] \frac{\sin^2 \vartheta}{\sin^2 \theta}. \tag{95}$$

Analog zu Gl. (87) gilt hier

$$\tan(\vartheta - \beta) = \frac{\varrho_0^*}{\varrho_s^*} \tan \vartheta. \tag{96}$$

Gl. (96) läßt sich nach $\tan \beta$ auflösen; man erhält, wenn man ϱ_s^* mittels
der Polytropengleichung (62) auf p_s^* zurückführt

$$\tan \beta = \frac{\left[1 - \dfrac{\varrho_0^*}{\varrho_{CJ}^*} \left(\dfrac{p_{CJ}^*}{p_s^*} \right)^{1/\gamma^*} \right] \tan \vartheta}{1 + \dfrac{\varrho_0^*}{\varrho_{CJ}^*} \left(\dfrac{p_{CJ}^*}{p_s^*} \right)^{1/\gamma^*} \tan^2 \vartheta}. \tag{97}$$

Elimination des Hilfswinkels ϑ aus den Gl. (95) und (97) führt auf die
gesuchte Reflexionskurve $p_s^* = p_s^*(\beta)$ der Schwaden. Deren Schnittpunkt
(p_g, β_g) mit der jeweiligen Hugoniot-Kurve $p_H = p_H(\beta)$ des Metalls liefert
den Gleichgewichtsdruck p_g. Wie in Fig. 13 angedeutet, verläuft die so
ermittelte Kurve $p_g = f(\theta)$ oberhalb derjenigen für die reguläre Stoßwellenreflexion. Dies bedeutet, daß sich die auf ihr liegenden Zustände
für $\theta < \theta'$ und $\theta > \theta''$ im allgemeinen nicht verwirklichen lassen, da dort
aus Stabilitätsgründen der niedrigere Druck der regulären Reflexion be-
vorzugt ist. Ob sich in Sonderfällen durch spezielle experimentelle Maß-
nahmen diese Werte dennoch erreichen lassen, ist ungeklärt. Im Winkelbereich $\theta' < \theta < \theta''$ jedoch, in dem überhaupt keine reguläre Reflexion
möglich ist, stellt die irreguläre die einzige mögliche Lösung dar. Der
sich einstellende Gleichgewichtsdruck p_g muß daher für $\theta = \theta'$ und $\theta = \theta''$
unstetig vom regulären Kurvenast auf den irregulären überspringen.
    Für die Metalle der Gruppe III ergibt sich damit ein Kurvenverlauf
$p_g = f(\theta)$, wie er in Fig. 11 für Kupfer dargestellt ist. Die errechneten
Kenngrößen der Sprungstellen sind in Tab. 11 zusammengestellt.
    Eine ähnliche Berechnung der Funktion $p_g = f(\theta)$ wurde von *Stern-
berg* u. *Piacesi* [37] für die Beschleunigung von Eisen durch Pentolit
(50:50) durchgeführt. Die genannten Autoren verwendeten keine formel-
mäßige Darstellung der Hugoniot-Kurve in Eisen, sondern gingen unmittelbar von den Meßdaten aus und führten die Integration im Fall
der Verdünnungswellenreflexion numerisch durch. Ihr Ergebnis steht
mit der vorliegenden Arbeit im Einklang.

Tabelle 11. *Kritische Daten der Metalle der Gruppe III*

Metall	θ' [°]	p'_{reg} [kbar]	p'_{irr} [kbar]	θ'' [°]	p''_{reg} [kbar]	p''_{irr} [kbar]
Ag	63,9	437,8	548,7	75,6	329,5	362,0
Au	57,1	548,4	720,2	82,2	299,8	308,9
Co	60,5	488,4	628,0	79,9	308,4	323,5
Cr	60,6	484,4	625,5	80,0	307,4	323,2
Cu	63,0	451,1	569,6	77,1	321,3	347,5
Fe	65,6	415,8	515,9	74,0	339,3	381,6
Mo	57,9	533,1	695,7	82,4	299,3	307,5
Ni	60,4	489,1	630,0	79,8	308,4	324,3
V	65,1	423,0	524,4	76,0	327,1	358,5
W	55,5	581,9	768,1	84,0	294,6	299,8

3.8 Klassifizierung der Metalle

Die bisherigen Diskussionen zur Funktion $p_g = f(\theta)$ haben gezeigt, daß man die betrachteten Metalle in mehrere Gruppen mit unterschiedlichen Eigenschaften unterteilen kann. Die Zugehörigkeit zu den einzelnen Gruppen wurde dabei lediglich als Tatsache registriert. Im folgenden werden physikalische Kriterien für die Zuordnung formuliert, die sich dann auch auf Fälle übertragen lassen, in denen andere Sprengstoffe als Composition B (60:40) verwendet werden.

Bei den Metallen der Gruppe I wird bei jedem Auftreffwinkel θ in $0° \leq \theta \leq 90°$ eine Verdünnungswelle in die Schwaden reflektiert; bei allen übrigen gibt es einen Winkelbereich $0 \leq \theta < \theta_{CJ}$, in dem eine Stoßwelle reflektiert wird. Grenzfall der Gruppe I ist $\theta_{CJ} = 0°$, d. h. $p_\perp = p_{CJ}^*$. Für diesen Grenzfall gilt einerseits nach Gl. (25) und (26): $p_{CJ}^* = \varrho_0 \cdot u_{CJ}^*$ $(C + S \cdot u_{CJ}^*)$ und andererseits nach Gl. (57b): $p_{CJ}^* = \varrho_0^* \cdot u_{CJ}^* \cdot D$. Daraus folgt als Kriterium für die Zugehörigkeit eines Metalls zur Gruppe I

$$\varrho_0 \cdot C \cdot \left(1 + \frac{S}{C} u_{CJ}^*\right) < \varrho_0^* \cdot D \qquad (98)$$

(Zugehörigkeit zu I).

Gl. (98) ist gleichbedeutend damit, daß das Impedanzverhältnis $\varrho_0^* \cdot D / \varrho_0 \cdot v$ für Metalle der Gruppe I kleiner als Eins ist. Nach dem Sprachgebrauch der Akustik sind die Sprengstoffschwaden „schallhärter" als die Metalle der Gruppe I.

Es ist einleuchtend, daß das Impedanzverhältnis $\varrho_0^* \cdot D / \varrho_0 \cdot v$ auch für die Beurteilung der Zugehörigkeit zur Gruppe II bzw. III von Bedeutung ist. In Fig. 15 sind die kritischen Winkel θ' und θ'' für die be-

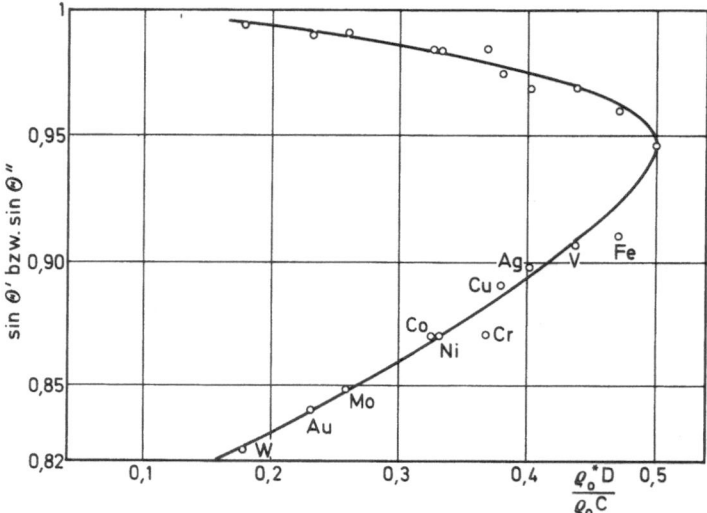

Fig. 15. Kritische Winkel für die Metalle der Gruppe III. Oberer Kurvenast: $\sin\theta''$. Unterer Kurvenast: $\sin\theta'$

rechneten Metalle der Gruppe III über diesem Impedanzverhältnis aufgetragen. Dabei wurde $\varrho_0^* \cdot D/\varrho_0 \cdot v$ näherungsweise durch $\varrho_0^* \cdot D/\varrho_0 \cdot C$ ersetzt, da für Auftreffwinkel θ in der Größenordnung θ' und θ'' im allgemeinen $u \ll v$, d. h. nach Gl. (26) $v \approx C$ ist. Man erkennt, daß die einzelnen Punkte in guter Näherung auf einem geschlossenen Kurvenzug liegen, wobei die beiden Äste für θ' und θ'' etwa bei $\varrho_0^* \cdot D/\varrho_0 \cdot C = 0{,}5$, entsprechend $\sin\theta = 0{,}947$, ineinander übergehen. Daraus läßt sich näherungsweise als Kriterium für die Zugehörigkeit eines Metalls zur Gruppe III folgern

$$\varrho_0 \cdot C > 2\varrho_0^* \cdot D \qquad\qquad (99)$$
(Zugehörigkeit zu III).

Alle Metalle, die keines der beiden Kriterien Gl. (98) und (99) erfüllen, gehören zur Gruppe II. Man erkennt aus Fig. 15, daß der Winkelbereich $\theta' < \theta < \theta''$ der irregulären Stoßwellenreflexion desto größer ist, je größer $\varrho_0 \cdot C$, d. h. je schallhärter das Metall ist.

Während für die Zugehörigkeit zu den Gruppen I, II und III die Schallhärte des Metalls maßgebend ist, ist dies bezüglich IV die plastische Schallgeschwindigkeit C. Die Gruppe IV fügt sich daher nicht in das Ordnungsschema der übrigen Gruppen ein und überlagert sich diesen teilweise. Bei ihr treten für Auftreffwinkel θ mit $\sin\theta > D/C$ nur Unterschallvorgänge im Metall auf; im restlichen Winkelbereich für θ lassen sich aber wiederum die Kriterien Gl. (98) und (99) anwenden. Für Be ergibt sich $\varrho_0^* \cdot D/\varrho_0 \cdot C = 0{,}92$; das bedeutet, daß sich Be für diese Winkel θ wie ein Metall der Gruppe II verhält.

4. Beschleunigung von Metallplatten durch Stoßwellenbelastung

4.1 Entlastungswellen in Metallen

Gelangt eine durch einen metallischen Körper hindurchlaufende Stoßwelle an die freie Oberfläche, so erfolgt dort eine Entspannung des Materials, die zu einer rücklaufenden Entlastungswelle führt. Die damit verbundenen Zustandsänderungen erfolgen nicht sprunghaft wie bei Stoßwellen, sondern stetig und können daher als isentrop angesehen werden.

Druck p und Dichte ϱ des Metalls seien im Ausgangszustand mit p_0, ϱ_0, hinter der Stoßfront mit p_1, ϱ_1 und hinter der Verdünnungswelle mit p_2, ϱ_2 bezeichnet. Die Zustandsänderung von p_0 nach p_1 erfolgt sprunghaft an der Stoßfront, diejenige von p_1 nach p_2 wegen der Abflachung der Entlastungswelle dagegen stetig über eine endliche Wegstrecke.

Zur Berechnung der isentropen Zustandsänderung bei der Entspannung auf den Druck $p = 0$ sei die Hilfsfunktion $y(V, T)$ betrachtet.

$$y(V, T) = p(V, T) - p_T(V) \tag{100}$$

betrachtet. Nach Gl. (11) und wegen der allgemein gültigen Beziehung $p = -(\partial U/\partial V)_S$ folgt aus Gl. (100) in Verbindung mit Gl. (31)

$$\frac{1}{y}\left(\frac{\partial y}{\partial \eta}\right)_S = \frac{1}{\gamma}\,\frac{\mathrm{d}\gamma}{\mathrm{d}\eta} + \frac{1+\gamma}{1+\eta}\,. \tag{101}$$

Diese Differentialgleichung für $y(\eta)$ auf der Isentropen läßt sich leicht integrieren, wenn man entsprechend der Vorgehensweise in Abschnitt 2.4 von dem Reihenansatz

$$y(\eta) = y_0 + y_1 \cdot \eta + y_2 \cdot \eta^2 + y_3 \cdot \eta^3 + \cdots \tag{102}$$

ausgeht und Gl. (34) in Gl. (101) einsetzt. Dabei kann man für die höheren Glieder in Gl. (34) willkürlich $\gamma_2 = 0$ und $\gamma_3 = 0$ einsetzen, da sie wegen der schon besprochenen Ungenauigkeiten bei der Bestimmung der Grüneisen-Konstante γ ohnehin keinen sinnvollen Beitrag mehr liefern. Man erhält so für die Koeffizienten in Gl. (102)

$$y_1 = y_0 \left\{ 1 + \gamma_0 + \frac{\gamma_1}{\gamma_0} \right\}; \qquad (103\,a)$$

$$y_2 = y_0 \frac{1}{2} \left\{ \left(\frac{y_1}{y_0}\right)^2 + \gamma_1 - 1 - \gamma_0 - \left(\frac{\gamma_1}{\gamma_0}\right)^2 \right\}; \qquad (103\,b)$$

$$y_3 = y_0 \frac{1}{3} \left\{ 3 \frac{y_2}{y_0} \frac{y_1}{y_0} - \left(\frac{y_1}{y_0}\right)^3 + \left(\frac{\gamma_1}{\gamma_0}\right)^3 + 1 + \gamma_0 - \gamma_1 \right\}. \qquad (103\,c)$$

y_0 ist der für die betreffende Isentrope charakteristische Kurvenparameter; er wird durch die Anfangsbedingung der Kurve festgelegt. Diese lautet hier nach Gl. (100): $y(V_1, T_1) = p_H(V_1) - p_T(V_1)$, was nach Gl. (32), (33), (38) und (40) auf

$$y_0 = \frac{1}{3} \varrho_0 \cdot C^2 \cdot S \cdot \gamma_0 \frac{\eta_1^3}{1 + \frac{y_1}{y_0}\eta_1 + \frac{y_2}{y_0}\eta_1^2 + \frac{y_3}{y_0}\eta_1^3} \qquad (104)$$

führt.

Hinter der Entlastungswelle, d. h. nach der Entspannung herrscht der Druck $p_2 = 0$. Nach Gl. (100), (102) und (33) ergibt sich damit als Bestimmungsgleichung für η_2

$$(y_3 + C_T)\eta_2^3 + (y_2 + B_T)\eta_2^2 + (y_1 + A_T)\eta_2 + y_0 = 0 \qquad (105)$$

In Tab. 12 sind die Zahlenwerte für η_2 zusammengestellt, die sich nach Gl. (105) für $p_1 = p_\perp$ und $\eta_1 = \eta_\perp$ mit den Daten in Tab. 8 ergeben. Sie repräsentieren denjenigen Metallzustand, der sich bei senkrecht auf eine ebene Platte auftreffender Detonationsfront hinter der erstmalig reflektierten Entlastungswelle an der Unterseite der Platte einstellt.

Man erkennt, daß in allen Fällen $\eta_2 < 0$ ist, was nach Gl. (31) einer thermischen Volumenausdehnung gegenüber dem Ausgangszustand $\eta_0 = 0$ entspricht. Setzt man den Wärmeausdehnungskoeffizienten α nach Gl. (6) im Temperaturintervall $T_0 \leqq T \leqq T_2$ als konstant voraus, so erhält man wegen $p_0 = p_2 = 0$ für T_2

$$T_2 = T_0 - \frac{\eta_2}{\alpha}. \qquad (106)$$

Die Temperaturwerte T_2 sind ebenfalls in Tab. 12 eingetragen; dabei wurde von $T_0 = 293\,°K$ ausgegangen. Bei Cd, Pb, Sn und Tl ist T_2 größer als die jeweilige Schmelztemperatur unter Normalbedingungen.

Tabelle 12. *Numerische Ergebnisse im Fall senkrecht auftreffender Detonationsfront*

Gruppe	Metall	η_2	T_2[°K]	$T_1 = T_\perp$[°K] nach Gl. 108	nach Tab. 8
II	Al	−0,01251	470	674	608
	Cd	−0,04494	877	(1737)	(1939)
	Pb	−0,06198	1077	(2133)	(3139)
	Sn	−0,04674	1041	(1904)	(2176)
	Th	−0,05693	1699	2895	2973
	Ti	−0,00903	641	829	617
	Tl	−0,07956	1213	(2558)	(3481)
	Zn	−0,02864	615	1067	1047
III	Ag	−0,01356	524	820	766
	Au	−0,00625	439	622	495
	Co	−0,00366	373	481	338
	Cr	−0,00402	501	669	426
	Cu	−0,00763	443	631	446
	Fe	−0,01388	670	(1121)	(706)
	Mo	−0,00179	416	504	319
	Ni	−0,00433	409	546	365
	V	−0,00457	476	609	377
	W	−0,00166	421	510	310
IV	Be	−0,00423	408	503	237

Diese Metalle liegen also nach der Entspannung zumindest partiell im geschmolzenen Zustand vor.

Aus T_2 läßt sich rückwärts wieder auf $T_1 = T_\perp$ schließen und so ein Vergleich mit den entsprechenden, direkt berechneten Werten in Tab. 8 durchführen. Wegen $(\partial S/\partial V)_T = (\partial p/\partial T)_V$ gilt für die Temperaturänderung längs einer Isentropen

$$C_V \cdot \left(\frac{\partial T}{\partial V}\right)_S + T\left(\frac{\partial p}{\partial T}\right)_V = 0 \,.$$

Nach Gl. (7) und (9) folgt damit

$$\frac{\mathrm{d}T}{T} = \gamma(\eta)\frac{\mathrm{d}\eta}{1+\eta} \tag{107}$$

und weiter nach Einsetzen der Gl. (34) durch Integration

$$T_1 = T_2\left(\frac{1+\eta_1}{1+\eta_2}\right)^{\gamma_0-\gamma_1} \cdot \mathrm{e}^{\gamma_1(\eta_1-\eta_2)} \,. \tag{108}$$

Man erkennt aus Tab. 12, daß die Übereinstimmung der nach Gl. (108) berechneten Zahlenwerte $T_1 = T_\perp$ mit denjenigen nach Tab. 8 nicht sonderlich gut ist. Das bestätigt die bereits in Abschnitt 3.3 getroffene Feststellung, daß die angewandten Näherungsverfahren zur Berechnung der Temperatur im komprimierten Zustand im allgemeinen nicht mehr genau genug sind.

Die Metalle der Gruppe I sind in Tab. 12 nicht aufgeführt. Bei ihnen ist η_1 nach Tab. 8 so groß, daß die dem Berechnungsverfahren zugrunde gelegten Reihenentwicklungen nach Abschnitt 2.4 nicht mehr anwendbar sind.

4.2 Plattenbeschleunigung bei senkrecht auftreffender Detonationsfront

Es sei zunächst die Beschleunigung einer ebenen Metallplatte endlicher Dicke für den Fall betrachtet, daß sich die Detonationsfront senkrecht auf ihre Oberfläche hin bewegt. Dabei sei die Sprengstoffbelegung wiederum als unendlich ausgedehnt vorausgesetzt.

Der Ablauf des Beschleunigungsvorgangs läßt sich am einfachsten überblicken, wenn man für den Augenblick annimmt, die Zustandsänderung in der Entlastungswelle erfolge wie die in der Stoßwelle sprunghaft an einer Diskontinuitätsfläche, einer „Verdünnungsfront".

In diesem Modell läuft zunächst eine Stoßfront von den Sprengstoffschwaden aus in das Metall hinein; der Zustand (1) hinter der Stoßfront entspricht im einzelnen den Zahlenwerten nach Tab. 8.

An der Plattenunterseite wird die Stoßfront als „Verdünnungsfront" reflektiert; der Zustand (2) hinter der „Verdünnungsfront" entspricht im einzelnen den Zahlenwerten nach Tab. 12. Beim Durchgang der „Verdünnungsfront" dehnt sich das Metall wieder aus, so daß seine vom Sprengstoff fort gerichtete Materialflußgeschwindigkeit u von u_1 auf einen höheren Wert u_2 ansteigt.

Wenn diese „Verdünnungsfront" an der Plattenoberseite ankommt, läuft von dort aus eine Verdünnungswelle in die Schwaden und eine Stoßwelle in das Metall. Der Zustand (3) hinter dieser Stoßfront bestimmt sich ähnlich wie der Zustand (1) in Abschnitt 3.3 aus den Gleichgewichtsbedingungen zwischen Metall und Schwaden.

Anschließend wird an der Plattenunterseite wieder eine „Verdünnungsfront" reflektiert, hinter der der Zustand (4) herrscht.

Diese Reflexionsvorgänge wiederholen sich nun ständig. Insgesamt ergibt sich so eine stufenweise Geschwindigkeitszunahme, wobei jeder von oben nach unten laufenden Stoßfront mit dem Zustand $(2n-1)$ dahinter eine von unten nach oben laufende „Verdünnungsfront" $(2n)$ und dann wieder eine Stoßfront $(2n+1)$ von oben nach unten folgt.

Dieser idealisierte Stufenprozeß wird in Wirklichkeit durch die Abflachung der Entlastungswellen verwischt. Unmittelbar nach ihrer Entstehung an der Plattenunterseite läßt sich die Entlastungswelle noch als Treppenfunktion ansehen: Überall im Platteninnern herrscht noch die Dichte ϱ_1. Lediglich an der Oberfläche selbst ist das Material bereits auf den Dichtewert $\varrho_2 < \varrho_1$ entspannt. Bei ihrem Hineinlaufen in das Platteninnere flacht sich die Welle dann ständig ab. In Fig. 16 stellt die gestrichelte Kurve den Verlauf der Funktion η nach Gl. (31) zu einem solchen späteren Zeitpunkt dar.

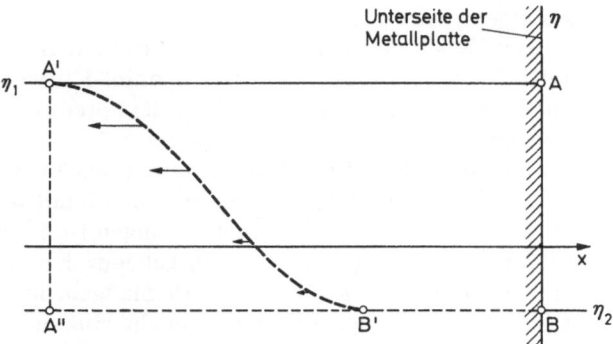

Fig. 16. Profil der Entlastungswelle in einem mit der Plattenunterseite mitbewegten Koordinatensystem

Die Abflachung der Entlastungswelle bewirkt, daß bei deren Eintreffen an der Plattenoberseite der Aufbau der reflektierten Verdichtungswelle nicht mehr momentan, sondern allmählich während einer endlichen Zeitspanne erfolgt. Beim Hineinlaufen in das Platteninnere steilt sich die Verdichtungswelle aber wieder auf, wodurch die vorherige Abflachung der Verdünnungswelle gewissermaßen zum Teil wieder kompensiert wird. Zum Zeitpunkt ihres Eintreffens an der Plattenunterseite herrscht daher im Platteninnern im wesentlichen doch derjenige Zustand, wie er sich nach dem idealisierten Stufenprozeß berechnet. Quantitative Angaben hierzu folgen in Abschnitt 4.3.

Der folgenden Berechnung des Beschleunigungsvorgangs wird der idealisierte Stufenprozeß zugrunde gelegt. Die Materialflußgeschwindigkeit u_2 hinter der „Verdünnungsfront" erhält man, indem man Gl. (68) sinngemäß auf die Zustandsgrößen des Metalls anwendet. Mit Gl. (31) und wegen $c^2 = (\partial p/\partial \varrho)_S$ ergibt sich so

$$u_2 = u_1 - \int_{\eta_1}^{\eta_2} \sqrt{\frac{1}{\varrho_0}\left(\frac{\partial p}{\partial \eta}\right)_s}\,\frac{\mathrm{d}\eta}{1+\eta}\,, \tag{109}$$

wobei sich $(\partial p / \partial \eta)_S$ nach Gl. (100), (102), (103) und (104) als Funktion von η ausdrücken läßt. Numerisch ergibt sich nach Gl. (109) $u_2 \approx 2 u_1$, wobei der relative Näherungsfehler in allen Fällen kleiner als 1 % ist. Hieraus läßt sich allgemein auch für die weiteren Beschleunigungsstufen folgern

$$u_{2n+2} = u_{2n} + 2(u_{2n+1} - u_{2n}); \quad n = 0, 1, 2, \dots . \tag{110}$$

Da sich Hugoniot-Kurven und Isentropen bei den höheren Beschleunigungsstufen immer mehr annähern, sollte der Näherungsfehler der Gl. (110) desto kleiner sein, je größer n ist.

Den jeweiligen Druck p_{2n+1} hinter den Stoßfronten erhält man wie p_1 in Abschnitt 3.3 dadurch, daß man die Hugoniot-Kurve $p_H = p_H(u_H)$ des Metalls mit der Reflexionskurve $p_s^* = p_s^*(u_s^*)$ der Sprengstoffschwaden zum Schnitt bringt. Für die Reflexionskurve der Schwaden gilt wieder Gl. (69), wobei lediglich überall der Index 1 durch den Index $2n + 1$ zu ersetzen ist. Hinsichtlich der Hugoniot-Kurve im Metall liegen keine unmittelbaren Meßwerte vor, da sich die Messungen [4, 17, 18, 23] alle auf die durch den Fußpunkt ($p_0 = 0$; $\eta_0 = 0$) gehende Kurve beziehen, während hier der Fußpunkt ($p_{2n} = 0$; $\eta_{2n} < 0$) maßgebend ist. Wegen $|\eta_{2n}| \ll 1$ ist es jedoch naheliegend, die ursprüngliche Hugoniot-Kurve im

Tabelle 13. *Daten der höheren Beschleunigungsstufen bei senkrechtem Auftreffen der Detonationsfront*

Gruppe	Metall	p_1 [kbar]	u_1 $\left[\dfrac{mm}{\mu s}\right]$	p_3 [kbar]	u_3 $\left[\dfrac{mm}{\mu s}\right]$	p_5 [kbar]	u_5 $\left[\dfrac{mm}{\mu s}\right]$	p_7 [kbar]	u_7 $\left[\dfrac{mm}{\mu s}\right]$
II	Al	353,2	1,659	104,2	3,932	52,3	4,875	32,5	5,413
	Cd	437,5	1,157	183,5	2,985	95,5	4,055	59,3	4,715
	Pb	450,9	1,083	198,9	2,835	105,6	3,908	65,9	4,579
	Sn	417,5	1,270	161,0	3,221	81,7	4,288	50,1	4,925
	Th	449,4	1,091	199,2	2,832	107,9	3,875	68,3	4,530
	Ti	396,1	1,395	145,6	3,394	78,6	4,343	50,6	4,914
	Tl	450,4	1,086	196,2	2,860	102,1	3,959	63,0	4,637
	Zn	427,6	1,212	174,8	3,074	92,7	4,105	58,2	4,739
III	Ag	475,2	0,953	239,4	2,476	143,2	3,417	96,2	4,048
	Au	530,7	0,670	330,0	1,799	225,2	2,592	164,4	3,175
	Co	482,2	0,916	253,2	2,364	158,6	3,243	110,3	3,840
	Cr	468,3	0,989	233,1	2,530	141,9	3,433	97,1	4,034
	Cu	472,1	0,969	237,0	2,496	143,4	3,415	97,4	4,028
	Fe	458,8	1,041	215,9	2,679	124,2	3,654	81,4	4,290
	Mo	504,0	0,803	287,7	2,098	189,3	2,924	135,9	3,503
	Ni	482,7	0,913	253,5	2,361	158,3	3,246	109,8	3,847
	V	443,5	1,124	200,5	2,821	117,4	3,745	78,7	4,338
	W	545,5	0,597	359,0	1,610	256,5	2,331	194,1	2,873

$p - \eta$-Diagramm einfach in η-Richtung so zu verschieben, daß sie durch den neuen Fußpunkt geht. Man erhält dann nach Gl. (39)

$$p_H = \varrho_0 C^2 \left\{ \frac{\eta(\eta + 1)}{[1 - \eta(S - 1)]^2} - \frac{\eta_{2n}(\eta_{2n} + 1)}{[1 - \eta_{2n}(S - 1)]^2} \right\}.$$ (111)

Aus den Gl. (22) und (23) folgt weiterhin wegen $p_{2n} = 0$

$$u_H = u_{2n} + \sqrt{\frac{p_H}{\varrho_0} \frac{\eta - \eta_{2n}}{(1 + \eta_{2n})(1 + \eta)}}.$$ (112)

Elimination von η aus den Gl. (111) und (112) liefert die gesuchte Hugoniot-Kurve $p_H = p_H(u_H)$ im Metall.

Im übrigen berechnen sich die höheren Beschleunigungsstufen analog der Vorgehensweise in den Abschnitten 3.3 und 4.1. In Tab. 13 sind die errechneten Daten zusammengestellt. Man erkennt, daß schon nach wenigen Reflexionen der Druckabbau in den Schwaden und damit die weitere Beschleunigung der Metallplatte nur noch langsam vonstatten geht.

4.3 Abflachung der Entlastungswelle

Es soll die Abflachung der an der Plattenunterseite reflektierten Entlastungswelle untersucht werden. Der Materialfluß gehorcht den Bewegungsgleichungen [22]

$$\text{a)} \ \frac{\partial \varrho}{\partial t} + \frac{\partial(\varrho w)}{\partial x} = 0; \quad \text{b)} \ \varrho\left(\frac{\partial w}{\partial t} + w \frac{\partial w}{\partial x}\right) + \frac{\partial p}{\partial x} = 0.$$ (113)

Dabei soll die negative x-Richtung die Laufrichtung der Welle in der Platte von unten nach oben sein. Die positive x-Richtung ist dann die Richtung der Materialflußgeschwindigkeit w. Diese hängt, abgesehen von den Randbedingungen, nur von den Zustandsvariablen p und ϱ ab. Andererseits liefert die Isentropiebedingung einen weiteren Zusammenhang zwischen den Zustandsgrößen, so daß im Endeffekt p und damit auch w nur noch von ϱ abhängen. Mit $w = w(\varrho)$ folgt aus den Gl. (113) nach einigen Umformungen

$$\frac{dw}{d\varrho} = (\pm) \frac{1}{\varrho} \sqrt{\left(\frac{\partial p}{\partial \varrho}\right)_s}.$$ (114)

In Gl. (114) muß das Minuszeichen vor der Wurzel gelten, da w desto größer ist, je stärker der Verdünnungseffekt, d. h. je kleiner ϱ ist.

Mittels der Gl. (31), (100), (102), (103) und (104) erhält man aus Gl. (114) durch Integration die Funktion $w = w(\eta)$. Diese lautet, wenn man sich auf ein Koordinatensystem bezieht, das mit der Geschwindigkeit u_2 der Plattenunterseite mitbewegt wird, in dem also $w_2 = 0$ ist

$$w(\eta) = \sqrt{\frac{y_1 + A_T}{\varrho_0}} \left\{ \omega(\eta_2) - \omega(\eta) \right\}, \tag{115a}$$

$$\omega(\eta) = \eta \left[1 + \frac{a}{2} \eta + \frac{b}{3} \eta^2 \right], \tag{115b}$$

$$a = \frac{y_2 + B_T}{y_1 + A_T} - 1, \tag{115c}$$

$$b = \frac{3}{2} \frac{y_3 + C_T}{y_1 + A_T} - \frac{1}{2} \left(\frac{y_2 + B_T}{y_1 + A_T} \right)^2 - a. \tag{115d}$$

Dabei bestimmen sich die Konstanten nach Gl. (38) und (40) bzw. nach Gl. (103) und (104).

Einsetzen der Funktion $w = w(\eta)$ nach Gl. (115) in Gl. (113a) führt auf eine partielle Differentialgleichung für $\eta = \eta(x, t)$

$$\frac{\partial \eta}{\partial t} + \frac{\partial \eta}{\partial x} \left\{ w + (1 + \eta) \frac{dw}{d\eta} \right\} = 0. \tag{116}$$

Diese läßt sich leicht integrieren, wenn man von $\eta = \eta(x, t)$ zur inversen Funktion $x = x(\eta, t)$ übergeht. Man erhält

a) $x(\eta, t) = -F(\eta) \cdot t$,

b) $F(\eta) = \sqrt{\frac{y_1 + A_T}{\varrho_0} \left\{ 1 + (2 + a)\eta + \left(\frac{3a}{2} + b \right) \eta^2 - \omega(\eta_2) \right\}}$, (117)

wenn man der Plattenunterseite die Koordinate $x = 0$ gibt und davon ausgeht, daß im Augenblick ($t = 0$) der Entstehung der Verdünnungswelle diese eine Treppenfunktion darstellt, d. h. $x(\eta, 0) = 0$ ist.

In Gl. (117) ist $F(\eta)$ die lokale Fortpflanzungsgeschwindigkeit der Welle. Speziell ist $F_1 = F(\eta_1)$ die Geschwindigkeit der Wellenspitze und $F_2 = F(\eta_2)$ diejenige des Wellenendes. Numerisch ergibt sich mit einem Fehler von wenigen Prozent: $F_2 = C$. Die Werte für F_1 sind in Tab. 14 eingetragen. Bei einem Vergleich mit den Zahlenwerten für C in Tab. 1 erkennt man, daß sich die Welle sehr schnell auseinanderzieht.

Damit läßt sich leicht die Dicke δ' der Entlastungswelle für den Zeitpunkt ausrechnen, in dem deren Spitze gerade an der Plattenoberseite eintrifft. Man erhält

$$\frac{\delta'}{d'} = 1 - \frac{C}{F_1}.$$ (118)

Dabei ist $d' = d/(1 + \eta_1)$ die Dicke der Metallplatte im komprimierten Zustand (1) und d die ursprüngliche Dicke im Ausgangszustand (0). Die anschließende Aufsteilung der durch Reflexion an der Plattenoberseite entstehenden Verdichtungswelle soll unter einigen vereinfachenden Voraussetzungen abgeschätzt werden. Insbesondere soll davon abgesehen werden, daß die hinteren Teile der ankommenden Verdünnungswelle sich den vorderen Teilen der zurücklaufenden Verdichtungswelle überlagern. Ferner sei $\eta_2 \approx 0$ gesetzt und in erster Näherung angenommen, daß sich die Spitze der Verdichtungswelle mit plastischer Schallgeschwindigkeit C und das Ende mit der sich nach Gl. (25) und (26) ergebenden Stoßwellengeschwindigkeit v_3

$$v_3 = \frac{C}{2} + \sqrt{\frac{C^2}{4} + \frac{p_3 \cdot S}{\varrho_0}}$$ (119)

fortpflanzen. (Dabei wurde $\varrho_0 \approx \varrho_2$ entsprechend $\eta_2 \approx 0$ eingesetzt.) Die Zeitspanne τ, die zwischen Start der Spitze und Start des Endes der Verdichtungswelle an der Plattenoberseite vergeht, ist unter den gemachten Voraussetzungen gerade diejenige Zeit, die das Ende der Entlastungswelle benötigt, um den Weg δ' zurückzulegen: $\tau = \delta'/C$. In dieser Zeit τ durchläuft die Spitze der Verdichtungswelle den Weg $C \cdot \tau$. Damit ergibt sich schließlich für die Dicke δ der Verdichtungswelle für den Zeitpunkt, in dem ihre Spitze gerade an der Plattenunterseite eintrifft

$$\frac{\delta}{d} = 1 - \frac{v_3}{C}\left[1 - \frac{1}{1+\eta_1}\left(1 - \frac{C}{F_1}\right)\right].$$ (120)

Die nach Gl. (118), (119) und (120) berechneten Zahlenwerte sind in Tab. 14 zusammengestellt. Man erkennt, daß bei Ankunft der Verdichtungswelle an der Plattenunterseite jeweils bereits 80 bis 90% der Platte denjenigen Zustand angenommen haben, der sich nach dem idealisierten Stufenprozeß errechnet; die übrigen 10 bis 20% zeigen graduelle Abweichungen. Hieraus folgt, daß der idealisierte Stufenprozeß hinsichtlich des Gesamtverhaltens der Metallplatte eine brauchbare Näherungsbasis darstellt. Insbesondere läßt sich nach ihm die zeitliche Zunahme des integralen Plattenimpulses in guter Näherung berechnen.

Tabelle 14. *Daten zur Abflachung der Entlastungswelle*

Gruppe	Metall	$F_1\left[\dfrac{mm}{\mu s}\right]$	$\dfrac{\delta'}{d'}$	$\dfrac{\delta}{d}$	$v_3\left[\dfrac{mm}{\mu s}\right]$
II	Al	9,17	0,40	0,22	6,25
	Cd	5,38	0,55	0,15	3,47
	Pb	4,45	0,54	0,11	2,93
	Sn	5,50	0,52	0,16	3,56
	Th	4,21	0,49	0,11	2,89
	Ti	7,37	0,35	0,18	5,43
	Tl	4,21	0,56	0,09	2,77
	Zn	6,07	0,50	0,18	4,00
III	Ag	5,80	0,44	0,18	4,12
	Au	4,92	0,37	0,16	3,78
	Co	6,94	0,32	0,16	5,45
	Cr	7,82	0,33	0,17	6,02
	Cu	6,53	0,38	0,18	4,83
	Fe	6,92	0,47	0,20	4,76
	Mo	6,99	0,26	0,14	5,76
	Ni	7,01	0,34	0,17	5,41
	V	7,52	0,32	0,17	5,79
	W	5,40	0,26	0,12	4,53

4.4 Plattenbeschleunigung bei schräg auftreffender Detonationsfront

Die von der freien Plattenunterseite aus rücklaufende Entlastungswelle läßt sich bei schräg auftreffender Detonationsfront als Prandtl-Meyer-Strömung berechnen. Die Vorgehensweise entspricht dabei derjenigen, die in Abschnitt 3.4 zur Berechnung der seitlichen Schwadenexpansion angewandt wurde. Fig. 17 zeigt die Verhältnisse schematisch. Es erübrigt sich jedoch, die auf dieser Basis entwickelten, teilweise recht komplizierten Gleichungen hier im einzelnen anzuführen. Die durchgeführte Rechnung zeigt nämlich, daß man mit hoher Genauigkeit von erheblich einfacheren Näherungsgleichungen ausgehen kann.

Man betrachte hierzu den Materialfluß nach Fig. 17 in einem mitbewegten Bezugssystem. Ein in A′ befindlicher Beobachter sieht das Metall längs der Plattenunterseite mit der Geschwindigkeit $\overline{EA'} = D/\sin\theta$ von E her auf sich zukommen und mit $\overline{A'F} = w_2$ nach F hin abfließen. Die Materialflußgeschwindigkeit u_2 im Laborsystem erhält man aus derjenigen (w_2) im mitbewegten System, indem man die Relativgeschwindigkeit $\overline{EA'}$ beider Systeme vektoriell von $\overline{A'F}$ subtrahiert: $u_2 = \overline{A'B}$. Zur Ermittlung von w_2 kann man wegen $\eta_2 \approx 0$ näherungsweise davon ausgehen, daß der Materialfluß längs der Stromlinie EA′F praktisch inkompressibel erfolgt, so daß man die Bernouillische Gleichung $\varrho w^2/2 + p = $ const. auf ihn anwenden kann. Daraus folgt, da an einer freien Oberfläche stets der Druck $p = 0$ herrscht: $w^2 = $ const. und somit $\overline{A'F} = \overline{EA'}$. Nach Fig. 17 ergibt sich damit

$$u_2 = 2D\,\frac{\sin(\beta_2/2)}{\sin\theta}. \tag{121}$$

Die der Näherung Gl. (121) äquivalente Beziehung $w_2 = D/\sin\theta$ wurde mit den numerischen Ergebnissen für die Prandtl-Meyer-Strömung verglichen. Dabei zeigte sich in allen gerechneten Fällen ein Näherungsfehler von nur wenigen Prozent.

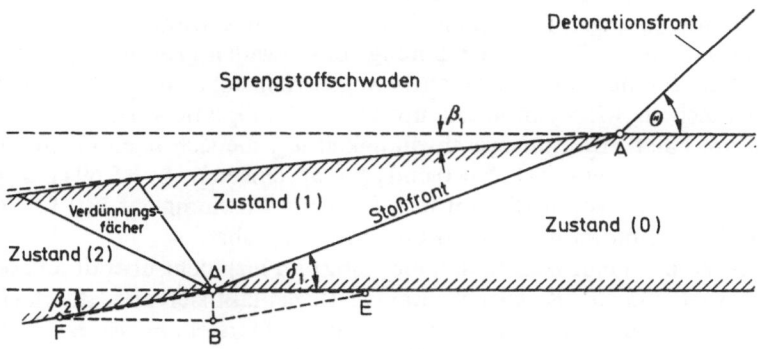

Fig. 17. Entlastungswellen-Reflexion an der freien Plattenunterseite

Ein weiteres Ergebnis nach den Gleichungen der Eckenströmung ist

$$\beta_2 = 2 \cdot \beta_1 \tag{122}$$

ebenfalls mit einem Näherungsfehler von nur wenigen Prozent in den überprüften Fällen. Es entspricht dem Ergebnis $u_2 = 2 \cdot u_1$ bei senkrecht auftreffender Detonationsfront in Abschnitt 4.2.

Die weitere Beschleunigung der Metallplatte läßt sich ähnlich wie in Abschnitt 4.2 aus einem idealisierten Stufenprozeß herleiten. Dabei müssen die Neigungswinkel der jeweiligen Fronten gegeneinander im einzelnen berechnet werden. Im übrigen lehnt sich die Vorgehensweise eng an diejenige in Abschnitt 4.2 an, so daß auf eine explizite Darstellung der Gleichungen verzichtet werden kann. Darüber hinaus ist es auch nicht sinnvoll, die numerischen Ergebnisse im einzelnen anzuführen; sie lassen sich übersichtlicher durch die Formulierung von Ähnlichkeitsgesetzen darstellen, wie dies im folgenden Abschnitt geschieht.

4.5 Ähnlichkeitsgesetz für die mittlere Plattengeschwindigkeit

Die numerischen Ergebnisse lassen sich näherungsweise in einem für alle Metalle gültigen Ähnlichkeitsgesetz zusammenfassen. Dieses bezieht sich auf die mittlere Plattengeschwindigkeit $\bar{u}$ in Abhängigkeit von der Zeit t.

Im Fall senkrecht auftreffender Detonationsfront ist $\bar{u}$ definiert durch

$$\varrho_0 \cdot \bar{u} = \frac{1}{d} \int_{x'}^{x''} \varrho(x) \cdot u(x) \cdot dx. \tag{123}$$

Dabei ist $x = x'$ die Koordinate der Plattenunterseite und $x = x''$ die der Plattenoberseite. Einsetzen der numerischen Ergebnisse nach Abschnitt 4.2 in Gl. (123) führt auf $\bar{u}(t)$.

Da die Sprengstoffbelegung als unendlich dick vorausgesetzt wurde, ist die kinetische Energie der Sprengstoffschwaden groß gegenüber der der Metallplatte. Die Schwaden bewegen sich daher nach hinreichend langer Zeit so, wie wenn überhaupt keine Metallplatte vorhanden wäre, d. h. mit einer Strömungsgeschwindigkeit u_∞^*, die sich nach Gl. (69) für u_s^* ergibt, wenn man dort $p_s^* = 0$ und $p_1^* = p_{CJ}^*$ einsetzt: $u_\infty^* = 8,694$ mm/µs. Für $t \to \infty$ geht somit die mittlere Plattengeschwindigkeit $\bar{u}(t)$ für alle Metalle asymptotisch in diesen Grenzwert u_∞^* über.

Es ist leicht einzusehen, daß die Funktion $\bar{u}(t)$ nicht überall differenzierbar ist, sondern Knickstellen aufweist. Sie läßt sich aber durch eine analytische Funktion annähern. Für deren Darstellung ist es zweckmäßig, zu den dimensionslosen Variablen ζ und τ

$$\text{a) } \zeta = \bar{u}\frac{1}{C}; \quad \text{b) } \tau = t\frac{1}{d}\frac{p_\perp}{\varrho_0 C} \tag{124}$$

und damit von $\bar{u}(t)$ zu $\zeta(\tau)$ überzugehen. Man erhält die für alle besprochenen Metalle der Gruppen II und III nachgeprüfte Näherungsbeziehung im Fall senkrecht auftreffender Detonationsfront ($\theta = 90°$)

$$\text{a) } \zeta(\tau) = \frac{\tau}{1 + \tau/\zeta_\infty}; \quad \text{b) } \zeta_\infty = \frac{u_\infty^*}{C} \quad u_\infty^* = 8,694 \text{ mm/µs}. \tag{125}$$

Für $\tau \to 0$ und für $\tau \to \infty$ gilt Gl. (125) exakt. Im übrigen ist der Näherungsfehler am größten ungefähr während der zweiten Beschleunigungsstufe ($v = 2$). Er ist absolut genommen kleiner als 10 % für folgende Beschleunigungsstufen v:

bei Au	für alle v,
bei Cd, Pb, Tl	für $v > 1$,
bei Ag, Cu, Fe, Mo, Ni, W, Zn	für $v > 2$,
bei Co, Cr, Sn, Th, V	für $v > 3$,
bei Al, Ti	für $v > 4$.

Hieraus folgt, daß man den Beginn der Plattenbewegung zweckmäßigerweise nicht nach der Näherung Gl. (125), sondern nach den vollständigen Gleichungen der Stoß- und Verdünnungswellenreflexion berechnet.

Die asymptotische Plattengeschwindigkeit u_∞^* läßt sich nur mit unendlich dicker Sprengstoffbelegung erreichen – ein Fall, der sich in der Praxis nicht verwirklichen läßt. Im folgenden soll abgeschätzt werden, welche Sprengstoffbelegung mindestens erforderlich ist, um eine Kupferplatte wenigstens auf 95 % ihrer asymptotischen Geschwindigkeit u_∞^* zu beschleunigen.

Mit $\xi = 0{,}95\,\xi_\infty$ folgt aus Gl. (125a) $\tau = 19 \cdot \xi_\infty$ und weiter nach Gl. (124b) und (125b): $t/d = 19 \cdot \varrho_0 u_\infty^*/p_\perp$. Mit den Zahlenwerten für Cu nach Tab. 1 und Tab. 8 führt dies auf $t/d = 31{,}25$ µs/mm. Bei einer 1 mm dicken Cu-Platte dauert es also 31,25 µs, bis sie auf die mittlere Geschwindigkeit $\bar{u} = 0{,}95\,u_\infty^*$ gebracht ist. Wenigstens so lange muß also an ihrer Oberfläche ein Schwadendruck aufrecht erhalten werden.

Nun erfolgt der Abbau der Schwaden von ihrem freien Ende her mit Schallgeschwindigkeit c^*, nach Gl. (63) also etwa mit $c^* = 5{,}8$ mm/µs. Die Mindestdicke der Schwadenschicht zu Beginn der Plattenbeschleunigung beträgt daher 181 mm. Die erforderliche Sprengstoffdicke ist das $D/(D - c^*)$fache hiervon und beträgt mithin 668 mm. Sie übertrifft die Dicke der zu beschleunigenden Metallplatte also ganz beträchtlich.

Im Fall schräg auftreffender Detonationsfront erhält man die asymptotische Grenzgeschwindigkeit $\lim_{t \to \infty} \bar{u}(t) = \bar{u}_\infty$ dadurch, daß man in den Gl. (87), (91) und (92) zu $p_s^* = 0$ übergeht. Man erhält für $\bar{u}_\infty$ eine etwas komplizierte Abhängigkeit von θ, die sich aber näherungsweise durch

$$\bar{u}_\infty(\theta) = 5{,}603 + 3{,}091 \cdot \cos\theta \;\; [\text{mm}/\,\text{µs}] \tag{126}$$

darstellen läßt. Der Näherungsfehler der Gl. (126) ist für alle θ kleiner als 1 %.

Erweitert man die Definition von τ nach Gl. (124b) für $\theta < 90°$ dadurch, daß man $p_1(\theta)$ statt $p_\perp$ einsetzt, so läßt sich Gl. (125) verallgemeinern zu

$$\text{a)}\;\; \xi(\tau, \theta) = \frac{\tau}{1 + \tau/\xi_\infty(\theta)} \cdot \sqrt{1 - \frac{C^2 \sin^2\theta}{D^2}}\,; \quad \text{b)}\;\; \xi_\infty(\theta) = \frac{\bar{u}_\infty(\theta)}{C}. \tag{127}$$

Gl. (127) wurde in einer Reihe von Einzelfällen numerisch überprüft. Sie geht für $\theta = 90°$ in Gl. (125) über.

5. Schlußfolgerungen

Das dynamische Verhalten von Metallen unter Stoßwellenbelastung läßt sich unter Verwendung experimenteller Daten nach den Gleichungen der Stoßwellentheorie und denen des thermodynamischen Zustands-verhaltens unter hohen Drücken berechnen. Da die Materialfestigkeit klein ist gegenüber den herrschenden Drücken, kann man das Metall in den meisten Fällen als ideal fluides Medium ansehen und daher bei der Aufstellung der Bewegungsgleichungen unmittelbar an die entsprechenden Gleichungen der Gasdynamik anknüpfen. Hinsichtlich der zu verwendenden Zustandsgleichung läßt sich auf der Basis der älteren Theorien von *Debye, Mie* und *Grüneisen* eine gut brauchbare Näherung entwickeln.

Der Bewegungsablauf errechnet sich nach den angegebenen Näherungsgleichungen hinreichend genau. Zur Berechnung der im Metall auftretenden Temperaturen reicht die Güte dieser Näherungen dagegen meist nicht aus.

Die numerischen Ergebnisse geben einerseits Hinweise auf die Einflußgrößen und die zu beachtenden Phänomene bei der verfahrenstechnischen Verwendung von Sprengstoffen zur Metallverformung. Andererseits zeigen sie eine Reihe noch nicht ganz geklärter Zusammenhänge zwischen den Kenngrößen der Metalle und werfen damit neue physikalische Fragestellungen auf.

Literatur

1. *Bridgman, P. W.:* The Physics of High Pressure. London: G. Bell & Sons, Ltd. 1952 (Nachdruck).
2. — Rev. Mod. Phys. **18**, 1–93 (1946).
3. *Swenson, C. A.:* Physics at High Pressures. In: Solid State Physics, Vol. 11, p. 41–147. F. Seitz, D. Turnbull (eds.). New York-London: Academic Press 1960.
4. *Rice, M. H., McQueen, R. G., Walsh, J. M.:* Compression of Solids by Strong Shock Waves. In: Solid State Physics, Vol. 6, p. 1–63. F. Seitz, D. Turnbull (eds.). New York-London: Academic Press 1958.
5. *Al'Tshuler, L. V.:* Soviet Phys. Usp. (engl.) **8**, 52–91 (1965).
6. *De Beaumont, Ph., Leygonie, J.:* Vaporizing of Uranium after Shock Loading, Fifth Symposium on Detonation, Pasadena 1970.
7. *Doran, D. G., Linde, R. K.:* Shock Effects in Solids. In: Solid State Physics, Vol. 19, p. 229–290. F. Seitz, D. Turnbull (eds.). New York-London: Academic Press 1966.
8. *Rinehart, J. S., Pearson, J.:* Explosive Working of Metals. Oxford: Pergamon Press 1963.
9. *Brush, S. G.:* Theories of the Equation of State of Matter at High Pressures and Temperatures. In: Progress in High Temperature Physics and Chemistry, p. 1–137. C. A. Rose (ed.). Oxford: Pergamon Press 1967.
10. *Knopoff, L.:* Equations of State of Solids at Moderately High Pressures. In: High Pressure Physics and Chemistry, Vol. 1, p. 227–245. R. S. Bradley (ed.). London-New York: Academic Press 1963.
11. vgl. z. B.: *Vogt, E.:* Physikalische Eigenschaften der Metalle. 1. Band. Leipzig: Akademische Verlagsgesellschaft Geest & Portig 1958.
12. *Geiger, W., Hornberg, H., Schramm, K. H.:* Zustand der Materie unter sehr hohen Drücken und Temperaturen. In: Springer Tracts in Modern Physics, Vol. 46, S. 1–52. Berlin-Heidelberg-New York: Springer 1968.

13. vgl. z. B.: *Sommerfeld, A.:* Mechanik der deformierbaren Medien. § 44. Leipzig: Akademische Verlagsgesellschaft Geest & Portig 1949.
14. *Huang, Y. K.:* J. Chem. Phys. **45**, 1979–1984 (1966).
15. *Schramm, K. H.:* Z. Physik **167**, 29–38 (1962).
16. *Dugdale, J. S., Mac Donald, D. K. C.:* Phys. Rev. **89**, 832–834 (1953).
17. *Mc Queen, R. G., Marsh, S. P.:* J. Appl. Phys. **31**, 1253–1269 (1960).
18. — — *Carter, W. J.:* The Determination of New Standards for Shock Wave Equation – of – State Work. In: Proc. I.U.T.A.M. Symposium H.D.P.: Behavior of Dense Media under High Dynamic Pressures, p. 67–83. Paris: Dunod 1967.
19. *Argous, J. P., Aveille, J.:* Adiabatique dynamique du cuivre a pression élevée. In: Behavior of Dense Media under High Dynamic Pressure, p. 173–178. Paris: Dunod 1967.
20. *Duvall, G. E.:* Some Properties and Applications of Shock Waves. In: Shewmon, P. G., Zackay, V. F.: Response of Metals to High Velocity Deformation, p. 165–203. New York: Interscience Publishers 1961.
21. *Zel'Dovich, Ya. B., Raizer, Yu. P.:* Physics of Shock Waves and High-Temperature Hydrodynamic Phenomena, Vol. 2, p. 685–784. New York-London: Academic Press 1967.
22. vgl. z. B.: *Oswatitsch, K.:* Gasdynamik. Wien: Springer 1952.
23. *Bakanova, A. A., Dudoladov, I. P., Trunin, R. F.:* Soviet Phys. Solid State (engl. Transl.) **7**, 1307–1313 (1965).
24. *Bancroft, D., Peterson, E. L., Minshall, S.:* J. Appl. Phys. **27**, 291–298 (1956).
25. *Johnson, P. C., Stein, B. A., Davis, R. S.:* J. Appl. Phys. **33**, 557–561 (1962).
26. *Jamieson, J. C., Lawson, A. W.:* J. Appl. Phys. **33**, 776–780 (1962).
27. *Duvall, G. E., Horie, Y.:* Shock Induced Phase Transitions. Proc. IV Symposium (int.) on Detonation. p. 248–257. Washington D.C.: Office of Naval Research, Dept. of Navy 1966.
28. *Kraut, E. A., Kennedy, G. C.:* Phys. Rev. **151**, 668–675 (1966).
29. *Kǫrmer, S. B., Sinitsyn, M. V., Kirillov, G. A., Urlin, V. D.:* Soviet Phys. JETP (engl. Transl.) **21**, 689–700 (1965).
30. *Horie, Y.:* J. Phys. Chem. Solids **28**, 1569–1577 (1967).
31. *D'Ans-Lax:* Taschenbuch für Chemiker und Physiker. 3. Aufl., Bd. 1. Berlin-Heidelberg-New York: Springer 1967.
32. American Institute of Physics Handbook. 2nd ed. 1963.
33. *Schramm, K. H.:* Z. Metallk. **53**, 729–735 (1962).
34. — Z. Metallk. **53**, 316–320 (1962).
35. *Deal, W. E.:* J. Chem. Phys. **27**, 796–800 (1957).
36. *Dunne, B. B.:* Phys. Fluids **7**, 1707–1712 (1964).
37. *Sternberg, H. M., Piacesi, D.:* Phys. Fluids **9**, 1307–1315 (1966).

Dr. *K. H. Schramm*
Battelle Institut e.V.
D – 6000 Frankfurt/Main
Deutschland

SPRINGER TRACTS IN MODERN PHYSICS

Ergebnisse
der exakten Natur-
wissenschaften

Volume **58**

Reprint

W. Kundt

Survey of Cosmology

Springer-Verlag Berlin Heidelberg New York 1971

SPRINGER-VERLAG
BERLIN·HEIDELBERG·NEW YORK

Springer Tracts
in Modern Physics

Ergebnisse der exakten Naturwissenschaften

Editor: **G. Höhler**
Editorial Board: P. Falk-Vairant, S. Flügge, J. Hamilton, F. Hund,
H. Lehmann, E. A. Niekisch, W. Paul

SPRINGER TRACTS IN MODERN PHYSICS

Ergebnisse
der exakten Natur-
wissenschaften

Volume **58**

Reprint

J. Feitknecht

Silicon Carbide as a Semiconductor

Springer-Verlag Berlin Heidelberg GmbH 1971

SPRINGER TRACTS
IN MODERN PHYSICS

Ergebnisse
der exakten Natur-
wissenschaften

Volume **58**

Reprint

K. Dettmann

High Energy Treatment of Atomic Collisions

Springer-Verlag Berlin Heidelberg GmbH 1971

SPRINGER TRACTS
IN MODERN PHYSICS

Ergebnisse
der exakten Natur-
wissenschaften

Volume 58

Reprint

K. H. Schramm

**Dynamisches Verhalten von Metallen
unter Stoßwellenbelastung**

Springer-Verlag Berlin Heidelberg GmbH 1971